STUDENT SOLUTIONS MANUAL

PAUL STANLEY
California Lutheran University

to accompany

Volumes One and Two

Physics

Fifth Edition

DAVID HALLIDAY
University of Pittsburgh

ROBERT RESNICK
Rensselaer Polytechnic Institute

KENNETH KRANE
Oregon State University

JOHN WILEY & SONS, INC.

COVER PHOTO: Courtesy IBM, Almaden Research Center.

To order books or for customer service call 1-800-CALL-WILEY (225-5945).

ISBN 0-471-39829-2

Printed in the United States of America.

10 9 8 7 6 5 4 3 2 1

Printed and bound by Bradford & Bigelow, Inc.

Here are the solutions to approximately 25% of the exercises and problems. Enjoy your reading, but remember that reading my solutions will make a poor substitute for deriving your own.

I have tried to be very consistent in my units, showing them at all times. After the first few chapters, however, I begin to assume that you have mastered some of the more common conversions, such as minutes to seconds or years to hours.

I have usually respected the rules for significant figures in calculations throughout; usually, but not always, this meant only two or three significant figures are shown. When intermediate calculations are done I used the significant figures from those calculations, so expect rounding to have occurred. The answers in the back of the book are also written to the correct number of significant digits, but sometimes the results from intermediate calculations were left at whatever the calculator came up with. Consequently, we don't always agree, and neither will you.

Each question has been answered by at least two people, and our answers agree within the errors expected from rounding of significant figures. There are, however, a few exceptions, because part of this text was written in Tabakea's mwaneaba in the hills of Delainavesi on Viti Levu (*ti aki toki ni moi te nangkona n te tairiki*), and I wasn't able to communicate discrepancies. I don't think it was me that made the mistake; however, if you do find a mistake, and let me know, you might be entered in a drawing which could have as a grand prize a monetary award in excess of 10,000,000 nano-dollars! At the very least, I'll probably acknowledge the first sender of each significant contribution which is incorporated into any revision.

At times I may have been too complete in my descriptions. Forgive my verbosity.

I want to give special thanks to Tebanimarawa Stanley for scanning in what must have felt like thousands of pages of text and helping to convert this text to LaTeX; Andrea Katz for checking my math; Jessica Helms, Alison Hill, and Nicole Imhof for punching holes in the rough drafts and keeping track of the numerous handwritten notes on the various edits.

Additionally I must thank Ken Krane and David Halliday for their positive comments and good suggestions, and also Aliza Atik and Stuart Johnson from John Wiley for relentlessly encouraging me to finish on time.

Paul Stanley
California Lutheran University
stanley@clunet.edu

Contents

Chapter 1

Measurement

E1-3 Multiply out the factors which make up a century. Note that each expression in parentheses is equal to one.

$$1 \text{ century} = 100 \text{ years} \left(\frac{365 \text{ days}}{1 \text{ year}} \right) \left(\frac{24 \text{ hours}}{1 \text{ day}} \right) \left(\frac{60 \text{ minutes}}{1 \text{ hour}} \right)$$

This gives 5.256×10^7 minutes in a century. The prefix *micro* means multiply by 10^{-6}, so a microcentury is $10^{-6} \times 5.256 \times 10^7$ or 52.56 minutes.

The percentage difference of x from y is defined as $(x - y)/y \times 100\%$. So the percentage difference from Fermi's approximation is $(2.56 \text{ min})/(50 \text{ min}) \times 100\%$ or 5.12%. Note that the percentage difference is dimensionless.

E1-7 The speed of a runner is given by the distance ran divided by the elapsed time. We don't know the actual distances, although they are probably very close to a mile. We'll assume, for convenience only, that the runner with the longer time ran *exactly* one mile. Let the speed of the runner with the shorter time be given by v_1, and call the distance actually ran by this runner d_1. Then $v_1 = d_1/t_1$. Remember, d_1 might not be a mile, it is instead 1 mile plus some error in measurement δ_1, which could be positive or negative. Similarly, $v_2 = d_2/t_2$ for the other runner, and $d_2 = 1$ mile.

We want to know when $v_1 > v_2$. Substitute our expressions for speed, and get $d_1/t_1 > d_2/t_2$. Rearrange, and $d_1/d_2 > t_1/t_2$ or $d_1/d_2 > 0.99937$. Remember to properly convert units when dividing the times! Then $d_1 > 0.99937$ mile $\times$ (5280 feet/1 mile) or $d_1 > 5276.7$ feet is the condition that the first runner was indeed faster. The first track can be no more than 3.3 feet too short to guarantee that the first runner was faster.

We originally assumed that the second runner ran on a perfectly measured track. You could solve the problem with the assumption that the first runner ran on the perfectly measured track, and find the error for the second runner. The answer will be slightly, but not significantly, different.

E1-9 First find the "logarithmic average" by

$$\log t_{\text{av}} = \frac{1}{2}\left(\log(5 \times 10^{17}) + \log(6 \times 10^{-15})\right),$$
$$= \frac{1}{2}\log\left(5 \times 10^{17} \times 6 \times 10^{-15}\right),$$
$$= \frac{1}{2}\log 3000 = \log\left(\sqrt{3000}\right).$$

Solve, and $t_{\text{av}} = 54.8$ seconds. This is about one minute. Note that we didn't need to specify *which* logarithm we were going to use, the answer would be the same with a base ten or a natural log! Not only that, you never needed to press the log button on your calculator to work out the answer.

E1-15 The volume of Antarctica is approximated by the area of the base time the height; the area of the base is the area of a semicircle. Then

$$V = Ah = \left(\frac{1}{2}\pi r^2\right)h,$$

where the factor of 1/2 comes from the *semi* in semicircle. The volume, keeping track of units, is

$$V = \frac{1}{2}(3.14)(2000 \times 1000 \text{ m})^2(3000 \text{ m}) = 1.88 \times 10^{16} \text{ m}^3$$
$$= 1.88 \times 10^{16} \text{ m}^3 \times \left(\frac{100 \text{ cm}}{1 \text{ m}}\right)^3 = 1.88 \times 10^{22} \text{ cm}^3.$$

Note that we needed to convert *each* factor of a meter in the answer, and not just one of them. So we needed to cube the expression in the parenthesis in order to get the correct answer.

E1-19 One light-year is the distance traveled by light in one year. Since distance is speed times time, one light-year $= (3 \times 10^8 \text{ m/s}) \times (1 \text{ year})$. Now we convert the units by multiplying through with appropriate factors of 1.

$$19,200\frac{\text{mi}}{\text{hr}}\left(\frac{\text{light-year}}{(3 \times 10^8 \text{ m/s}) \times (1 \text{ year})}\right)\left(\frac{1609 \text{ m}}{1 \text{ mi}}\right)\left(\frac{1 \text{ hr}}{3600 \text{ s}}\right)\left(\frac{100 \text{ year}}{1 \text{ century}}\right),$$

which is equal to 0.00286 light-year/century.

E1-23 1.0 kg of hydrogen atoms is equal to the number of atoms times the mass of one atom. Table 1-6 shows that one hydrogen atom has a mass of $1.00783u$, where $u = 1.661 \times 10^{-27}$ kg. Then the number of atoms is given by $(1 \text{ kg})/(1.00783 \times 1.661 \times 10^{-27} \text{ kg})$, or 5.974×10^{26} atoms.

6

E1-27 One sugar cube has a volume of 1.0 cm³, so a mole of sugar cubes would have a volume of $N_A \times 1.0$ cm³, where N_A is the Avogadro constant. Since the volume of a cube is equal to the length cubed, $V = l^3$, then $l = \sqrt[3]{N_A}$ cm $= 8.4 \times 10^7$ cm. With an edge length equal to 844 kilometers, the top of such a cube would be higher than the orbit of the International Space Station.

E1-29 The definition of the meter was wavelengths per meter; the question asks for meters per wavelength, so we want to take the reciprocal. The definition is accurate to 9 figures, so the reciprocal should be written as $1/1,650,763.73 = 6.05780211 \times 10^{-7}$ m. A *nano* is 10^{-9}. Dividing our answer by 10^{-9} will then give 605.780211 nm.

E1-31 The easiest approach is to first solve Darcy's Law for K, and then substitute the known SI units for the other quantities. Then

$$ K = \frac{VL}{AHt} \text{ has units of } \frac{(\text{m}^3)\,(\text{m})}{(\text{m}^2)\,(\text{m})\,(\text{s})} $$

which can be simplified to m/s.

P1-1 There are $24 \times 60 = 1440$ traditional minutes in a day, which is equivalent to the 1000 decimal minutes of metric clock. The conversion plan is then fairly straightforward

$$ 822.8 \text{ dec. min} \left(\frac{1440 \text{ trad. min}}{1000 \text{ dec. min}} \right) = 1184.8 \text{ trad. min.} $$

This is traditional minutes since midnight, the time in traditional hours can be found by dividing by 60 min/hr, the integer part of the quotient is the hours, while the remainder is the minutes. So the time is 19 hours, 45 minutes, which would be 7:45 pm.

P1-7 Break the problem down into parts. Some of the questions that need to be answered are (1) what is the surface area of a sand grain of radius 50 μm? (2) what is volume of this sand grain? It might be tempting to calculate the numerical value of each quantity, but it is more instructive to keep the expressions symbolic. Let the radius of the grain be given by r_g. Then the surface area of the grain is $A_g = 4\pi r_g^2$, and the volume is given by $V_g = (4/3)\pi r_g^3$.

If N grains of sand have a total surface area equal to that of a cube 1 m on a edge, then $NA_g = 6$ m², since the cube has six sides each with an area of 1 m². The total volume V_t of this number of grains of sand is NV_g. We can eliminate N from these two expressions and get

$$ V_t = NV_g = \frac{(6 \text{ m}^2)}{A_g} V_g = \frac{(6 \text{ m}^2) r_g}{3} $$

where the last step involved substituting the expressions for A_g and V_g. We haven't really started using numbers yet, and our expressions have simplified as a result. Now is, however, a good time to put in the numbers. Then $V_t = (2 \text{ m}^2)(50 \times 10^{-6} \text{ m}) = 1 \times 10^{-4} \text{ m}^3$.

All that is left is to find the mass. We were given that 2600 kg occupies a volume of 1 m^3, so the mass of the volume V_t is given by

$$1 \times 10^{-4} \text{ m}^3 \left(\frac{2600 \text{ kg}}{1 \text{ m}^3} \right) = 0.26 \text{ kg},$$

about the mass of two quarter-pound hamburgers.

Chapter 2

Motion in One Dimension

E2-1 There are two ways of solving this particular problem.

Method I Add the vectors as is shown in Fig. 2-4. If $\vec{a}$ has length $a = 4$ m and $\vec{b}$ has length $b = 3$ m then the sum is given by $\vec{s}$. The cosine law can be used to find the magnitude s of $\vec{s}$,

$$s^2 = a^2 + b^2 - 2ab\cos\theta,$$

where θ is the angle between sides a and b in the figure. Put in the given numbers for each instance and solve for the angle.

(a) $(7\text{ m})^2 = (4\text{ m})^2 + (3\text{ m})^2 - 2(4\text{ m})(3\text{ m})\cos\theta$, so $\cos\theta = -1.0$, and $\theta = 180°$. This means that $\vec{a}$ and $\vec{b}$ are pointing in the same direction.

(b) $(1\text{ m})^2 = (4\text{ m})^2 + (3\text{ m})^2 - 2(4\text{ m})(3\text{ m})\cos\theta$, so $\cos\theta = 1.0$, and $\theta = 0°$. This means that $\vec{a}$ and $\vec{b}$ are pointing in the opposite direction.

(c) $(5\text{ m})^2 = (4\text{ m})^2 + (3\text{ m})^2 - 2(4\text{ m})(3\text{ m})\cos\theta$, so $\cos\theta = 0$, and $\theta = 90°$. This means that $\vec{a}$ and $\vec{b}$ are pointing at right angles to each other.

Method II You might have been able to just look at the numbers and "guess" the answers. This is a perfectly acceptable method, as long as you recognize the limitations and still verify your initial assumptions with concrete calculations. The verification is simple enough: for the vectors pointing in the same direction, add the magnitudes $(4 + 3 = 7)$; for vectors pointing in opposite directions, subtract the magnitudes $(|4-3| = 1)$; for vectors which meet at right angles, apply the Pythagoras relation $(4^2 + 3^2 = 5^2)$. If none of these approaches works then you need to solve the problem with the first method.

E2-5 We'll solve this problem in two steps. First, we find the components of the displacement vector along the north-south and east-west street system. Then we'll show that the sum of these components is actually the shortest distance.

The components are given by the trigonometry relations $O = H \sin \theta = (3.42$ km$) \sin 35.0° = 1.96$ km and $A = H \cos \theta = (3.42$ km$) \cos 35.0° = 2.80$ km. The stated angle is measured from the east-west axis, counter clockwise from east. So O is measured against the north-south axis, with north being positive; A is measured against east-west with east being positive.

Now we find the shortest distance by considering that the person can only walk east-west or north-south. Since her individual steps are displacement vectors which are only north-south or east-west, she must eventually take enough north-south steps to equal 1.96 km, and enough east-west steps to equal 2.80 km. Any individual step can only be along one or the other direction, so the minimum total will be 4.76 km.

E2-7 (a) In unit vector notation we need only add the components; $\vec{a} + \vec{b} = (5\hat{i} + 3\hat{j}) + (-3\hat{i} + 2\hat{j}) = (5 - 3)\hat{i} + (3 + 2)\hat{j} = 2\hat{i} + 5\hat{j}$.

(b) The magnitude of the sum is found from Pythagoras' theorem, because these components are at right angles. If we define $\vec{c} = \vec{a} + \vec{b}$ and write the magnitude of $\vec{c}$ as c, then $c = \sqrt{c_x^2 + c_y^2} = \sqrt{2^2 + 5^2} = 5.39$. The 2 and the 5 under the square root sign were the components found in part (a). We use those same components to find the direction, according to $\tan \theta = c_y/c_x$ which gives an angle of 68.2°, measured counterclockwise from the positive x-axis.

E2-13 Displacement vectors are given by the final position minus the initial position. Eventually we need to represent the position in each of the three positions where the minute hand is. Our axes will be chosen so that $\hat{i}$ points toward 3 O'clock and $\hat{j}$ points toward 12 O'clock.

(a) The two relevant positions are $\vec{r}_i = (11.3$ cm$)\hat{i}$ and $\vec{r}_f = (11.3$ cm$)\hat{j}$. The displacement in the interval is $\Delta \vec{r} = \vec{r}_f - \vec{r}_i$; we can evaluate this expression by looking at the components, then

$$
\begin{aligned}
\Delta \vec{r} &= \vec{r}_f - \vec{r}_i \\
&= (11.3 \text{ cm})\hat{j} - (11.3 \text{ cm})\hat{i} \\
&= -(11.3 \text{ cm})\hat{i} + (11.3 \text{ cm})\hat{j},
\end{aligned}
$$

where in the last line we wrote the answer in the more traditional ordering of unit vectors. *But line 2 should be a perfectly adequate answer.*

(b) The two relevant positions are now $\vec{r}_i = (11.3 \text{ cm})\hat{j}$ and $\vec{r}_f = (-11.3 \text{ cm})\hat{j}$. Note that the 6 O'clock position for the minute hand has a negative sign. As before we can evaluate this expression by looking at the components, so

$$
\begin{aligned}
\Delta\vec{r} &= \vec{r}_f - \vec{r}_i \\
&= (11.3 \text{ cm})\hat{j} - (-11.3 \text{ cm})\hat{j} \\
&= (22.6 \text{ cm})\hat{j}.
\end{aligned}
$$

There's a double negative in the second line that is often missed by students of introductory physics classes.

(c) The two relevant positions are now $\vec{r}_i = (-11.3 \text{ cm})\hat{j}$ and $\vec{r}_f - (-11.3 \text{ cm})\hat{j}$. As before we can evaluate this expression by looking at the components, so

$$
\begin{aligned}
\Delta\vec{r} &= \vec{r}_f - \vec{r}_i \\
&= (-11.3 \text{ cm})\hat{j} - (-11.3 \text{ cm})\hat{j} \\
&= (0 \text{ cm})\hat{j}.
\end{aligned}
$$

The displacement is zero, since we have started and stopped in the same position!

E2-17 As always, remember to take the time derivatives *before* you substitute in for the time!

(a) Evaluate $\vec{r}$ when $t = 2$ s.

$$
\begin{aligned}
\vec{r} &= [(2 \text{ m/s}^3)t^3 - (5 \text{ m/s})t]\hat{i} + [(6 \text{ m}) - (7 \text{ m/s}^4)t^4]\hat{j} \\
&= [(2 \text{ m/s}^3)(2 \text{ s})^3 - (5 \text{ m/s})(2 \text{ s})]\hat{i} + [(6 \text{ m}) - (7 \text{ m/s}^4)(2 \text{ s})^4]\hat{j} \\
&= [(16 \text{ m}) - (10 \text{ m})]\hat{i} + [(6 \text{ m}) - (112 \text{ m})]\hat{j} \\
&= [(6 \text{ m})]\hat{i} + [-(106 \text{ m})]\hat{j}.
\end{aligned}
$$

(b) Take the derivative of $\vec{r}$ with respect to time, using the full form of $\vec{r}$ from the first line of the equations above.

$$
\begin{aligned}
\vec{v} = \frac{d\vec{r}}{dt} &= [(2 \text{ m/s}^3)3t^2 - (5 \text{ m/s})]\hat{i} + [-(7 \text{ m/s}^4)4t^3]\hat{j} \\
&= [(6 \text{ m/s}^3)t^2 - (5 \text{ m/s})]\hat{i} + [-(28 \text{ m/s}^4)t^3]\hat{j}.
\end{aligned}
$$

Into this last expression we now evaluate $\vec{v}(t = 2$ s$)$ and get

$$
\begin{aligned}
\vec{v} &= [(6 \text{ m/s}^3)(2 \text{ s})^2 - (5 \text{ m/s})]\hat{i} + [-(28 \text{ m/s}^4)(2 \text{ s})^3]\hat{j} \\
&= [(24 \text{ m/s}) - (5 \text{ m/s})]\hat{i} + [-(224 \text{ m/s})]\hat{j} \\
&= [(19 \text{ m/s})]\hat{i} + [-(224 \text{ m/s})]\hat{j},
\end{aligned}
$$

for the velocity $\vec{v}$ when $t = 2$ s.

(c) We'll take the time derivative of $\vec{v}$ to find $\vec{a}$, making sure that we use the expression for $\vec{v}$ before we substituted for t.

$$\vec{a} = \frac{d\vec{v}}{dt} = [(6 \text{ m/s}^3)2t]\hat{\mathbf{i}} + [-(28 \text{ m/s}^4)3t^2]\hat{\mathbf{j}}$$
$$= [(12 \text{ m/s}^3)t]\hat{\mathbf{i}} + [-(84 \text{ m/s}^4)t^2]\hat{\mathbf{j}}.$$

Into this last expression we now evaluate $\vec{a}(t = 2 \text{ s})$ and get

$$\vec{a} = [(12 \text{ m/s}^3)(2 \text{ s})]\hat{\mathbf{i}} + [-(84 \text{ m/s}^4)(2\ 2)^2]\hat{\mathbf{j}}$$
$$= [(24 \text{ m/s}^2)]\hat{\mathbf{i}} + [-(336 \text{ m/s}^2)]\hat{\mathbf{j}}.$$

However tempting it might be, it makes no physical sense to compare the acceleration with either the velocity or position at $t = 2$ s, or any other time, except to maybe note that one or more quantities might be zero.

E2-21 For the record, *Namulevu* and *Vanuavinaka* are perfectly good words in some language. See if your instructor will give you extra credit for translating the meaning. Of course, knowing the language would help, but *au na sega ni tukuna vei kemuni!*

Let the actual flight time, as measured by the passengers, be T. There is some time difference between the two cities, call it ΔT = Namulevu time - Los Angeles time. The ΔT will be positive if Namulevu is east of Los Angeles. The time in Los Angeles can then be found from the time in Namulevu by subtracting ΔT.

The actual time of flight from Los Angeles to Namulevu is then the difference between when the plane lands (LA times) and when the plane takes off (LA time):

$$T = (18{:}50 - \Delta T) - (12{:}50)$$
$$= 6{:}00 - \Delta T,$$

where we have written times in 24 hour format to avoid the AM/PM issue. The return flight time can be found from

$$T = (18{:}50) - (1{:}50 - \Delta T)$$
$$= 17{:}00 + \Delta T,$$

where we have again changed to LA time for the purpose of the calculation.

Now we just need to solve the two equations and two unknowns. The way we have written it makes it easier to solve for ΔT first by setting the two expressions for T equal to each other:

$$17{:}00 + \Delta T = 6{:}00 - \Delta T$$
$$2\Delta T = 6{:}00 - 17{:}00$$
$$\Delta T = -5{:}30,$$

and yes, there are a number of places in the world with time zones that differ by half an hour. Since this is a negative number, Namulevu is located *west* of Los Angeles.

(a) Choose either the outbound or the inbound flight to find T. If we choose the outbound flight, $T = 6:00 - \Delta T = 11:30$, or eleven and a half hours. We've already found the time difference, so we move straight to (c).

(c) The distance traveled by the plane is given by $d = vt = (520 \text{ mi/hr})(11.5 \text{ hr}) = 5980$ mi. We'll draw a circle around Los Angeles with a radius of 5980 mi, and then we look for where it intersects with longitudes that would belong to a time zone ΔT away from Los Angeles. Since the Earth rotates once every 24 hours and there are 360 longitude degrees, then each hour corresponds to 15 longitude degrees, and then Namulevu must be located approximately $15° \times 5.5 = 83°$ west of Los Angeles, or at about longitude 160 east. The location on the globe is then latitude 5°, in the vicinity of Vanuatu.

When this exercise was originally typeset the times for the outbound and the inbound flights were inadvertently switched. I suppose that we could blame this on the airlines; nonetheless, when the answers were prepared for the back of the book the reversed numbers put Namulevu *east* of Los Angeles. That would put it in either the North Atlantic or Brazil.

E2-25 Speed is distance traveled divided by time taken; this is equivalent to the inverse of the slope of the line in Fig. 2-32. The line appears to pass through the origin and through the point $(1600 \text{ km}, 80 \times 10^6 \text{ y})$, so the speed is $v = 1600 \text{ km}/80 \times 10^6$ y$= 2 \times 10^{-5}$ km/y. The answer requests units of centimeters per year, so we convert units by

$$v = 2 \times 10^{-5}\text{km/y} \left(\frac{1000 \text{ m}}{1 \text{ km}}\right) \left(\frac{100 \text{ cm}}{1 \text{ m}}\right) = 2 \text{ cm/y}$$

E2-29 Don't fall into the trap of assuming that the average speed is the average of the speeds. We instead need to go back to the definition: speed is distance traveled divided by time taken. It might look as if there isn't enough information to solve this problem, since we weren't given the distance or the time. We can solve the problem, however, with some algebra. Let $v_1 = 40$ km/hr be the speed up the hill, t_1 be the time taken, and d_1 be the distance traveled in that time. We similarly define $v_2 = 60$ km/hr for the down hill trip, as well as t_2 and d_2. Note that $d_2 = d_1$, because the car drove down the same hill it drove up.

Now for the algebra. $v_1 = d_1/t_1$ or $t_1 = d_1/v_1$; $v_2 = d_2/t_2$ or $t_2 = d_2/v_2$. The *average* speed will be $v_{av} = d/t$, where d total distance and t is the total time. But the total distance is $d_1 + d_2 = 2d_1$ because the up distance is same as the down distance. The total time t is just the sum of t_1 and t_2, so

$$
\begin{aligned}
v_{\text{av}} &= \frac{d}{t} \\
&= \frac{2d_1}{t_1 + t_2}
\end{aligned}
$$

$$= \frac{2d_1}{d_1/v_1 + d_2/v_2}$$

$$= \frac{2}{1/v_1 + 1/v_2},$$

where in the last line we used $d_2 = d_1$ and then factored out d_1. So, as expected, we never needed to know the height of the hill, or the time. The last expression looks a little nasty; but we can take the reciprocal of both sides to get a simpler looking expression

$$\frac{2}{v_{av}} = \frac{1}{v_1} + \frac{1}{v_2}.$$

In either case, the average speed is 48 km/hr.

E2-33 The initial velocity is $\vec{v}_i = (18 \text{ m/s})\hat{i}$, the final velocity is $\vec{v}_f = (-30 \text{ m/s})\hat{i}$. Negative signs *can't* be ignored in this problem; acceleration and velocity are both vectors and require some indication of direction. The average acceleration is then

$$\vec{a}_{av} = \frac{\Delta \vec{v}}{\Delta t} = \frac{\vec{v}_f - \vec{v}_i}{\Delta t} = \frac{(-30 \text{ m/s})\hat{i} - (18 \text{ m/s})\hat{i}}{2.4 \text{ s}},$$

which gives $\vec{a}_{av} = (-20.0 \text{ m/s}^2)\hat{i}$.

E2-37 When the displacement-time graph is at a maximum or minimum the velocity should be zero, meaning the velocity-time graph will pass through the time axis. There are no straight line segments in the distance-time graph, so there are no constant velocity segments for the velocity-time graph.

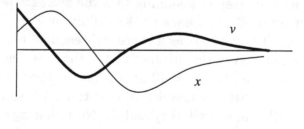

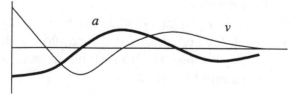

E2-41 This one dimensional, constant acceleration problem states the acceleration, $a_x = 9.8 \text{ m/s}^2$, the initial velocity, $v_{0x} = 0$, and the final velocity $v_x = 0.1c = 3.0 \times 10^7$ m/s.

(a)   We are then asked for the time it will take for the space ship to acquire the final velocity. Applying Eq. 2-26,

$$
\begin{aligned}
v_x &= v_{0x} + a_x t, \\
(3.0 \times 10^7 \text{ m/s}) &= (0) + (9.8 \text{ m/s}^2)t, \\
3.1 \times 10^6 \text{ s} &= t.
\end{aligned}
$$

This is about one month. Although accelerations of this magnitude are well within the capability of modern technology, we are unable to sustain such accelerations for even several hours, much less a month.

(b)   If, however, the acceleration could be sustained, how far would the rocket ship travel? We apply Eq. 2-28 using an initial position of $x_0 = 0$,

$$
\begin{aligned}
x &= x_0 + v_{0x} + \frac{1}{2}a_x t^2, \\
x &= (0) + (0) + \frac{1}{2}(9.8 \text{ m/s}^2)(3.1 \times 10^6 \text{ s})^2, \\
x &= 4.7 \times 10^{13} \text{ m}.
\end{aligned}
$$

This distance is 8000 times farther than Pluto, and considerably farther than any spacecraft has ever traveled; it is also about the same distance that light travels in a day and a half, and only a small fraction of the distance to the nearest star to the sun.

E2-45   Given in this problem are the initial velocity, $v_{0x} = 1020$ km/hr; the final velocity, $v_x = 0$, and the time taken to stop the sled, $t = 1.4$ s. It will be easier to solve the problem if we change the units for the initial velocity,

$$
v_{0x} = 1020\frac{\text{km}}{\text{hr}} \left(\frac{1000 \text{ m}}{\text{km}}\right) \left(\frac{\text{hr}}{3600 \text{ s}}\right) = 283\frac{\text{m}}{\text{s}},
$$

and then applying Eq. 2-26,

$$
\begin{aligned}
v_x &= v_{0x} + a_x t, \\
(0) &= (283 \text{ m/s}) + a_x(1.4 \text{ s}), \\
-202 \text{ m/s}^2 &= a_x.
\end{aligned}
$$

The negative sign reflects the fact that the rocket sled is slowing down. The problem asks for this in terms of g, so

$$
-202 \text{ m/s}^2 \left(\frac{g}{9.8 \text{ m/s}^2}\right) = 21g.
$$

E2-49 The problem will be somewhat easier if the units are consistent, so we'll write the maximum speed as

$$1000\frac{\text{ft}}{\text{min}}\left(\frac{\text{min}}{60\text{ s}}\right) = 16.7\frac{\text{ft}}{\text{s}}.$$

(a) The distance traveled during acceleration can't be found directly from the information given (although some textbooks do introduce a *third* kinematic relationship in addition to Eq. 2-26 and Eq. 2-28 that would make this possible.) We can, however, easily find the time required for the acceleration from Eq. 2-26,

$$\begin{aligned}
v_x &= v_{0x} + a_x t,\\
(16.7\text{ ft/s}) &= (0) + (4.00\text{ ft/s}^2)t,\\
4.18\text{ s} &= t.
\end{aligned}$$

And from this and Eq 2-28 we can find the distance

$$\begin{aligned}
x &= x_0 + v_{0x} + \frac{1}{2}a_x t^2,\\
x &= (0) + (0) + \frac{1}{2}(4.00\text{ ft/s}^2)(4.18\text{ s})^2,\\
x &= 34.9\text{ ft}.
\end{aligned}$$

This is considerably less than 624 ft; this means that through most of the journey the elevator is traveling at the maximum speed.

(b) The motion of the elevator is divided into three parts: acceleration from rest, constant speed motion, and deceleration to a stop. The total distance is given at 624 ft and in part (a) we found the distance covered during acceleration was 34.9 ft. By symmetry, the distance traveled during deceleration should also be 34.9 ft, but it would be good practice to verify this last assumption with calculations. The distance traveled at constant speed is then $(624 - 34.9 - 34.9)$ ft $= 554$ ft. The time required for the constant speed portion of the trip is found from Eq. 2-22, rewritten as

$$\Delta t = \frac{\Delta x}{v} = \frac{554\text{ ft}}{16.7\text{ ft/s}} = 33.2\text{ s}.$$

The total time for the trip is the sum of times for the three parts: accelerating (4.18 s), constant speed (33.2 s), and decelerating (4.18 s). The total is 41.6 seconds.

E2-53 The initial velocity of the "dropped" wrench would be zero. The acceleration would be 9.8 m/s^2. Although we can orient our coordinate system any way we want, I choose vertical to be along the y axis with up as positive, which is the convention of Eq. 2-29 and Eq. 2-30. Note that Eq. 2-29 is equivalent to Eq. 2-26, we could have used either. The same is true for Eq. 2-30 and Eq. 2-28.

It turns out that it is much easier to solve part (b) before solving part (a). So that's what we'll do.

(b) We solve Eq. 2-29 for the time of the fall.

$$v_y = v_{0y} - gt,$$
$$(-24.0 \text{ m/s}) = (0) - (9.8 \text{ m/s}^2)t,$$
$$2.45 \text{ s} = t.$$

We manually insert the minus sign for the final velocity because the object is moving *down*. We don't insert an extra minus sign in front of the 9.8 because it is explicit in Eq. 2-29; we would have, however, needed to insert it if we had used Eq. 2-26.

(a) Now we can easily use Eq. 2-30 to find the height from which the wrench fell.

$$y = y_0 + v_{0y}t - \frac{1}{2}gt^2,$$
$$(0) = y_0 + (0)(2.45 \text{ s}) - \frac{1}{2}(9.8 \text{ m/s}^2)(2.45 \text{ s})^2,$$
$$0 = y_0 - 29.4 \text{ m}$$

We have set $y = 0$ to correspond to the final position of the wrench: on the ground. This results in an initial position of $y_0 = 29.4$ m; it is positive because the wrench was dropped from a point *above* where it landed.

We could instead have chosen the initial point to correspond to $y_0 = 0$, where we make our measurements from the point where the wrench was dropped. If we do this, we find $y = -29.4$ m. The negative sign indicates that the wrench landed *below* the point from which it was dropped.

E2-57 Don't assume that 36.8 m corresponds to the highest point.

(a) Solve Eq. 2-30 for the initial velocity. Let the distances be measured from the ground so that $y_0 = 0$.

$$y = y_0 + v_{0y}t - \frac{1}{2}gt^2,$$
$$(36.8 \text{ m}) = (0) + v_{0y}(2.25 \text{ s}) - \frac{1}{2}(9.8 \text{ m/s}^2)(2.25 \text{ s})^2,$$
$$36.8 \text{ m} = v_{0y}(2.25 \text{ s}) - 24.8 \text{ m},$$
$$27.4 \text{ m/s} = v_{0y}.$$

(b)   Solve Eq. 2-29 for the velocity, using the result from part (a).

$$\begin{aligned} v_y &= v_{0y} - gt, \\ v_y &= (27.4 \text{ m/s}) - (9.8 \text{ m/s}^2)(2.25 \text{ s}), \\ v_y &= 5.4 \text{ m/s}. \end{aligned}$$

(c)   We need to solve Eq. 2-30 to find the height to which the ball rises, but we don't know how long it takes to get there. So we first solve Eq. 2-29, because we do know the velocity at the highest point ($v_y = 0$).

$$\begin{aligned} v_y &= v_{0y} - gt, \\ (0) &= (27.4 \text{ m/s}) - (9.8 \text{ m/s}^2)t, \\ 2.8 \text{ s} &= t. \end{aligned}$$

And then we find the height to which the object rises,

$$\begin{aligned} y &= y_0 + v_{0y}t - \frac{1}{2}gt^2, \\ y &= (0) + (27.4 \text{ m/s})(2.8 \text{ s}) - \frac{1}{2}(9.8 \text{ m/s}^2)(2.8 \text{ s})^2, \\ y &= 38.3\text{m}. \end{aligned}$$

This is the height as measured from the ground; so the ball rises $38.3 - 36.8 = 1.5$ m above the point specified in the problem.

E2-61   The total time the pot is visible is 0.54 s; the pot is visible for 0.27 s on the way down. We'll define the initial position as the highest point and make our measurements from there. Then $y_0 = 0$ and $v_{0y} = 0$. Define t_1 to be the time at which the *falling* pot passes the top of the window y_1, then $t_2 = t_1 + 0.27$ s is the time the pot passes the bottom of the window $y_2 = y_1 - 1.1$ m. We have two equations we can write, both based on Eq. 2-30,

$$\begin{aligned} y_1 &= y_0 + v_{0y}t_1 - \frac{1}{2}gt_1^2, \\ y_1 &= (0) + (0)t_1 - \frac{1}{2}gt_1^2, \end{aligned}$$

and

$$\begin{aligned} y_2 &= y_0 + v_{0y}t_2 - \frac{1}{2}gt_2^2, \\ y_1 - 1.1 \text{ m} &= (0) + (0)t_2 - \frac{1}{2}g(t_1 + 0.27 \text{ s})^2, \end{aligned}$$

Isolate y_1 in this last equation and then set the two expressions equal to each other so that we can solve for t_1,

$$-\frac{1}{2}gt_1^2 = 1.1 \text{ m} - \frac{1}{2}g(t_1 + 0.27 \text{ s})^2,$$

$$-\frac{1}{2}gt_1^2 = 1.1 \text{ m} - \frac{1}{2}g(t_1^2 + [0.54 \text{ s}]t_1 + 0.073 \text{ s}^2),$$

$$0 = 1.1 \text{ m} - \frac{1}{2}g([0.54 \text{ s}]t_1 + 0.073 \text{ s}^2).$$

This last line can be directly solved to yield $t_1 = 0.28$ s as the time when the falling pot passes the top of the window. Use this value in the first equation above and we can find $y_1 = -\frac{1}{2}(9.8 \text{ m/s}^2)(0.28 \text{ s})^2 = -0.38$ m. The negative sign is because the top of the window is beneath the highest point, so the pot must have risen to 0.38 m above the top of the window.

P2-3 We align the coordinate system so that the origin corresponds to the starting position of the fly and that all positions inside the room are given by positive coordinates.

(a) The displacement vector can just be written,

$$\Delta \vec{\mathbf{r}} = (10 \text{ ft})\hat{\mathbf{i}} + (12 \text{ ft})\hat{\mathbf{j}} + (14 \text{ ft})\hat{\mathbf{k}}.$$

(b) The magnitude of the displacement vector is the square root of the sum of the squares of the components, a generalization of Pythagoras' Theorem. $|\Delta \vec{\mathbf{r}}| = \sqrt{10^2 + 12^2 + 14^2}$ ft$= 21$ ft.

(c) The straight line distance between two points is the shortest possible distance, so the length of the path taken by the fly must be greater than or equal to 21 ft.

(d) If the fly walks it will need to cross two faces. The shortest path will be the diagonal across these two faces. If the lengths of sides of the room are l_1, l_2, and l_3, then the diagonal length across two faces will be given by

$$\sqrt{(l_1 + l_2)^2 + l_3^2},$$

where we want to choose the l_i from the set of 10 ft, 12 ft, and 14 ft that will minimize the length. Trial and error works (there are only three possibilities), or we can reason out that we want the squares to be minimal, in particular we want the largest square to be the smallest possible. This will happen if $l_1 = 10$ ft, $l_2 = 12$ ft, and $l_3 = 14$. Then the minimal distance the fly would *walk* is 26 ft.

P2-7 (a) Don't try to calculate this by brute force, unless you are a glutton for punishment. Assume the bird has no size, the trains have some separation, and the bird is just leaving one of the trains. The bird will be able to fly from one train to the other *before* the two trains collide, regardless of how close together the trains are. After doing so, the bird is now on the other train, the trains are still separated, so once again the bird can fly between the trains before they collide. This process can be repeated every time the bird touches one of the trains, so the bird will make an infinite number of trips between the trains. But it makes this infinite number of trips in a finite time, because eventually the trains do collide.

(b) The trains collide in the middle; using a simple application of distance equals speed times time we find that the trains collide after $(51 \text{ km})/(34 \text{ km/hr}) = 1.5 \text{ hr}$. The bird was flying with constant speed this entire time, so the distance flown by the bird is $(58 \text{ km/hr})(1.5 \text{ hr}) = 87 \text{ km}$. This apparent paradox of an infinite number of trips summing to a finite length was investigated by Zeno quite a number of years ago.

P2-11 (a) *easy* The average velocity is displacement divided by change in time,

$$v_{av} = \frac{(2.0 \text{ m/s}^3)(2.0 \text{ s})^3 - (2.0 \text{ m/s}^3)(1.0 \text{ s})^3}{(2.0 \text{ s}) - (1.0 \text{ s})} = \frac{14.0 \text{ m}}{1.0 \text{ s}} = 14.0 \text{ m/s}.$$

The average acceleration is the change in velocity. So we need an expression for the velocity, which is the time derivative of the position,

for a_{av} take derivative of

$$v = \frac{dx}{dt} = \frac{d}{dt}(2.0 \text{ m/s}^3)t^3 = (6.0 \text{ m/s}^3)t^2.$$

From this we find average acceleration

a_{av} = derivative of V_{avg}

$$\frac{}{\Delta t} \qquad a_{av} = \frac{(6.0 \text{ m/s}^3)(2.0 \text{ s})^2 - (6.0 \text{ m/s}^3)(1.0 \text{ s})^2}{(2.0 \text{ s}) - (1.0 \text{ s})} = \frac{18.0 \text{ m/s}}{1.0 \text{ s}} = 18.0 \text{ m/s}^2.$$

(b) The instantaneous velocities can be found directly from $v = (6.0 \text{ m/s}^2)t^2$, so $v(2.0 \text{ s}) = 24.0 \text{ m/s}$ and $v(1.0 \text{ s}) = 6.0 \text{ m/s}$. We can get an expression for the instantaneous acceleration by taking the time derivative of the velocity

$$a = \frac{dv}{dt} = \frac{d}{dt}(6.0 \text{ m/s}^3)t^2 = (12.0 \text{ m/s}^3)t.$$

Then the instantaneous accelerations are $a(2.0 \text{ s}) = 24.0 \text{ m/s}^2$ and $a(1.0 \text{ s}) = 12.0 \text{ m/s}^2$

(c) Since the motion is monotonic we expect the average quantities to be somewhere between the instantaneous values at the endpoints of the time interval. Indeed, that is the case.

P2-17 The runner covered a distance d_1 in a time interval t_1 during the acceleration phase and a distance d_2 in a time interval t_2 during the constant speed phase. Since the runner started from rest we know that the constant speed is given by $v = at_1$, where a is the runner's acceleration.

The distance covered during the acceleration phase is given by Eq. 2-28 with $v_{0x} = 0$,

$$d_1 = \frac{1}{2}at_1^2.$$

The distance covered during the constant speed phase can also be found from Eq. 2-28 except now with $a = 0$,

$$d_2 = vt_2 = at_1t_2.$$

We want to use these two expressions, along with $d_1 + d_2 = 100$ m and $t_2 = (12.2\ \text{s}) - t_1$, to get

$$
\begin{aligned}
100\ \text{m} &= d_1 + d_2 = \frac{1}{2}at_1^2 + at_1(12.2\ \text{s} - t_1), \\
&= -\frac{1}{2}at_1^2 + a(12.2\ \text{s})t_1, \\
&= -(1.40\ \text{m/s}^2)t_1^2 + (34.2\ \text{m/s})t_1.
\end{aligned}
$$

This last expression is quadratic in t_1, and is solved to give $t_1 = 3.40\,\text{s}$ or $t_1 = 21.0\,\text{s}$. Since the race only lasted 12.2 s we can ignore the second answer.

(b)   The distance traveled during the acceleration phase is then

$$d_1 = \frac{1}{2}at_1^2 = (1.40\ \text{m/s}^2)(3.40\,\text{s})^2 = 16.2\,\text{m}.$$

P2-21   The rocket travels a distance $d_1 = \frac{1}{2}at_1^2 = \frac{1}{2}(20\ \text{m/s}^2)(60\ \text{s})^2 = 36{,}000$ m during the acceleration phase; the rocket velocity at the end of the acceleration phase is $v = at = (20\ \text{m/s}^2)(60\ \text{s}) = 1200$ m/s. The second half of the trajectory can be found from Eqs. 2-29 and 2-30, with $y_0 = 36{,}000$ m and $v_{0y} = 1200$ m/s.

(a)   The highest point of the trajectory occurs when $v_y = 0$, so we solve Eq. 2-29 for time.

$$
\begin{aligned}
v_y &= v_{0y} - gt, \\
(0) &= (1200\ \text{m/s}) - (9.8\ \text{m/s}^2)t, \\
122\ \text{s} &= t.
\end{aligned}
$$

This time is used in Eq. 2-30 to find the height to which the rocket rises,

$$
\begin{aligned}
y &= y_0 + v_{0y}t - \frac{1}{2}gt^2, \\
&= (36000\ \text{m}) + (1200\ \text{m/s})(122\text{s}) - \frac{1}{2}(9.8\ \text{m/s}^2)(122\ \text{s})^2 = 110000\ \text{m}.
\end{aligned}
$$

(b) The easiest way to find the total time of flight is to solve Eq. 2-30 for the time when the rocket has returned to the ground. Then

$$y = y_0 + v_{0y}t - \frac{1}{2}gt^2,$$

$$(0) = (36000 \text{ m}) + (1200 \text{ m/s})t - \frac{1}{2}(9.8 \text{ m/s}^2)t^2.$$

This quadratic expression has two solutions for t; one is negative so we don't need to worry about it, the other is $t = 270$ s. This is the free-fall part of the problem, to find the total time we need to add on the 60 seconds of accelerated motion. The total time is then 330 seconds.

P2-25 This is a problem that is best solved backwards, then forwards. We want to find the deceleration of the woman. We know the distance through which she decelerated (18 in) and her final velocity (0), but not the time taken nor the initial velocity at the start of the deceleration phase. So neither Eq. 2-26 nor Eq. 2-28 is of much use. However, since her deceleration is assumed uniform, we could apply Eq. 2-27 to find her average velocity (if we knew her initial velocity), and then use Eq. 2-22 to find the time elapsed during deceleration since we know the distance, and then use Eq. 2-26 to find the acceleration. So all that we need to find is the initial velocity at the start of the deceleration phase.

But this initial velocity is the same as the final velocity at the end of the freely falling part of the motion. We would need to use Eq. 2-29 to find this final velocity, but we don't know the time to fall. However, we could get that time from Eq. 2-30, since we know the initial velocity at the start of the fall (0) and the distance through which she fell (144 ft). So now we are prepared to solve the problem. Since inches and feet are used throughout, I'm going to use $g = 32$ ft/s^2 for the acceleration of free-fall.

Now we reverse our approach and work forwards through the problem and find the time she fell from Eq. 2-30. I've written this equation a number of times in the past few pages, so I'll just substitute the variables in directly.

$$(0 \text{ ft}) = (144 \text{ ft}) + (0)t - \frac{1}{2}(32 \text{ ft/s}^2)t^2,$$

which is a simple quadratic with solutions $t = \pm 3.0$ s. Only the positive solution is of interest, since we assume she was falling forward in time. Use this time in Eq. 2-29 to find her speed when she hit the ventilator box,

$$v_y = (0) - (32 \text{ ft/s}^2)(3.0 \text{ s}) = -96 \text{ ft/s}.$$

This becomes the initial velocity for the deceleration motion, so her average speed during deceleration is given by Eq. 2-27,

$$v_{\text{av},y} = \frac{1}{2}(v_y + v_{0y}) = \frac{1}{2}((0) + (-96 \text{ ft/s})) = -48 \text{ ft/s}.$$

This average speed, used with the distance of 18 in (1.5 ft), can be used to find the time of deceleration

$$v_{\mathrm{av}, y} = \Delta y / \Delta t,$$

and putting numbers into the expression gives $\Delta t = 0.031$ s. We actually used $\Delta y = -1.5$ ft, where the negative sign indicated that she was still moving downward. Finally, we use this in Eq. 2-26 to find the acceleration,

$$(0) = (-96 \text{ ft/s}) + a(0.031 \text{ s}),$$

which gives $a = +3100 \text{ ft/s}^2$. The *important* positive sign is because she is accelerating upward when she stops. In terms of g this is $a = 97g$, which can be found by multiplying through by $1 = g/(32 \text{ ft/s}^2)$.

P2-31 Assume each hand can toss n objects per second. Let τ be the amount of time that any one object is in the air. Then $2n\tau$ is the number of objects that are in the air at any time, where the "2" comes from the fact that (most?) jugglers have two hands. We'll estimate n, but τ can be found from Eq. 2-30 for an object which falls a distance h from rest:

$$0 = h + (0)t - \frac{1}{2}gt^2,$$

solving, $t = \sqrt{2h/g}$. But τ is twice this, because the object had to go up before it could come down. So the number of objects that can be juggled is

$$4n\sqrt{2h/g}$$

We need to estimate n. Do this by standing in front of a bathroom mirror and flapping a hand up and down as fast as you can. I can manage to simulate 10 tosses in 5 seconds while frantic, which means $n = 2$ tosses/second. So the maximum number of objects I could juggle to a height h would be

$$3.6\sqrt{h/\text{meters}}.$$

I doubt I could toss objects higher than 4 meters, so my absolute maximum would be about 7 objects. In reality I almost juggled one object once.

Chapter 3

Force and Newton's Laws

E3-3 We want to find the force on the electron; we do this by first finding the acceleration; that's actually the most involved part of the problem. We are given the distance through which the electron accelerates and the final speed. Assuming constant acceleration we can find the average speed during the interval from Eq. 2-27

$$v_{\text{av},x} = \frac{1}{2}(v_x + v_{0x}) = \frac{1}{2}\left((5.8 \times 10^6 \text{ m/s}) + (0)\right) = 2.9 \times 10^6 \text{ m/s}.$$

From this we can find the time spent accelerating from Eq. 2-22, since $\Delta x = v_{\text{av},x}\Delta t$. Putting in the numbers, we find that the time is $\Delta t = 5.17 \times 10^{-9}$s. This can be used in component form of Eq. 2-14 to find the acceleration,

$$a_x = \frac{\Delta v_x}{\Delta t} = \frac{(5.8 \times 10^6 \text{ m/s}) - (0)}{(5.17 \times 10^{-9}\text{s})} = 1.1 \times 10^{15} \text{m/s}^2.$$

We are not done yet. The net force on the electron is from Eq. 3-5,

$$\sum F_x = ma_x = (9.11 \times 10^{-31}\text{kg})(1.1 \times 10^{15}\text{m/s}^2) = 1.0 \times 10^{-15} \text{ N}.$$

E3-5 The *net* force on the sled is 92 N−90 N= 2 N; we subtract because the forces are in opposite directions. This net force is used with Newton's Second law to find the acceleration; $\sum F_x = ma_x$, so

$$a_x = \frac{\sum F_x}{m} = \frac{(2\,\text{N})}{(25\,\text{kg})} = 8.0 \times 10^{-2}\text{m/s}^2.$$

E3-9 There are too many unknowns to find a numerical value for the force or for either mass. So don't try. Write the expression for the motions of the first object as $\sum F_x = m_1 a_{1x}$ and that of the second object as $\sum F_x = m_2 a_{2x}$. In both cases there is only one force, F, on the object, so $\sum F_x = F$. We will solve these for the mass as $m_1 = F/a_1$ and $m_2 = F/a_2$. Since $a_1 > a_2$ we can conclude that $m_2 > m_1$

(a) The acceleration of and object with mass $m_2 - m_1$ under the influence of a single force of magnitude F would be

$$a = \frac{F}{m_2 - m_1} = \frac{F}{F/a_2 - F/a_1} = \frac{1}{1/(3.30\,\mathrm{m/s^2}) - 1/(12.0\,\mathrm{m/s^2})},$$

which has a numerical value of $a = 4.55$ m/s^2.

(b) Similarly, the acceleration of an object of mass $m_2 + m_1$ under the influence of a force of magnitude F would be

$$a = \frac{1}{1/a_2 + 1/a_1} = \frac{1}{1/(3.30\,\mathrm{m/s^2}) + 1/(12.0\,\mathrm{m/s^2})},$$

which is the same as part (a) except for the sign change. Then $a = 2.59$ m/s^2.

E3-11 The existence of the spring has little to do with the problem except to "connect" the two blocks; the consequence of this connection is that the force of block 1 on block 2 is equal in magnitude to the force of block 2 on block 1.

(a) The net force on the second block is given by

$$\sum F_x = m_2 a_{2x} = (3.8\,\mathrm{kg})(2.6\,\mathrm{m/s^2}) = 9.9\,\mathrm{N}.$$

There is only one (relevant) force on the block, the force of block 1 on block 2.

(b) There is only one (relevant) force on block 1, the force of block 2 on block 1. By Newton's third law this force has a magnitude of 9.9 N. Then Newton's second law gives $\sum F_x = -9.9$ N$= m_1 a_{1x} = (4.6$ kg$)a_{1x}$. So $a_{1x} = -2.2$ m/s^2 at the instant that $a_{2x} = 2.6$ m/s^2. Note the minus sign, it isn't frivolous; it reflects the fact the two block are necessarily accelerating in opposite directions. We could have instead defined the direction of acceleration of block 2 to be negative, then the acceleration of block 1 would be positive.

E3-15 The numerical weight of an object is given by Eq. 3-7, $W = mg$. If $g = 9.81$ m/s^2, then $m = W/g = (26.0$ N$)/(9.81$ m/s$^2) = 2.65$ kg.

(a) Apply $W = mg$ again, but now $g = 4.60$ m/s^2, so at this point $W = (2.65$ kg$)(4.60$ m/s$^2) = 12.2$ N. Just a reminder; the mass didn't change between these two points, only the *weight* did.

(b) If there is no gravitational force, there is no weight, because $g = 0$. There is still mass, however, and that mass is still 2.65 kg.

E3-19 We'll assume the net force in the x direction on the plane as it accelerates down the runway is from the two engines, so $\sum F_x = 2(1.4 \times 10^5$ N$) = ma_x$. Then $m = 1.22 \times 10^5$ kg. We want the *weight* of the plane, so

$$W = mg = (1.22 \times 10^5 \text{kg})(9.81 \text{ m/s}^2) = 1.20 \times 10^6 \text{ N}.$$

E3-23 Look back at Problem 2-25 for a detailed description of solving the first part of this exercise. We won't go through all of the reasoning here.

(a) Find the time during the "jump down" phase from Eq. 2-30. I'll substitute the variables in directly.

$$(0 \text{ m}) = (0.48 \text{ m}) + (0)t - \frac{1}{2}(9.8 \text{ m/s}^2)t^2,$$

which is a simple quadratic with solutions $t = \pm 0.31$ s. Only the positive solution is of interest. Use this time in Eq. 2-29 to find his speed when he hit ground,

$$v_y = (0) - (9.8 \text{ m/s}^2)(0.31 \text{ s}) = -3.1 \text{ m/s}.$$

This becomes the initial velocity for the deceleration motion, so his average speed during deceleration is given by Eq. 2-27,

$$v_{\text{av},y} = \frac{1}{2}(v_y + v_{0y}) = \frac{1}{2}((0) + (-3.1 \text{ m/s})) = -1.6 \text{ m/s}.$$

This average speed, used with the distance of -2.2 cm (-0.022 m), can be used to find the time of deceleration
$$v_{\text{av},y} = \Delta y / \Delta t,$$

and putting numbers into the expression gives $\Delta t = 0.014$ s. Finally, we use this in Eq. 2-26 to find the acceleration,

$$(0) = (-3.1 \text{ m/s}) + a(0.014 \text{ s}),$$

which gives $a = 220 \text{ m/s}^2$.

(b) The average *net* force on the man is

$$\sum F_y = ma_y = (83 \text{ kg})(220 \text{ m/s}^2) = 1.8 \times 10^4 \text{N}.$$

This *isn't* the force of the ground on the man, and it isn't the force of gravity on the man; it is the *vector sum* of these two forces. That the net force is positive means that it is directed up; a direct consequence is that the *upward* force from the ground must have a larger magnitude than the *downward* force of gravity.

E3-25 Remember that pounds are a measure of force, not a measure of mass. From appendix G we find $1\ \text{lb} = 4.448\ \text{N}$; so the weight is $(100\ \text{lb})(4.448\ \text{N}/1\ \text{lb}) = 445\ \text{N}$; similarly the cord will break if it pulls upward on the object with a force greater than $387\ \text{N}$. It will be necessary to know the mass of the object sooner or later, using Eq. 3-7, $m = W/g = (445\ \text{N})/(9.8\ \text{m/s}^2) = 45\ \text{kg}$.

There are two vertical forces on the 45 kg object, an upward force from the cord F_{OC} (which has a maximum value of 387 N) and a downward force from gravity F_{OG}. Since the objective is to *gently* lower the object we will assume the upward force is as large as it can be. Then $\sum F_y = F_{OC} - F_{OG} = (387\ \text{N}) - (445\ \text{N}) = -58\ \text{N}$. Since the net force is negative, the object must be accelerating downward according to

$$a_y = \sum F_y/m = (-58\,\text{N})/(45\,\text{kg}) = -1.3\,\text{m/s}^2.$$

So long as you lower the cord with this acceleration (or greater), the upward force on the object from the cable will be *less* than the breaking strength. But don't stop! The instant that you feed the cord out with an acceleration of less than $-1.3\ \text{m/s}^2$ the cord will snap, and the object will fall with an acceleration equal to g.

E3-31 (a) The vertical (upward) force from the air on the blades, F_{BA}, can be considered to act on a system consisting of the helicopter alone, or the helicopter + car (or is it a Hummer?). We choose the latter; the *total* mass of this system is 19,500 kg; and the only other force acting on the system is the force of gravity, which is

$$W = mg = (19,500\,\text{kg})(9.8\,\text{m/s}^2) = 1.91 \times 10^5\,\text{N}.$$

The force of gravity is directed down, so the net force on the system is $\sum F_y = F_{BA} - (1.91 \times 10^5\ \text{N})$. The net force can also be found from Newton's second law: $\sum F_y = ma_y = (19,500\ \text{kg})(1.4\ \text{m/s}^2) = 2.7 \times 10^4\ \text{N}$. The positive sign for the acceleration was important; the object was accelerating *up*. Equate the two expression for the net force, $F_{BA} - (1.91 \times 10^5\ \text{N}) = 2.7 \times 10^4\ \text{N}$, and solve; $F_{BA} = 2.2 \times 10^5\ \text{N}$.

(b) We basically repeat the above steps except: (1) the system will consist only of the car, and (2) the upward force on the car comes from the supporting cable only F_{CC}. Then the weight of the car is $W = mg = (4500\ \text{kg})(9.8\ \text{m/s}^2) = 4.4 \times 10^4\ \text{N}$. The net force is $\sum F_y = F_{CC} - (4.4 \times 10^4\ \text{N})$, it can also be written as $\sum F_y = ma_y = (4500\ \text{kg})(1.4\ \text{m/s}^2) = 6300\ \text{N}$. Equating, $F_{CC} = 50,000\ \text{N}$.

P3-3 (a) Start with block one. It starts from rest, accelerating through a distance of 16 m in a time of 4.2 s. Applying Eq. 2-28,

$$x = x_0 + v_{0x}t + \frac{1}{2}a_x t^2,$$

$$-16\ \text{m} = (0) + (0)(4.2\ \text{s}) + \frac{1}{2}a_x(4.2\ \text{s})^2$$

we find the acceleration to be $a_x = -1.8$ m/s^2. The negative sign is because I choose the convention that lower down the ramp is negative.

Now for the second block. The acceleration of the second block is identical to the first for much the same reason that all objects fall with approximately the same acceleration. See the statement at the end of the problem.

(b) The second block is projected up the plane with some initial velocity, rises to some highest point, and then slides back down. Since the acceleration while the block moves up the plane is the same as the acceleration while the block moves down the plane, it is reasonable to assume that the motion is symmetric: the magnitude of the initial velocity at the bottom of the incline is the same as the magnitude of the final velocity on the way down; the time it takes to go up the ramp is the same as the time it takes to come back down.

If the initial and final velocities are related by a sign, then $v_x = -v_{0x}$ and Eq. 2-26 would become

$$
\begin{aligned}
v_x &= v_{0x} + a_x t, \\
-v_{0x} &= v_{0x} + a_x t, \\
-2v_{0x} &= (-1.8 \text{ m/s}^2)(4.2 \text{ s}).
\end{aligned}
$$

which gives an initial velocity of $v_{0x} = 3.8$ m/s.

(c) The time it takes the second block to go up the ramp is the same as the time it takes to come back down. This means that half of the time is spent coming down from the highest point, so the time to "fall" is 2.1 s. The distance traveled is found from Eq. 2-28,

$$
x = (0) + (0)(2.1\,\text{s}) + \frac{1}{2}(-1.8\,\text{m/s}^2)(2.1\,\text{s})^2 = -4.0\,\text{m}.
$$

The negative sign is because it ended up *beneath* the starting point.

P3-7 This problem requires repeated, but careful, application of Newton's second law.

(a) Consider all three carts as one system. There is one (relevant) force $P = 6.5$ N on this system. Then $\sum F_x = P = 6.5$ N. It is the total mass of the system that matters, so Newton's second law will be applied as

$$
\begin{aligned}
\sum F_x &= m_{\text{total}} a_x, \\
6.5 \text{ N} &= (3.1\,\text{kg} + 2.4\,\text{kg} + 1.2\,\text{kg}) a_x, \\
0.97 \text{ m/s}^2 &= a_x.
\end{aligned}
$$

(b) Now choose your system so that it only contains the third car. There is one force on the third car, the pull from car two F_{23} directed to the right, so $\sum F_x = F_{23}$ if we choose the convention that right is positive. We know the acceleration of the car from part (a), so our application of Newton's second law will be

$$\sum F_x = F_{23} = m_3 a_x = (1.2\,\text{kg})(0.97\,\text{m/s}^2).$$

The unknown can be solved to give $F_{23} = 1.2\,\text{N}$ directed to the right.

(c) We can either repeat part (b) except apply it to the second and third cart combined, or we can just look at the second cart. Since looking at the second and third cart combined involves fewer forces, we'll do it that way. There is one (relevant) force on our system, F_{12}, the force of the first cart on the second. The $\sum F_x = F_{12}$, so Newton's law applied to the system gives

$$F_{12} = (m_2 + m_3)a_x = (2.4\,\text{kg} + 1.2\,\text{kg})(0.97\,\text{m/s}^2) = 3.5\,\text{N}.$$

The system contained two masses, so we need to add them in the above expression.

P3-11 This problem is really no different than Problem 3-7, except that there are no numbers here. The horizontal force $\vec{P}$ is a vector of magnitude P, and since the only relevant quantities in this problem are directed along what I'll conveniently choose to call the x-axis, we'll restrict ourselves to a scalar presentation.

(a) Treat the system as including both the block and the rope, so that the mass of the system is $M + m$. There is one (relevant) force which acts on the system, so $\sum F_x = P$. Then Newton's second law would be written as $P = (M + m)a_x$. Solve this for a_x and get $a_x = P/(M + m)$.

(b) Now consider only the block. The horizontal force doesn't act on the block; instead, there is the force of the rope on the block. We'll assume that force has a magnitude R, and this is the *only* (relevant) force on the block, so $\sum F_x = R$ for the net force on the block.. In this case Newton's second law would be written $R = Ma_x$. Yes, a_x is the same in part (a) and (b); the acceleration of the block is the same as the acceleration of the block + rope. Substituting in the results from part (a) we find

$$R = \frac{M}{M + m}P.$$

Chapter 4

Motion in Two and Three Dimensions

E4-3 The derivative is a linear operator, so it operates on each component individually.

(a) The velocity is given by

$$
\begin{aligned}
\frac{d\vec{\mathbf{r}}}{dt} &= \frac{d}{dt}\left(A\hat{\mathbf{i}} + Bt^2\hat{\mathbf{j}} + Ct\hat{\mathbf{k}}\right), \\
&= \frac{d}{dt}\left(A\hat{\mathbf{i}}\right) + \frac{d}{dt}\left(Bt^2\hat{\mathbf{j}}\right) + \frac{d}{dt}\left(Ct\hat{\mathbf{k}}\right), \\
\vec{\mathbf{v}} &= (0) + 2Bt\hat{\mathbf{j}} + C\hat{\mathbf{k}}.
\end{aligned}
$$

(b) The acceleration is given by

$$
\begin{aligned}
\frac{d\vec{\mathbf{v}}}{dt} &= \frac{d}{dt}\left(2Bt\hat{\mathbf{j}} + C\hat{\mathbf{k}}\right), \\
&= \frac{d}{dt}\left(2Bt\hat{\mathbf{j}}\right) + \frac{d}{dt}\left(C\hat{\mathbf{k}}\right), \\
\vec{\mathbf{v}} &= (0) + 2B\hat{\mathbf{j}} + (0).
\end{aligned}
$$

(c) Nothing exciting happens in the x direction, so we will focus on the yz plane. The trajectory in this plane is a parabola.

E4-7 The x and y components of the force $\vec{\mathbf{P}}$ are given by the trigonometry relations $P_x = P\cos\theta$ and $P_y = P\sin\theta$, where $\theta = 25°$ and P is the magnitude of $\vec{\mathbf{P}}$. The block has a weight $W = mg = (5.1\text{ kg})(9.8\text{ m/s}^2) = 50\text{ N}$.

(a) Initially $P = 12$ N, so $P_y = 5.1$ N and $P_x = 11$ N. Since the upward component is less than the weight, the block doesn't leave the floor, and a normal force will be present which will make $\sum F_y = 0$. There is only one contribution to the horizontal force, so $\sum F_x = P_x$. Newton's second law then gives $a_x = P_x/m = (11$ N$)/(5.1$ kg$) = 2.2$ m/s^2.

(b) As P is increased, so is P_y; eventually P_y will be large enough to overcome the weight of the block. This happens just after $P_y = W = 50$ N, which occurs when $P = P_y/\sin\theta = 120$ N.

(c) Repeat part (a), except now $P = 120$ N. Then $P_x = 110$ N, and the acceleration of the block is $a_x = P_x/m = 22$ m/s^2.

E4-11 If the barge is headed straight down the canal then that is also the way it is accelerating. If the x axis is parallel to the river and the y axis is perpendicular, then $\vec{a} = 0.12\hat{i}$ m/s^2. The net force on the barge is $\sum \vec{F} = m\vec{a} = (9500$ kg$)(0.12\hat{i}$ m/s$^2) = 1100\hat{i}$ N.

The force exerted on the barge by the horse has components in both the x and y direction. If $P = 7900$ N is the magnitude of the pull and $\theta = 18°$ is the direction, then $\vec{P} = P\cos\theta\hat{i} + P\sin\theta\hat{j} = (7500\hat{i} + 2400\hat{j})$ N.

Let the force exerted on the barge by the water be $\vec{F_w} = F_{w,x}\hat{i} + F_{w,y}\hat{j}$. Then $\sum F_x = (7500$ N$) + F_{w,x}$ and $\sum F_y = (2400$ N$) + F_{w,y}$. But we already found $\sum \vec{F}$, so

$$F_x = 1100 \text{ N} = 7500 \text{ N} + F_{w,x},$$
$$F_x = 0 = 2400 \text{ N} + F_{w,y}.$$

Solving, $F_{w,x} = -6400$ N and $F_{w,y} = -2400$ N. The magnitude is found by $F_w = \sqrt{F_{w,x}^2 + F_{w,y}^2} = 6800$ N.

E4-13 It is easiest to solve this problem as separate x and y components using Eq. 4-9 and Eq. 4-10.

(a) First we solve the vertical problem for the time the ball was in the air. Since it rolled off horizontally, $v_{0y} = 0$. Then applying Eq. 4-10(b),

$$y = v_{0y}t - \frac{1}{2}gt^2,$$

$$(-4.23 \text{ ft}) = (0)t - \frac{1}{2}(32.2 \text{ ft/s}^2)t^2,$$

which can be solved to yield $t = \pm 0.514$ s. The negative sign in front of the distance is because the ball landed at a point beneath where it started. We only care about the positive answer.

(b) The initial velocity in the x direction can be found from Eq. 4-10, $x = v_{0x}t$; rearranging, $v_{0x} = x/t = (5.11 \text{ ft})/(0.514 \text{ s}) = 9.94 \text{ ft/s}$. Since there is no y component to the velocity, then by Eq. 4-11 the initial speed is $v_0 = 9.94 \text{ ft/s}$.

E4-17 The highest point occurs when $v_y = 0$, so use Eq. 4-9(b) to find the time in terms of the the initial y component of the velocity:

$$v = v_0 - gt$$

$$
\begin{aligned}
v_y &= v_{0y} - gt, \\
(0) &= v_{0y} - gt, \\
t &= v_{0y}/g.
\end{aligned}
$$

Use this time in Eq. 4-10(b) to find the highest point:

$$
\begin{aligned}
y &= v_{0y}t - \frac{1}{2}gt^2, \\
y_{\text{max}} &= v_{0y}\left(\frac{v_{0y}}{g}\right) - \frac{1}{2}g\left(\frac{v_{0y}}{g}\right)^2, \\
&= \frac{v_{0y}^2}{2g}.
\end{aligned}
$$

Finally, we know the initial y component of the velocity from Eq. 2-6, so $y_{\text{max}} = (v_0 \sin \phi_0)^2 / 2g$.

E4-25 We use Eqs. 4-9 and 4-10, and we define the point the ball leaves the racquet as $\vec{r} = 0$.

(a) The initial conditions are given as $v_{0x} = 23.6 \text{ m/s}$ and $v_{0y} = 0$, since the ball was launched horizontally. We can use Eq. 4-10(a) to find out how long it takes for the ball to reach the horizontal location of the net:

$$
\begin{aligned}
x &= v_{0x}t, \\
(12 \text{ m}) &= (23.6 \text{ m/s})t, \\
0.51 \text{ s} &= t,
\end{aligned}
$$

From Eq. 4-10(b) we can find how far the ball has moved horizontally in this time:

$$y = v_{0y}t - \frac{1}{2}gt^2 = (0)(0.51 \text{ s}) - \frac{1}{2}(9.8 \text{ m/s}^2)(0.51 \text{ s})^2 = -1.3 \text{ m}.$$

Did the ball clear the net? The ball started 2.37 m above the ground (must have been an overhand serve!) and "fell" through a distance of 1.3 m by the time it arrived at the net. So it is still 1.1 m above the ground and 0.2 m above the net.

(b) The initial conditions are now given by $v_{0x} = (23.6 \text{ m/s})(\cos[-5.0°]) = 23.5$ m/s and $v_{0y} = (23.6 \text{ m/s})(\sin[-5.0°]) = -2.1$ m/s, found by applying Eq. 4-6. Now we find the time to reach the net just as done in part (a):

$$t = x/v_{0x} = (12.0 \text{ m})/(23.5 \text{ m/s}) = 0.51 \text{ s}.$$

It actually takes slightly more time than in part (a), but this difference is lost at our level of significant figures. Now find the vertical position of the ball when it arrives at the net:

$$y = v_{0y}t - \frac{1}{2}gt^2 = (-2.1 \text{ m/s})(0.51 \text{ s}) - \frac{1}{2}(9.8 \text{ m/s}^2)(0.51 \text{ s})^2 = -2.3 \text{ m}.$$

Did the ball clear the net? Not this time; it started 2.37 m above the ground and then passed the net 2.3 m lower, or only 0.07 m above the ground. Since the net is 0.9 m high, the player goofed.

E4-29 That the speed of the pebble is constant means that the pebble is *not* accelerating. Then the net force on the pebble must be zero, so $\sum F_y = 0$. There are only two forces on the pebble, the force of gravity W and the force of the water on the pebble F_{PW}. These point in opposite directions, so $0 = F_{PW} - W$. But $W = mg = (0.150 \text{ kg})(9.81 \text{ m/s}^2) = 1.47 \text{ N}$. Since $F_{PW} = W$ in this problem, the force of the water on the pebble must also be 1.47 N.

No, we didn't need any of the other information; and it is important that you discern what is and is not important for yourself.

E4-31 Eq. 4-22 is
$$v_y(t) = v_T \left(1 - e^{-bt/m}\right),$$
where we have used Eq. 4-24 to substitute for the terminal speed. We want to solve this equation for time when $v_y(t) = v_T/2$, so

$$\tfrac{1}{2}v_T = v_T \left(1 - e^{-bt/m}\right),$$
$$\tfrac{1}{2} = \left(1 - e^{-bt/m}\right),$$
$$e^{-bt/m} = \tfrac{1}{2}$$
$$bt/m = -\ln(1/2)$$
$$t = \tfrac{m}{b}\ln 2$$

Although we are finished, it is neater to make use of Eq. 4-24 again, and let $m/b = v_T/g$ so that the time for the speed to reach one-half the terminal velocity is given by $t = v_T \ln 2/g$.

E4-35 (a) The speed can be found from Eq. 4-29; rearrange this equation to get $v = \sqrt{ra_c} = \sqrt{(5.2 \text{ m})(6.8)(9.8 \text{ m/s}^2)} = 19 \text{ m/s}$. This is only about 40 miles/hour, so really isn't that fast.

(b) This is mostly a unit conversion problem, but you do need to use the fact that one revolution corresponds to a length of $2\pi r$:

$$19\frac{\text{m}}{\text{s}}\left(\frac{1 \text{ rev}}{2\pi(5.2 \text{ m})}\right)\left(\frac{60 \text{ s}}{1 \text{ min}}\right) = 35\frac{\text{rev}}{\text{min}}.$$

E4-39 What a fun problem! Since the last part of the question implies that we might not even need to know the length of the Escalator, we are going to try not to use it. So let $\Delta x = 15$ m be the length, and see if it cancels out. Other useful variables will be $t_w = 90$ s, the time to walk the stalled Escalator; $t_s = 60$ s, the time to ride the moving Escalator; and t_m, the time to walk up the moving Escalator.

The walking speed of the person relative to a fixed Escalator is $v_{we} = \Delta x/t_w$; the speed of the Escalator relative to the ground is $v_{eg} = \Delta x/t_s$; and the speed of the walking person relative to the ground on a moving Escalator is $v_{wg} = \Delta x/t_m$. But Eq. 4-32 shows that these three speeds are related by $v_{wg} = v_{we} + v_{eg}$. Combine all the above:

$$\begin{aligned} v_{wg} &= v_{we} + v_{eg}, \\ \frac{\Delta x}{t_m} &= \frac{\Delta x}{t_w} + \frac{\Delta x}{t_s}, \\ \frac{1}{t_m} &= \frac{1}{t_w} + \frac{1}{t_s}. \end{aligned}$$

Yes, the length Δx did drop out of the expression; we never need to know it. Putting in the numbers, $t_m = 36$ s.

E4-43 (a) This is a relative motion problem, although it is subtle. The position of the bolt relative to the elevator is y_{be}, the position of the bolt relative to the shaft is y_{bs}, and the position of the elevator relative to the shaft is y_{es}. We zero all three positions at $t = 0$; at this time we also have $v_{0,bs} = v_{0,es} = 8.0$ ft/s.

The three equations describing the positions are

$$\begin{aligned} y_{bs} &= v_{0,bs}t - \frac{1}{2}gt^2, \\ y_{es} &= v_{0,es}t + \frac{1}{2}at^2, \\ y_{be} + r_{es} &= r_{bs}, \end{aligned}$$

where $a = 4.0$ m/s^2 is the upward acceleration of the elevator. Rearrange the last equation and solve for y_{be}; we soon get $y_{be} = -\frac{1}{2}(g+a)t^2$, where advantage was taken of the fact that the initial velocities are the same.

Solve for the time, then

$$t = \sqrt{-2y_{be}/(g+a)} = \sqrt{-2(-9.0 \text{ ft})/(32 \text{ ft/s}^2 + 4 \text{ ft/s}^2)} = 0.71 \text{ s}$$

The double negative inside the square root *is* important; the bolt was falling down relative to the elevator.

(b) We then use the expression for y_{bs} to find how the bolt moved relative to the shaft:

$$y_{bs} = v_{0,bs}t - \frac{1}{2}gt^2 = (8.0 \text{ ft})(0.71 \text{ s}) - \frac{1}{2}(32 \text{ ft/s}^2)(0.71 \text{ s})^2 = -2.4 \text{ ft}.$$

P4-1 Let $\vec{r}_A$ be the position of particle of particle A, and $\vec{r}_B$ be the position of particle B. The equations for the motion of the two particles are then

$$
\begin{aligned}
\vec{r}_A &= \vec{r}_{0,A} + \vec{v}t, \\
&= d\hat{j} + vt\hat{i}; \\
\vec{r}_B &= \frac{1}{2}\vec{a}t^2, \\
&= \frac{1}{2}a(\sin\theta\hat{i} + \cos\theta\hat{j})t^2.
\end{aligned}
$$

A collision will occur if, and only if, there is a time when $\vec{r}_A = \vec{r}_B$. So we equate,

$$d\hat{j} + vt\hat{i} = \frac{1}{2}a(\sin\theta\hat{i} + \cos\theta\hat{j})t^2,$$

but this is really *two* equations: $d = \frac{1}{2}at^2\cos\theta$ and $vt = \frac{1}{2}at^2\sin\theta$. That's a good thing, as we have two unknowns, t and θ.

We'll solve the second one for t and get $t = 2v/(a\sin\theta)$. Substitute that into the first equation, and then rearrange,

$$
\begin{aligned}
d &= \frac{1}{2}at^2\cos\theta, \\
d &= \frac{1}{2}a\left(\frac{2v}{a\sin\theta}\right)^2\cos\theta, \\
\sin^2\theta &= \frac{2v}{ad}\cos\theta, \\
1 - \cos^2\theta &= \frac{2v^2}{ad}\cos\theta, \\
0 &= \cos^2\theta + \frac{2v^2}{ad}\cos\theta - 1.
\end{aligned}
$$

This last expression is quadratic in $\cos\theta$. It simplifies the solution if we define $b = 2v/(ad) = 2(3.0 \text{ m/s})^2/([0.4 \text{ m/s}^2][30 \text{ m}]) = 1.5$, then

$$\cos\theta = \frac{-b \pm \sqrt{b^2 + 4}}{2} = -0.75 \pm 1.25.$$

Only the answer which has a magnitude less than 1 is physically relevant, so $\cos\theta = 0.5$ and $\theta = 60°$.

P4-7 The components of the initial velocity are given by $v_{0x} = v_0 \cos\theta = 56$ ft/s and $v_{0y} = v_0 \sin\theta = 106$ ft/s where we used $v_0 = 120$ ft/s and $\theta = 62°$.

(a) To find h we need only find out the vertical position of the stone when $t = 5.5$ s. Using Eq. 4-10(b),

$$y = v_{0y}t - \frac{1}{2}gt^2 = (106 \text{ ft/s})(5.5 \text{ s}) - \frac{1}{2}(32 \text{ ft/s}^2)(5.5 \text{ s})^2 = 99 \text{ ft.}$$

(b) The speed of the stone upon reaching point A is the magnitude of the velocity, for a mild change of pace, let's look at this as a vector problem and use Eq. 4-1:

$$
\begin{aligned}
\vec{v} &= \vec{v}_0 + \vec{a}t, \\
&= \left(v_{0x}\hat{\mathbf{i}} + v_{0y}\hat{\mathbf{j}}\right) - g\hat{\mathbf{j}}t, \\
&= v_{0x}\hat{\mathbf{i}} + (v_{0y} - gt)\hat{\mathbf{j}}, \\
&= (56 \text{ ft/s})\hat{\mathbf{i}} + \left((106 \text{ ft/s} - (32 \text{ ft/s}^2)(5.5 \text{ s})\right)\hat{\mathbf{j}}, \\
&= (56 \text{ ft/s})\hat{\mathbf{i}} + (-70.0 \text{ ft/s})\hat{\mathbf{j}}.
\end{aligned}
$$

The magnitude of this vector gives the speed when $t = 5.5$ s; $v = \sqrt{56^2 + (-70)^2}$ ft/s$= 90$ ft/s.

(c) Highest point occurs when $v_y = 0$. Solving Eq. 4-9(b) for time; $v_y = 0 = v_{0y} - gt = (106 \text{ ft/s}) - (32 \text{ ft/s}^2)t$; $t = 3.31$ s. Use this time in Eq. 4-10(b),

$$y = v_{0y}t - \frac{1}{2}gt^2 = (106 \text{ ft/s})(3.31 \text{ s}) - \frac{1}{2}(32 \text{ ft/s}^2)(3.31 \text{ s})^2 = 176 \text{ ft.}$$

P4-9 To score the ball must pass the horizontal distance of 50 m with an altitude of no less than 3.44 m. The initial velocity components are $v_{0x} = v_0 \cos\theta$ and $v_{0y} = v_0 \sin\theta$ where $v_0 = 25$ m/s, and θ is the unknown.

We'll first find an expression for the time to the goal post from Eq. 4-10(a), which describes the horizontal motion, by a quick rearrangement: $t = x/v_{0x} = x/(v_0 \cos\theta)$.

And then we'll take this time and substitute is into Eq. 4-10(b) for the vertical motion:

$$
\begin{aligned}
y &= v_{0y}t - \frac{1}{2}gt^2 = (v_0 \sin\theta)\left(\frac{x}{v_0 \cos\theta}\right) - \frac{1}{2}g\left(\frac{x}{v_0 \cos\theta}\right)^2, \\
&= x\frac{\sin\theta}{\cos\theta} - \frac{gx^2}{2v_0^2}\frac{1}{\cos^2\theta}.
\end{aligned}
$$

In this last expression y needs to be greater than 3.44 m; the problem would be *much* easier if this were 0 instead of 3.44! The mix of sin and cos terms will require some inventive efforts. Try this:

$$\frac{1}{\cos^2\theta} - 1 + 1 = \frac{1}{\cos^2\theta} - \frac{\cos^2\theta}{\cos^2\theta} + 1 = \frac{1-\cos^2\theta}{\cos^2\theta} + 1 = \frac{\sin^2\theta}{\cos^2\theta} + 1 = \tan^2\theta + 1.$$

How were you supposed to know that this was the trick? You weren't; I didn't know until after I had tried a few false starts myself. Now we both know. This gives for our y expression

$$y = x\tan\theta - \frac{gx^2}{2v_0}\left(\tan^2\theta + 1\right),$$

which can be combined with numbers and constraints to give

$$(3.44 \text{ m}) \;\le\; (50 \text{ m})\tan\theta - \frac{(9.8 \text{ m/s}^2)(50 \text{ m})^2}{2(25 \text{ m/s})^2}\left(\tan^2\theta + 1\right),$$

$$3.44 \;\le\; 50\tan\theta - 20\left(\tan^2\theta + 1\right),$$

$$0 \;\le\; -20\tan^2\theta + 50\tan\theta - 23$$

This is a quadratic we can solve; assuming equality, we find $\tan\theta = 1.25 \pm 0.65$, so the allowed kicking angles are between $\theta = 31°$ and $\theta = 62°$. Notice that these two angles are *not* complimentary; this happens because the ball didn't land at the same level from which it took off.

P4-13 When the problem starts the balloon is moving down with a constant speed of $v_1 = 1.88$ m/s. There is a downward force on the balloon of 10.8 kN from gravity and an upward force of 10.3 kN from the buoyant force of the air (read chapter 15 for more exciting details on the buoyant force). The resultant of these two forces is 500 N down, but since the balloon is descending at constant speed, the acceleration of, and hence net force on, the balloon must be zero.

This is possible because there is a drag force on the balloon of $D = bv^2$, this force is directed upward. The magnitude must be 500 N, so the constant b is

$$b = \frac{(500 \text{ N})}{(1.88 \text{ m/s})^2} = 141 \text{ kg/m}.$$

If the crew drops 26.5 kg of ballast they are "lightening" the balloon by

$$(26.5 \text{ kg})(9.81 \text{ m/s}^2) = 260 \text{ N}.$$

This reduced the weight, but not the buoyant force, so the drag force at constant speed will now be $500 \text{ N} - 260 \text{ N} = 240 \text{ N}$.

The new constant downward speed will be

$$v = \sqrt{D/b} = \sqrt{(240 \text{ N})/(141 \text{ kg/m})} = 1.30 \text{ m/s}.$$

P4-17 (a) The acceleration is the time derivative of the velocity, so starting with Eq. 4-22,

$$a_y = \frac{dv_y}{dt} = \frac{d}{dt}\left(\frac{mg}{b}\left(1 - e^{-bt/m}\right)\right) = \frac{mg}{b}\frac{b}{m}e^{-bt/m},$$

which can be simplified as $a_y = ge^{-bt/m}$. For large t this expression approaches 0; for small t the exponent can be expanded to give

$$a_y \approx g\left(1 - \frac{bt}{m}\right) = g - v_{\mathrm{T}}t,$$

where in the last line we made use of Eq. 4-24.

(b) The position is the integral of the velocity, so we want to integrate Eq. 4-22 with respect to time

$$
\begin{aligned}
\int_0^t v_y\, dt &= \int_0^t \left(\frac{mg}{b}\left(1 - e^{-bt/m}\right)\right)\, dt, \\
\int_0^t \frac{dy}{dt}\, dt &= \frac{mg}{b}\left(t - (-m/b)e^{-bt/m}\right)\Big|_0^t, \\
\int_0^y dy &= v_{\mathrm{T}}\left(t + \frac{v_{\mathrm{T}}}{g}\left(e^{-v_{\mathrm{T}}t/g} - 1\right)\right), \\
y &= v_{\mathrm{T}}\left(t + \frac{v_{\mathrm{T}}}{g}\left(e^{-v_{\mathrm{T}}t/g} - 1\right)\right).
\end{aligned}
$$

P4-21 This is a problem best solved backwards. Start from where the stone lands; in order to get there the stone fell through a vertical distance of 1.9 m while moving 11 m horizontally. We can use Eq. 4-10(b) to find out what length of time the stone was in the air, assuming that it was launched horizontally. Then $v_{0y} = 0$, so

$$y = -\frac{1}{2}gt^2 \text{ which can be written as } t = \sqrt{\frac{-2y}{g}}.$$

Putting in the numbers, $t = 0.62\,\mathrm{s}$ is the time of flight from the moment the string breaks.

From this time we can use Eq. 4-10(a) to find the horizontal velocity,

$$v_x = \frac{x}{t} = \frac{(11\,\mathrm{m})}{(0.62\,\mathrm{s})} = 18\,\mathrm{m/s}.$$

Since the stone moves off in a tangent when the string breaks the speed of the stone in the circle must have been the same as the initial speed when the string breaks. Then the centripetal acceleration is

$$a_c = \frac{v^2}{r} = \frac{(18\,\mathrm{m/s})^2}{(1.4\,\mathrm{m})} = 230\,\mathrm{m/s^2}.$$

P4-23 (a) A cycloid looks something like this:

(b) The position of the particle is given by

$$\vec{\mathbf{r}} = (R\sin\omega t + \omega Rt)\hat{\mathbf{i}} + (R\cos\omega t + R)\hat{\mathbf{j}}.$$

The maximum value of y occurs whenever $\cos\omega t = 1$. The minimum value of y occurs whenever $\cos\omega t = -1$. At either of those times $\sin\omega t = 0$.

The velocity is the derivative of the displacement vector,

$$\vec{\mathbf{v}} = (R\omega\cos\omega t + \omega R)\hat{\mathbf{i}} + (-R\omega\sin\omega t)\hat{\mathbf{j}}.$$

When y is a maximum the velocity simplifies to

$$\vec{\mathbf{v}} = (2\omega R)\hat{\mathbf{i}} + (0)\hat{\mathbf{j}}.$$

When y is a minimum the velocity simplifies to

$$\vec{\mathbf{v}} = (0)\hat{\mathbf{i}} + (0)\hat{\mathbf{j}}.$$

The acceleration is the derivative of the velocity vector,

$$\vec{\mathbf{a}} = (-R\omega^2\sin\omega t)\hat{\mathbf{i}} + (-R\omega^2\cos\omega t)\hat{\mathbf{j}}.$$

When y is a maximum the acceleration simplifies to

$$\vec{\mathbf{a}} = (0)\hat{\mathbf{i}} + (-R\omega^2)\hat{\mathbf{j}}.$$

When y is a minimum the acceleration simplifies to

$$\vec{\mathbf{a}} = (0)\hat{\mathbf{i}} + (R\omega^2)\hat{\mathbf{j}}.$$

P4-27 The velocity of the police car with respect to the ground is $\vec{\mathbf{v}}_{pg} = -76\text{km/h}\hat{\mathbf{i}}$. The velocity of the motorist with respect the ground is $\vec{\mathbf{v}}_{mg} = -62 \text{ km/h}\hat{\mathbf{j}}$.

The velocity of the motorist with respect to the police car is given by solving

$$\vec{\mathbf{v}}_{mg} = \vec{\mathbf{v}}_{mp} + \vec{\mathbf{v}}_{pg},$$

so $\vec{\mathbf{v}}_{mp} = 76\text{km/h}\hat{\mathbf{i}} - 62 \text{ km/h}\hat{\mathbf{j}}$. This velocity has magnitude

$$v_{mp} = \sqrt{(76\text{km/h})^2 + (-62 \text{ km/h})^2} = 98 \text{ km/h}.$$

The direction is

$$\theta = \arctan(-62 \text{ km/h})/(76\text{km/h}) = -39°,$$

but that is relative to $\hat{\mathbf{i}}$. We want to know the direction relative to the line of sight. The line of sight is

$$\alpha = \arctan(57\,\text{m})/(41\,\text{m}) = -54°$$

relative to $\hat{\mathbf{i}}$, so the answer must be 15°.

Chapter 5

Applications of Newton's Laws

E5-1 There are three forces which act on the charged sphere— an electric force, F_E, the force of gravity, W, and the tension in the string, T. All arranged as shown in the figure on the right below. Since the sphere isn't moving, we can assume that the acceleration is zero, and consequently that the net force is zero, so that $\sum \vec{F} = 0$.

(a) We can solve this problem with components, but it is slightly more elegant to write the vectors so that they geometrically show that the sum is zero, as in the figure on the left below. Now $W = mg = (2.8 \times 10^{-4} \text{ kg})(9.8 \text{ m/s}^2) = 2.7 \times 10^{-3} \text{ N}$. The magnitude of the electric force can be found from the tangent relationship, so $F_E = W \tan \theta = (2.7 \times 10^{-3} \text{ N}) \tan(33°) = 1.8 \times 10^{-3} \text{ N}$.

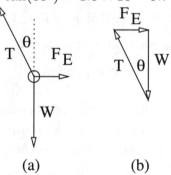

(a)   (b)

(b) The tension can be found from the cosine relation, so $T = W/\cos \theta = (2.7 \times 10^{-3} \text{ N})/\cos(33°) = 3.2 \times 10^{-3} \text{ N}$.

E5-5 This exercise is Atwood's machine in disguise, so read Sample Problem 5-5. Although it is *very* worthwhile to be able to duplicate the derivations (it makes for a common midterm exam question), we won't do that here, we'll instead start from Eq. 5-4.

(a) The magnitude of the man's acceleration is given by

$$a = \frac{m_2 - m_1}{m_2 + m_1}g = \frac{(110 \text{ kg}) - (74 \text{ kg})}{(110 \text{ kg}) + (74 \text{ kg})}g = 0.2g,$$

and is directed down. The time which elapses while he falls is found by solving $y = v_{0y}t + \frac{1}{2}a_y t^2$, or, with numbers, $(-12 \text{ m}) = (0)t + \frac{1}{2}(-0.2g)t^2$ which has the solutions $t = \pm 3.5$ s. The velocity with which he hits the ground is then $v = v_{0y} + a_y t = (0) + (-0.2g)(3.5 \text{ s}) = -6.9$ m/s. Not very fast, but fast enough to hurt yourself.

(b) Reducing the speed can be accomplished by reducing the acceleration. We can't change Eq. 5-4 without also changing one of the assumptions that went into it. Since the man is hoping to reduce the speed with which he hits the ground, it makes sense that he might want to climb up the rope.

E5-9 See Sample Problem 5-8. We need only apply the (unlabeled!) equation

$$\mu_s = \tan \theta$$

to find the egg angle. In this case $\theta = \tan^{-1}(0.04) = 2.3°$. I wish that I had one of these frying pans; the eggs don't slide out of mine even when inverted.

E5-13 The use of strings at angle doesn't affect the basic concern of the problem. A 75 kg mass has a weight of $W = (75 \text{ kg})(9.8 \text{ m/s}^2) = 735$ N, so the force of friction on each end of the bar must be 368 N.
  Using Eq. 5-7,

$$F \geq \frac{f_s}{\mu_s} = \frac{(368 \text{ N})}{(0.41)} = 900 \text{ N}.$$

E5-17 I encourage you to consider parts (b) and (c) before calculating anything. In which case do you expect the second worker will need to exert the greater force?

(a) The force of static friction is less than $\mu_s N$, where N is the normal force. Since the crate isn't moving up or down, $\sum F_y = 0 = N - W$. So in this case $N = W = mg = (136 \text{ kg})(9.81 \text{ m/s}^2) = 1330$ N; however, don't be fooled into thinking $N = W$ always! The force of static friction is less than or equal to $(0.37)(1330 \text{ N}) = 492$ N; moving the crate will require a force greater than or equal to 492 N.

(b) The second worker could lift upward with a force L, reducing the normal force, and hence reducing the force of friction. If the first worker can move the block with a 412 N force, then $412 \geq \mu_s N$. Solving for N, the normal force needs to be less than 1110 N. The crate doesn't move off the table, so then $N + L = W$, or $L = W - N = (1330 \text{ N}) - (1110 \text{ N}) = 220$ N.

(c) Or the second worker can help by adding a push so that the total force of both workers is equal to 492 N. If the first worker pushes with a force of 412 N, the second would need to push with a force of 80 N.

Did you "guess" right? Could there ever be a case where it is easier to lift than to help push? When?

E5-21 This exercise has a few parts; I don't know that it matters where we start, so we'll start with accelerations. Let a_1 be acceleration down frictionless incline of length l, and t_1 the time taken. The a_2 is acceleration down "rough" incline, and $t_2 = 2t_1$ is the time taken. Then

$$l = \frac{1}{2}a_1 t_1^2 \text{ and } l = \frac{1}{2}a_2(2t_1)^2.$$

Equate and we'll find $a_1/a_2 = 4$.

There are two force which act on the ice when it sits on the frictionless incline. The normal force acts perpendicular to the surface, so it doesn't contribute any components parallel to the surface. The force of gravity has a component parallel to the surface, given by

$$W_{\parallel} = mg\sin\theta,$$

and a component perpendicular to the surface given by

$$W_{\perp} = mg\cos\theta.$$

The acceleration down the frictionless ramp is then

$$a_1 = \frac{W_{\parallel}}{m} = g\sin\theta.$$

When friction is present the force of kinetic friction is $f_k = \mu_k N$; since the ice doesn't move perpendicular to the surface we also have $N = W_{\perp}$; and finally the acceleration down the ramp is

$$a_2 = \frac{W_{\parallel} - f_k}{m} = g(\sin\theta - \mu\cos\theta).$$

Previously we found the ratio of a_1/a_2, so we now have

$$\sin\theta = 4\sin\theta - 4\mu\cos\theta,$$
$$\sin 33° = 4\sin 33° - 4\mu\cos 33°,$$
$$\mu = 0.49.$$

E5-23 We want to find the force of friction on the block. There are four forces on the block— the force of gravity, $W = mg$; the normal force, N; the horizontal push, P, and the force of friction, f. Since the block is moving parallel to the plane, it makes sense to choose our coordinate system so that components are either parallel (x-axis) to the plane or perpendicular (y-axis) to it. $\theta = 39°$. Refer to the figure below.

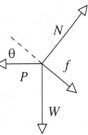

The magnitudes of the x components of the forces are $W_x = W \sin\theta$, $P_x = P \cos\theta$ and f; the magnitudes of the y components of the forces are $W_y = W \cos\theta$, $P_y = P \sin\theta$.

(a) We consider the first the case of the block moving up the ramp; then f is directed down as is shown in the figure below. Newton's second law for each set of components then reads as

$$\sum F_x = P_x - f - W_x = P \cos\theta - f - W \sin\theta = ma_x,$$
$$\sum F_y = N - P_y - W_y = N - P \sin\theta - W \cos\theta = ma_y$$

Since the block never leaves the table we can safely assume $a_y = 0$. Then the second equation is easy to solve for N

$$N = P \sin\theta + W \cos\theta = (46\text{ N}) \sin(39°) + (4.8\text{ kg})(9.8\text{ m/s}^2) \cos(39°) = 66\text{ N}.$$

The force of friction is found from $f = \mu_k N = (0.33)(66\text{ N}) = 22\text{ N}$. This is directed down the incline while the block is moving up. We can now find the acceleration in the x direction.

$$ma_x = P \cos\theta - f - W \sin\theta,$$
$$= (46\text{ N}) \cos(39°) - (22\text{ N}) - (4.8\text{ kg})(9.8\text{ m/s}^2) \sin(39°) = -16\text{ N}.$$

So the block is slowing down, with an acceleration of magnitude 3.3 m/s^2.

(b) The block has an initial speed of $v_{0x} = 4.3$ m/s; it will rise until it stops; so we can use $v_y = 0 = v_{0y} + a_y t$ to find the time to the highest point. Then $t = (v_y - v_{0y})/a_y = -(-4.3\text{ m/s})/(3.3\text{ m/s}^2 = 1.3\text{ s}$. Now that we know the time we can use the other kinematic relation to find the distance

$$y = v_{0y}t + \frac{1}{2}a_y t^2 = (4.3\text{ m/s})(1.3\text{ s}) + \frac{1}{2}(-3.3\text{ m/s}^2)(1.3\text{ s})^2 = 2.8\text{ m}$$

(c)   When the block gets to the top it *might* slide back down. But in order to do so the frictional force, which is now directed up the ramp, must be sufficiently small so that $f + P_x \leq W_x$. Solving for f we find $f \leq W_x - P_x$ or, using our numbers from above, $f \leq -6$ N. Is this possible? No, so the block will not slide back down the ramp, *even if the ramp were frictionless,* while the horizontal force is applied.

E5-29   This problem is similar to Sample Problem 5-7, except now there is friction which can act on block B. The relevant equations are now for block B

$$N - m_B g \cos \theta = 0$$

and

$$T - m_B g \sin \theta \pm f = m_B a,$$

where the sign in front of f depends on the direction in which block B is moving. If the block is moving up the ramp then friction is directed down the ramp, and we would use the negative sign. If the block is moving down the ramp then friction will be directed up the ramp, and then we will use the positive sign. Finally, if the block is stationary then friction we be in such a direction as to make $a = 0$.

For block A the relevant equation is

$$m_A g - T = m_A a.$$

We can combine the first two equations with $f = \mu N$ to get

$$T - m_B g \sin \theta \pm \mu m_B g \cos \theta = m_B a,$$

where we will need to take some care when interpreting friction for the static case, since the static value of μ yield the maximum possible static friction force, which is not necessarily the actual static frictional force.

We can combine this last equation with the block A equation,

$$m_A g - m_A a - m_B g \sin \theta \pm \mu m_B g \cos \theta = m_B a,$$

and then rearrange to get

$$a = g \frac{m_A - m_B \sin \theta \pm \mu m_B \cos \theta}{m_A + m_B}.$$

Now we can answer the questions. For convenience we will use metric units; then the masses are $m_A = 13.2$ kg and $m_B = 42.6$ kg. In addition, $\sin 42° = 0.669$ and $\cos 42° = 0.743$.

(a) If the blocks are originally at rest then

$$m_A - m_B \sin\theta = (13.2\,\text{kg}) - (42.6\,\text{kg})(0.669) = -15.3\,\text{kg}$$

where the negative sign indicates that block B would slide downhill if there were no friction.

If the block are originally at rest we need to consider static friction, so the last term can be as large as

$$\mu m_B \cos\theta = (.56)(42.6\,\text{kg})(0.743) = 17.7\,\text{kg}.$$

Since this quantity is larger than the first static friction would be large enough to stop the blocks from accelerating if they are at rest.

(b) If block B is moving up the ramp we use the negative sign, and the acceleration is

$$a = (9.81\,\text{m/s}^2)\frac{(13.2\,\text{kg}) - (42.6\,\text{kg})(0.669) - (.25)(42.6\,\text{kg})(0.743)}{(13.2\,\text{kg}) + (42.6\,\text{kg})} = -4.08\,\text{m/s}^2.$$

where the negative sign means down the ramp. The block, originally moving up the ramp, will slow down and stop. Once it stops the static friction takes over and the results of part (a) become relevant.

(c) If block B is moving down the ramp we use the positive sign, and the acceleration is

$$a = (9.81\,\text{m/s}^2)\frac{(13.2\,\text{kg}) - (42.6\,\text{kg})(0.669) + (.25)(42.6\,\text{kg})(0.743)}{(13.2\,\text{kg}) + (42.6\,\text{kg})} = -1.30\,\text{m/s}^2.$$

where the negative sign means down the ramp. This means that if the block is moving down the ramp it will continue to move down the ramp, faster and faster.

Note that my answer disagrees with the back of the book. One of us must have made a mistake. I think I'm right.

E5-33 There are, despite popular opinion, only three forces acting on the car: the force of gravity $W = mg$, the normal force of the road on the car N, and the force of friction f. Since the force of friction is the only force with a horizontal component, it *must* be directed toward the center of the turn, and is the only non-canceling contribution to the net force. Sooner or later we'll need to know the mass of the car, $m = W/g = (10700\ \text{N})/(9.8\ \text{m/s}^2) = 1100$ kg.

(a) The magnitude of net force on the car is given by Eq. 5-10, but this force is only from friction, so

$$f = \frac{mv^2}{r} = \frac{(1100\ \text{kg})(13.4\ \text{m/s})^2}{(61\ \text{m})} = 3200\ \text{N}.$$

(b) The coefficient of static friction is used here because the tires should *not* be skidding on the road. Then $f \leq \mu_s N$, or $\mu_s \geq f/N$. It is *so* tempting to say that the normal force is the weight. *Don't do it!* Instead look at the vertical motion; since the car stays on the road $a_y = 0$, and then we can write

$$\sum F_y = N - W = ma_y = 0,$$

from which we can conclude that $N = W$. Now don't think this means the normal force is the weight; it means that the magnitude of the normal force is the same as the magnitude of the weight. Finally we can find μ_s from $\mu_s \geq f/N = (3200 \text{ N})/(10700 \text{ N}) = 0.3$. This value is well below the average value for a rubber tire on a road.

E5-39 There are two forces on the hanging cylinder: the force of the cord pulling up T and the force of gravity $W = Mg$. The cylinder is at rest, so these two forces must balance, or $T = W$. There are three forces on the disk, but only the force of the cord on the disk T is relevant here, since there is no friction or vertical motion.

The disk undergoes circular motion, so $T = mv^2/r$. We want to solve this for v and then express the answer in terms of m, M, r, and G.

$$v = \sqrt{\frac{Tr}{m}} = \sqrt{\frac{Mgr}{m}}.$$

E5-41 A banked turn with friction requires a little more force to solve than the banked curve in Section 5-4. There are three forces to consider: the normal force of the road on the car N; the force of gravity on the car W; and the frictional force on the car f. The acceleration of the car in circular motion is toward the center of the circle; this means the *net* force on the car is horizontal, toward the center. We will arrange our coordinate system so that r is horizontal and z is vertical. Then the components of the normal force are $N_r = N \sin\theta$ and $N_z = N \cos\theta$; the components of the frictional force are $f_r = f \cos\theta$ and $f_z = f \sin\theta$.

The direction of the friction depends on the speed of the car. Think about it: if the car is moving too slowly it will tend to slide down the incline, so friction would need to be directed up the incline if it were to hold the car; the reverse would be true if the car were moving too fast. The figure below shows the two force diagrams.

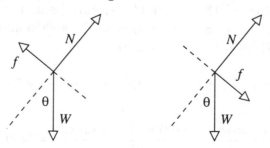

The turn is designed for 95 km/hr, at this speed a car should require *no* friction to stay on the road. Using Eq. 5-17 we find that the banking angle is given by

$$\tan \theta_b = \frac{v^2}{rg} = \frac{(26 \text{ m/s})^2}{(210 \text{ m})(9.8 \text{ m/s}^2)} = 0.33,$$

for a bank angle of $\theta_b = 18°$. I changed the units of the speed for you. You should check my work.

(a)   On the rainy day traffic is moving at 14 m/s. This is slower than the rated speed, so any frictional force must be directed up the incline. Newton's second law is then

$$\sum F_r = N_r - f_r = N \sin \theta - f \cos \theta = \frac{mv^2}{r},$$
$$\sum F_z = N_z + f_z - W = N \cos \theta + f \sin \theta - mg = 0.$$

We can substitute $f = \mu_s N$ to find the minimum value of μ_s which will keep the cars from slipping. There will then be two equations and two unknowns, μ_s and N. Solving for N,

$$N (\sin \theta - \mu_s \cos \theta) = \frac{mv^2}{r} \quad \text{and} \quad N (\cos \theta + \mu_s \sin \theta) = mg.$$

Combining,

$$(\sin \theta - \mu_s \cos \theta) mg = (\cos \theta + \mu_s \sin \theta) \frac{mv^2}{r}$$

The mass cancels out of this expression; we can move r to the left hand side; then rearrange to solve for μ_s.

$$\mu_s = \frac{gr \sin \theta - v^2 \cos \theta}{gr \cos \theta + v^2 \sin \theta}.$$

We know all the numbers. Put them in and we'll find $\mu_s = 0.22$

(b)   How fast can the car go without sliding off the top of the incline? Now the frictional force will point the other way, so Newton's second law is now

$$\sum F_r = N_r + f_r = N \sin \theta + f \cos \theta = \frac{mv^2}{r},$$
$$\sum F_z = N_z - f_z - W = N \cos \theta - f \sin \theta - mg = 0.$$

The bottom equation can be rearranged to show that

$$N = \frac{mg}{\cos \theta - \mu_s \sin \theta}.$$

47

This can be combined with the top equation to give

$$mg\frac{\sin\theta + \mu_s\cos\theta}{\cos\theta - \mu_s\sin\theta} = \frac{mv^2}{r}.$$

Once again, the mass of the car is irrelevant! we can solve this final expression for v using all our previous numbers and get $v = 35$ m/s. That's about 130 km/hr, which seems way too fast for a rainy day, but if cars can go safely at 95 km/hr, they can go safely at 130 km/hr.

E5-49 The force only has an x component, so we can use Eq. 5-19 to find the velocity.

$$v_x = v_{0x} + \frac{1}{m}\int_0^t F_x\, dt = v_0 + \frac{F_0}{m}\int_0^t (1 - t/T)\ dt$$

Even though we want to find the velocity at $t = T$ don't make the mistake of making this the upper limit of integration. To find the position we need to know the velocity for all times, not just one. Integrating,

$$v_x = v_0 + a_0\left(t - \frac{1}{2T}t^2\right)$$

Now put this expression into Eq. 5-20 to find the position as a function of time

$$x = x_0 + \int_0^t v_x\, dt = \int_0^t \left(v_{0x} + a_0\left(t - \frac{1}{2T}t^2\right)\right)\ dt$$

We *can* integrate this expression from 0 to T, because we only need to know $x(T)$. We find

$$x = v_0 T + a_0\left(\frac{1}{2}T^2 - \frac{1}{6T}T^3\right) = v_0 T + a_0\frac{T^2}{3}.$$

Now we can put $t = T$ into the expression for v. You'll get the answer in the book. Really.

P5-3 As the string is pulled the two masses will move together so that the configuration will look like the figure below. The point where the force is applied to the string is massless, so $\sum F = 0$ at that point. We can take advantage of this fact and the figure below to find the tension in the cords, $F/2 = T\cos\theta$. The factor of $1/2$ occurs because only $1/2$ of F is contained in the right triangle that has T as the hypotenuse. From the figure we can find the x component of the force on one mass to be $T_x = T\sin\theta$. Combining,

$$T_x = \frac{F}{2}\frac{\sin\theta}{\cos\theta} = \frac{F}{2}\tan\theta.$$

This doesn't look too much like the expression in the book, but if we look again at the figure and we remember that the tangent is the opposite over the adjacent side, then

$$\tan\theta = \frac{\text{Opposite}}{\text{Adjacent}} = \frac{x}{\sqrt{L^2 - x^2}}$$

And now we have the answer in the book.

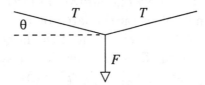

What happens when $x = L$? Well, a_x is infinite according to this expression. Since that could only happen if the tension in the string were infinite (I imagine the string would snap at that), then there must be some other physics that we had previously ignored. But what? The mass of the string, for one.

P5-7   There are four forces on the broom: the force of gravity $W = mg$; the normal force of the floor N; the force of friction f; and the applied force from the person P (the book calls it F). We'll choose y to be straight up and down and x horizontal. Then Newton's second law can be written

$$\sum F_x = P_x - f = P\sin\theta - f = ma_x,$$
$$\sum F_y = N - P_y - W = N - P\cos\theta - mg = ma_y = 0$$

We solve the second equation for N, with the goal of using it to find the friction,

$$N = P\cos\theta + mg.$$

(a)   If the mop slides at constant speed $a_x = 0$, and $f = \mu_k N$. Then Newton's second law for the x components is

$$P\sin\theta - f = P\sin\theta - \mu_k\left(P\cos\theta + mg\right) = 0.$$

We can solve this for P (which was called F in the book);

$$P = \frac{\mu mg}{\sin\theta - \mu_k\cos\theta}.$$

This is the force required to push the broom at constant speed. Push harder and it will accelerate, push less hard and it will eventually stop.

(b)   Note that P becomes negative (or infinite) if $\sin\theta \le \mu_k\cos\theta$. This occurs when $\tan\theta_c \le \mu_k$. If this happens the mop stops moving, to get it started again you must overcome the static friction, but this is impossible if $\tan\theta_0 \le \mu_s$

P5-9 This is another one of my favorite problems. To hold up the smaller block the frictional force between the larger block and smaller block must be as large as the weight of the smaller block. This can be written as $f = mg$. The normal force of the larger block on the smaller block is N, and the frictional force is given by $f \le \mu_s N$. So the smaller block won't fall if $mg \le \mu_s N$.

There are two systems we want to consider: (1) the large block only, and (2) both blocks. For the system there is only one horizontal force on the large block, which is the normal force of the small block on the large block. Newton's third law says this force has a magnitude N, so the acceleration of the large block is $N = Ma$.

There is only one horizontal force on the second system, the force F. So the acceleration of the second system is given by $F = (M + m)a$. The two accelerations are equal, otherwise the blocks won't stick together. Equating, then, gives $N/M = F/(M + m)$.

We can combine this last expression with $mg \le \mu_s N$ and get

$$mg \le \mu_s F \frac{M}{M + m}$$

or, solving for F,

$$F \ge \frac{g(M + m)m}{\mu_s M} = \frac{(9.81\,\mathrm{m/s^2})(88\,\mathrm{kg} + 16\,\mathrm{kg})(16\,\mathrm{kg})}{(0.38)(88\,\mathrm{kg})} = 490\,\mathrm{N}$$

P5-11 The rope wraps around the dowel and there is a contribution to the frictional force Δf from each small segment of the rope where it touches the dowel. There is also a normal force ΔN at each point where the contact occurs. We can find ΔN much the same way that we solve the circular motion problem.

In the figure on the left below we see that we can form a triangle with long side T and short side ΔN. In the figure on the right below we see a triangle with long side r and short side $r\Delta\theta$. These triangles are similar, so $r\Delta\theta/r = \Delta N/T$.

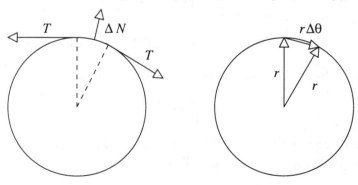

Now $\Delta f = \mu \Delta N$ and $T(\theta) + \Delta f \approx T(\theta + \Delta\theta)$. Combining, and taking the limit as $\Delta\theta \to 0$, $dT = df$

$$\int \frac{1}{\mu}\frac{dT}{T} = \int d\theta$$

integrating both sides of this expression,

$$\int \frac{1}{\mu}\frac{dT}{T} = \int d\theta,$$

$$\frac{1}{\mu}\ln T\Big|_{T_1}^{T_2} = \pi,$$

$$T_2 = T_1 e^{\pi\mu}.$$

In this case T_1 is the weight and T_2 is the downward force.

P5-17 (a) This is *not* a statics problem— the net force on the ball is *not* zero. In order to keep the ball moving in a circle there must be a net centripetal force F_c directed horizontally to toward the rod. There are only *three* forces which act on the ball: the force of gravity, $W = mg = (1.34\,\text{kg})(9.81\,\text{m/s}^2) = 13.1\,\text{N}$; the tension in the top string $T_1 = 35.0\,\text{N}$, and the tension in the bottom string, T_2.

The components of the force from the tension in the top string are

$$T_{1,x} = (35.0\,\text{N})\cos 30° = 30.3\,\text{N} \text{ and } T_{1,y} = (35.0\,\text{N})\sin 30° = 17.5\,\text{N}.$$

The vertical components *do* balance, so

$$T_{1,y} + T_{2,y} = W,$$

or $T_{2,y} = (13.1\,\text{N}) - (17.5\,\text{N}) = -4.4\,\text{N}$. From this we can find the tension in the bottom string,

$$T_2 = T_{2,y}/\sin(-30°) = 8.8\,\text{N}.$$

(b) The net force on the object will be the sum of the two horizontal components,

$$F_c = (30.3\,\text{N}) + (8.8\,\text{N})\cos 30° = 37.9\,\text{N}.$$

(c) The speed will be found from

$$v = \sqrt{a_c r} = \sqrt{F_c r/m},$$

$$= \sqrt{(37.9\,\text{m})(1.70\,\text{m})\sin 60°/(1.34\,\text{kg})} = 6.45\,\text{m/s}.$$

Chapter 6

Momentum

E6-3 The figure below shows the initial and final momentum vectors arranged to geometrically show $\vec{p}_f - \vec{p}_i = \Delta\vec{p}$. We can use the cosine law to find the length of $\Delta\vec{p}$.

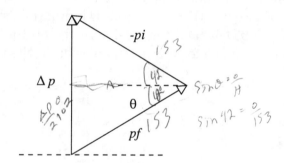

The angle $\alpha = 42° + 42°$, $p_i = mv = (4.88\,\text{kg})(31.4\,\text{m/s}) = 153\,\text{kg·m/s}$. Then the magnitude of $\Delta\vec{p}$ is

$$\Delta p = \sqrt{(153\,\text{kg·m/s})^2 + (153\,\text{kg·m/s})^2 - 2(153\,\text{kg·m/s})^2 \cos(84°)} = 205\,\text{kg·m/s},$$

directed up from the plate. By symmetry it must be perpendicular.

E6-7 Choose the coordinate system so that the ball is only moving along the x axis, with away from the batter as positive. Then $p_{fx} = mv_{fx} = (0.150\,\text{kg})(61.5\,\text{m/s}) = 9.23\,\text{kg·m/s}$ and $p_{ix} = mv_{ix} = (0.150\,\text{kg})(-41.6\,\text{m/s}) = -6.24\,\text{kg·m/s}$. The impulse is given by $J_x = p_{fx} - p_{ix} = 15.47\,\text{kg·m/s}$. We can find the average force by application of Eq. 6-7:

$$F_{\text{av},x} = \frac{J_x}{\Delta t} = \frac{(15.47\,\text{kg · m/s})}{(4.7 \times 10^{-3}\,\text{s})} = 3290\,\text{N}.$$

E6-11 The impulse is the area under the force curve on a force-time graph. We can (1) count the boxes, (2) approximate the curve as a polynomial, and exactly integrate

52

the polynomial, or (3) numerically integrate the curve with something like Simpson's rule.

The number of boxes *completely* inside the curve is 188. The number of boxes which *touch* the curve is 52. If we assume on that half of the area of these border boxes are inside the curve and half are outside, then the area of the curve is 214 boxes. Each box is 100 N by 0.2 ms or 0.02 kg·m/s, so the impulse is 4.28 kg·m/s.

Or we could use Simpson's rule. Then the area is given by

$$J_x = \frac{1}{3}h\left(f_0 + 4f_1 + 2f_2 + 4f_3 + \ldots + 4f_13 + f_14\right),$$

$$= \frac{1}{3}(0.2\,\text{ms})\left(200 + 4 \cdot 800 + 2 \cdot 1200\ldots\text{N}\right)$$

which gives $J_x = 4.28$ kg·m/s. Hey, look, we got the same answer!

Since the impulse is the change in momentum, and the ball started from rest, $p_{fx} = J_x + p_{ix} = 4.28$ kg·m/s. The final velocity is then found from $v_x = p_x/m = 8.6$ m/s.

E6-15 This problem is based on the somewhat embarrassing fiasco which cost NASA a Martian probe when someone forgot to convert between metric and English units in 1999.

A transverse direction means at right angles, so the thrusters have imparted a momentum sufficient to direct the spacecraft $100+3400 = 3500$ km to the side of the original path. The spacecraft is half-way through the six-month journey, so it has three months to move the 3500 km to the side. This corresponds to a transverse speed of
$$v = (3500 \times 10^3\text{m})/(3 \times 30 \times 24 \times 3600\,\text{s}) = 0.45\,\text{m/s}.$$
The required time for the rocket to fire is
$$\Delta t = (5400\,\text{kg})(0.45\,\text{m/s})/(1200\,\text{N}) = 2.0\,\text{s}.$$

E6-17 All of the motion happens along one dimension; we can set up the coordinate system so that all motion happens along the x axis. Since there are no other axes to consider we will drop the x subscript to cut down on confusion. Conservation of momentum is used to solve the problem:

$$P_f = P_i,$$
$$p_{f,m} + p_{f,c} = p_{i,m} + p_{i,c},$$
$$m_m v_{f,m} + m_c v_{f,c} = m_m v_{i,m} + m_c v_{i,c},$$
$$v_{f,c} - v_{i,c} = \frac{m_m v_{i,m} - m_m v_{f,m}}{m_c},$$
$$\Delta v_c = \frac{(75.2\,\text{kg})(2.33\,\text{m/s}) - (75.2\,\text{kg})(0)}{(38.6\,\text{kg})},$$
$$= 4.54\,\text{m/s}.$$

The answer is positive; the cart speed *increases*.

E6-23 As with the Exercise 6-17, all of the motion happens along one dimension; since there are no other axes to consider we will drop the x subscript to cut down on confusion. Conservation of momentum is used to solve the problem:

$$P_f = P_i,$$
$$p_{f,bl} + p_{f,bu} = p_{i,bl} + p_{i,bu},$$
$$m_{bl}v_{f,bl} + m_{bu}v_{f,bu} = m_{bl}v_{i,bl} + m_{bu}v_{i,bu},$$
$$(715 \text{ g})v_{f,bl} + (5.18 \text{ g})(428 \text{ m/s}) = (715 \text{ g})(0) + (5.18 \text{ g})(672 \text{ m/s}),$$

which has solution $v_{f,bl} = 1.77 \, \text{m/s}$.

E6-29 We want to solve Eq. 6-24 for m_2 given that $v_{1,f} = 0$ and $v_{1,i} = -v_{2,i}$. Making these substitutions

$$(0) = \frac{m_1 - m_2}{m_1 + m_2}v_{1,i} + \frac{2m_2}{m_1 + m_2}(-v_{1,i}),$$
$$0 = (m_1 - m_2)v_{1,i} - (2m_2)v_{1,i},$$
$$3m_2 = m_1$$

so $m_2 = 100$ g.

E6-33 Let the initial momentum of the first object be $\vec{p}_{1,i} = m\vec{v}_{1,i}$, that of the second object be $\vec{p}_{2,i} = m\vec{v}_{2,i}$, and that of the combined final object be $\vec{p}_f = 2m\vec{v}_f$. Then

$$\vec{p}_{1,i} + \vec{p}_{2,i} = \vec{p}_f,$$

implies that we can find a triangle with sides of length $p_{1,i}$, $p_{2,i}$, and p_f. These lengths are

$$p_{1,i} = mv_i,$$
$$p_{2,i} = mv_i,$$
$$p_f = 2mv_f = 2mv_i/2 = mv_i,$$

so this is an equilateral triangle. This means the angle between the initial velocities is 120°.

P6-1 The force is the change in momentum over change in time; the momentum is the mass time velocity, so

$$F = \frac{\Delta p}{\Delta t} = \frac{m\Delta v}{\Delta t} = \Delta v \frac{m}{\Delta t} = 2u\mu,$$

since μ is the mass per unit time.

P6-7 The reading on the scale will be the force on the box; there will be a contribution from the weight of the marbles in the box and the force required to stop the marbles as they fall.

The weight of the marbles in the box after a time t is $mgRt$ because Rt is the number of marbles in the box.

The marbles fall a distance h from rest; the time required to fall this distance is $t = \sqrt{2h/g}$, the speed of the marbles when they strike the box is $v = gt = \sqrt{2gh}$. The momentum each marble imparts on the box is then $m\sqrt{2gh}$. If the marbles strike at a rate R then the force required to stop them is $Rm\sqrt{2gh}$.

The reading on the scale is then

$$W = mR(\sqrt{2gh} + gt).$$

This will give a numerical result of

$$(4.60 \times 10^{-3}\text{kg})(115\,\text{s}^{-1}) \left(\sqrt{2(9.81\,\text{m/s}^2)(9.62\,\text{m})} + (9.81\,\text{m/s}^2)(6.50\,\text{s}) \right) = 41.0\,\text{N}.$$

P6-11 We align the coordinate system so that west is $+x$ and south is $+y$. The each car contributes the following to the initial momentum

$$A \;:\; (2720\,\text{lb}/g)(38.5\,\text{mi/h})\hat{\mathbf{i}} = 1.05 \times 10^5\,\text{lb} \cdot \text{mi/h}/g\,\hat{\mathbf{i}},$$
$$B \;:\; (3640\,\text{lb}/g)(58.0\,\text{mi/h})\hat{\mathbf{j}} = 2.11 \times 10^5\,\text{lb} \cdot \text{mi/h}/g\,\hat{\mathbf{j}}.$$

Yes, these are weird looking units. Don't worry; it will simplify to something nice. These become the components of the final momentum. The direction is then

$$\theta = \arctan \frac{2.11 \times 10^5\,\text{lb} \cdot \text{mi/h}/g}{1.05 \times 10^5\,\text{lb} \cdot \text{mi/h}/g} = 63.5°,$$

south of west. The magnitude is the square root of the sum of the squares,

$$2.36 \times 10^5\,\text{lb} \cdot \text{mi/h}/g,$$

and we divide this by the mass $(6360\,\text{lb}/g)$ to get the final speed after the collision: 37.1 mi/h.

P6-15 The equations for elastic collisions with an object at rest are Eq. 6-27, so

$$v_{2,\text{f}} = \frac{2m_1}{m_1 + m_2} v_{1,\text{i}}.$$

(a) We get

$$v_{2,\text{f}} = \frac{2(220\,\text{g})}{(220\,\text{g}) + (46.0\,\text{g})}(45.0\,\text{m/s}) = 74.4\,\text{m/s}.$$

(a) Doubling the mass of the clubhead we get

$$v_{2,f} = \frac{2(440\text{ g})}{(440\text{ g}) + (46.0\text{ g})}(45.0\,\text{m/s}) = 81.5\,\text{m/s}.$$

(a) Tripling the mass of the clubhead we get

$$v_{2,f} = \frac{2(660\text{ g})}{(660\text{ g}) + (46.0\text{ g})}(45.0\,\text{m/s}) = 84.1\,\text{m/s}.$$

Although the heavier club helps some, the maximum speed to get out of the ball will be less than twice the speed of the club.

P6-19 (a) The speed of the bullet after leaving the first block but before entering the second can be determined by momentum conservation.

$$
\begin{aligned}
P_f &= P_i, \\
p_{f,bl} + p_{f,bu} &= p_{i,bl} + p_{i,bu}, \\
m_{bl}v_{f,bl} + m_{bu}v_{f,bu} &= m_{bl}v_{i,bl} + m_{bu}v_{i,bu}, \\
(1.78\text{kg})(1.48\,\text{m/s}) + (3.54\times10^{-3}\text{kg})(1.48\,\text{m/s}) &= (1.78\text{kg})(0) + (3.54\times10^{-3}\text{kg})v_{i,bu},
\end{aligned}
$$

which has solution $v_{i,bl} = 746\,\text{m/s}$.

(b) We do the same steps again, except applied to the first block,

$$
\begin{aligned}
P_f &= P_i, \\
p_{f,bl} + p_{f,bu} &= p_{i,bl} + p_{i,bu}, \\
m_{bl}v_{f,bl} + m_{bu}v_{f,bu} &= m_{bl}v_{i,bl} + m_{bu}v_{i,bu}, \\
(1.22\text{kg})(0.63\,\text{m/s}) + (3.54\times10^{-3}\text{kg})(746\,\text{m/s}) &= (1.22\text{kg})(0) + (3.54\times10^{-3}\text{kg})v_{i,bu},
\end{aligned}
$$

which has solution $v_{i,bl} = 963\,\text{m/s}$.

Chapter 7

Systems of Particles

E7-3 The center of mass velocity is given by Eq. 7-1,

$$\vec{v}_{cm} = \frac{m_1\vec{v}_1 + m_2\vec{v}_2}{m_1 + m_2},$$

$$= \frac{(2210 \text{ kg})(105 \text{ km/h}) + (2080 \text{ kg})(43.5 \text{ km/h})}{(2210 \text{ kg}) + (2080 \text{ kg})} = 75.2 \text{ km/h}.$$

It does make a difference which direction the cars are moving; since we are told they are moving down the same road in the same direction, we only had a single component for the velocity vectors.

E7-7 The center of mass of the boat + dog doesn't move because there are no external forces on the system. Define the coordinate system so that distances are measured from the shore, so toward the shore is in the negative x direction. The *change* in position of the center of mass is given by

$$\Delta x_{cm} = \frac{m_d\Delta x_d + m_b\Delta x_b}{m_d + m_b} = 0,$$

and as pointed out, this equals zero. Both Δx_d and Δx_b are measured with respect to the shore; we are given $\Delta x_{db} = -8.50$ ft, the displacement of the dog with respect to the boat. But

$$\Delta x_d = \Delta x_{db} + \Delta x_b$$

an expression that can be substituted into the center of mass expression. Since we want to find out about the dog, we'll substitute for the boat's displacement,

$$0 = \frac{m_d\Delta x_d + m_b\left(\Delta x_d - \Delta x_{db}\right)}{m_d + m_b}.$$

Rearrange and solve for Δx_d. Now the problem gives the *weights* not the *masses*. We could spend the time to find the mass, but instead we'll use $W = mg$ and multiply the top and bottom of the following expression by g. Then we'll use the weights.

$$\Delta x_d = \frac{m_b \Delta x_{db}}{m_d + m_b} \frac{g}{g} = \frac{(46.4 \text{ lb})(-8.50 \text{ ft})}{(10.8 \text{ lb}) + (46.4 \text{ lb})} = -6.90 \text{ ft}.$$

The dog is now $21.4 - 6.9 = 14.5$ feet from shore.

E7-11 The center of mass of the three hydrogen atoms will be at the center of the pyramid base. We solve this with symmetry arguments; rotating the triangle through $120°$ about the dotted line in Fig. 7-27 returns the same triangle, so the center of mass shouldn't have moved. That will only happen if it is in the center. The problem is then reduced to finding the center of mass of the nitrogen atom and the three hydrogen atom triangle. This molecular center of mass must lie on the dotted line in Fig. 7-27, and because one nitrogen is considerably more massive than three hydrogens, the center of mass should be closer to the nitrogen.

Choose the coordinate system so that the y axis is aligned with the dotted line in the figure. Set the location of the nitrogen atom at $y = 0$. Then the location of the plane of the hydrogen atoms can be found from Pythagoras theorem

$$y_h = \sqrt{(10.14 \times 10^{-11} \text{m})^2 - (9.40 \times 10^{-11} \text{m})^2} = 3.8 \times 10^{-11} \text{m}.$$

This distance can be used to find the center of mass of the molecule. From Eq. 7-2,

$$y_{cm} = \frac{m_n y_n + m_h y_h}{m_n + m_h} = \frac{(13.9 m_h)(0) + (3 m_h)(3.8 \times 10^{-11} \text{m})}{(13.9 m_h) + (3 m_h)} = 6.75 \times 10^{-12} \text{m}.$$

E7-13 The center of mass should lie on the perpendicular bisector of the rod of mass $3M$. We can deduce this by symmetry; flip the picture over (left/right) and the picture looks the same, so the center of mass shouldn't have moved. We can view the system as having two parts: the heavy rod of mass $3M$ and the two light rods each of mass M. The two light rods have a center of mass at the center of the square.

Both of these center of masses are located along the vertical line of symmetry for the object. The center of mass of the heavy bar is at $y_{h,cm} = 0$, while the *combined* center of mass of the two light bars is at $y_{l,cm} = L/2$, where down is positive. The center of mass of the system is then at

$$y_{cm} = \frac{2M y_{l,cm} + 3M y_{h,cm}}{2M + 3M} = \frac{2(L/2)}{5} = L/5.$$

E7-19 Label the velocities of the various containers as $\vec{v}_k$ where k is an integer between one and twelve. Don't assume these are Initially zero, because they don't have to be. The mass of each container is m. The subscript "g" refers to the goo; the subscript k refers to the kth container.

The total momentum before the collision is given by

$$\vec{P} = \sum_k m\vec{v}_{k,i} + m_g\vec{v}_{g,i} = 12m\vec{v}_{\text{cont.,cm}} + m_g\vec{v}_{g,i}.$$

We are told, however, that the initial velocity of the center of mass of the containers is at rest, so the initial momentum simplifies to $\vec{P} = m_g\vec{v}_{g,i}$, and has a magnitude of 4000 kg·m/s.

(a) The *total* final momentum is the same as the initial momentum. Now there is a problem: if the question wants the center of mass velocity for *just* the cylinders but not the space goo, we can't answer it. If instead the question meant to ask for the center of mass velocity of the entire system, then

$$v_{\text{cm}} = \frac{P}{12m + m_g} = \frac{(4000\,\text{kg·m/s})}{12(100.0\,\text{kg}) + (50\,\text{kg})} = 3.2\,\text{m/s}.$$

(b) Assuming that our interpretation of the question in part (a) is correct, then it doesn't matter if the cord breaks, we'll get the same answer for the motion of the center of mass.

E7-21 Use Eq. 7-32. The initial velocity of the rocket is 0. The mass ratio can then be found from a minor rearrangement;

$$\frac{M_i}{M_f} = e^{|v_f/v_{\text{rel}}|}$$

The "flipping" of the left hand side of this expression is possible because the exhaust velocity is *negative* with respect to the rocket. For part (a) $M_i/M_f = e = 2.72$. For part (b) $M_i/M_f = e^2 = 7.39$.

E7-25 We'll use Eq. 7-4 to solve this problem, but since we are given *weights* instead of *mass* we'll multiply the top and bottom by g like we did in Exercise 7-7. Then

$$\vec{v}_{\text{cm}} = \frac{m_1\vec{v}_1 + m_2\vec{v}_2}{m_1 + m_2}\frac{g}{g} = \frac{W_1\vec{v}_1 + W_2\vec{v}_2}{W_1 + W_2}.$$

Now for the numbers

$$v_{\text{cm}} = \frac{(9.75\,\text{T})(1.36\,\text{m/s}) + (0.50\,\text{T})(0)}{(9.75\,\text{T}) + (0.50\,\text{T})} = 1.29\,\text{m/s}.$$

P7-3 This is a glorified Atwood's machine problem. The total mass on the right side is the mass per unit length times the length, $m_r = \lambda x$; similarly the mass on the left is given by $m_l = \lambda(L - x)$. We know from Eq. 5-4 the acceleration of two different masses in Atwood's machine, so

$$a = \frac{m_2 - m_1}{m_2 + m_1}g = \frac{\lambda x - \lambda(L - x)}{\lambda x + \lambda(L - x)}g = \frac{2x - L}{L}g$$

which solves the problem. The acceleration is in the direction of the side of length x if $x > L/2$.

P7-5 By symmetry, the center of mass of the empty storage tank should be in the very center, along the axis at a height $y_{t,cm} = H/2$. We can pretend that the entire mass of the tank, $m_t = M$, is located at this point.

The center of mass of the gasoline is also, by symmetry, located along the axis at half the height of the gasoline, $y_{g,cm} = x/2$. But the mass of the gasoline is more complicated to find. The mass, if the tank were filled to a height H, is m, assuming a uniform density for the gasoline, we can conclude that the mass present when the level of gas reaches a height x is $m_g = mx/H$.

(a) The center of mass of the entire system is at the center of the cylinder when the tank is full and when the tank is empty. When the tank is half full the center of mass is below the center. So as the tank changes from full to empty the center of mass drops, reaches some lowest point, and then rises back to the center of the tank.

(b) The center of mass of the entire system is found from

$$y_{cm} = \frac{m_g y_{g,cm} + m_t y_{t,cm}}{m_g + m_t} = \frac{(mx/H)(x/2) + (M)(H/2)}{(mx/H) + (M)} = \frac{mx^2 + MH^2}{2mx + 2MH}.$$

We want to find the minimum of y_{cm} with respect to changes in x. Take the derivative; you can do it by hand (I used Maple) and get

$$\frac{dy_{cm}}{dx} = \frac{m\left(mx^2 + 2xMH - MH^2\right)}{(mx + MH)^2}$$

Set this equal to zero to find the minimum, this means we want the numerator to vanish, or $mx^2 + 2xMH - MH^2 = 0$. There are two solutions,

$$x = \frac{-M \pm \sqrt{M^2 + mM}}{m}H.$$

Only the positive answer is of interest here.

P7-11 Consider Eq. 7-31. We want the barges to continue at constant speed, so the left hand side of that equation vanishes. Then

$$\sum \vec{F}_{ext} = -\vec{v}_{rel} \frac{dM}{dt}.$$

We are told that the frictional force is independent of the weight, since the speed doesn't change the frictional force should be constant and equal in magnitude to the force exerted by the engine *before* the shoveling happens. Then $\sum \vec{F}_{ext}$ is equal to the additional force required from the engines. We'll call it $\vec{P}$.

There is a sideways component to $\vec{v}_{rel}$; otherwise the coal can't get over to the other barge. We can't determine this sideways component from the information given, but it will probably be small, so we will ignore it, except to warn the pilots of each boat that they will need to turn in toward each other *slightly* to counteract the effect.

So we are now in a position to find the magnitude of $\vec{P}$ for the barge which has coal being shoveled into it. The relative speed of the coal to the faster moving cart has magnitude: $21.2 - 9.65 = 11.6$ km/h= 3.22 m/s. The mass flux is 15.4 kg/s, so $P = (3.22 \text{ m/s})(15.4 \text{kg/s}) = 49.6$ N. Not a big force. The faster moving cart will need to *increase* the engine force by 49.6 N. The slower cart won't need to do anything, because the coal left the slower barge with a relative speed of *zero* according to our approximation.

Chapter 8

Rotational Kinematics

E8-1   An n-dimensional object can be oriented by stating the position of n different *carefully chosen* points P_i inside the body. Since each point has n coordinates, one might think there are n^2 coordinates required to completely specify the position of the body. But if the body is rigid then the distances between the points are fixed. There is a distance d_{ij} for every pair of points P_i and P_j. For each distance d_{ij} we need one fewer coordinate to specify the position of the body. It is useful to write this in a table.

Dimension n	Number of P_i	n^2	Number of d_{ij}	Coordinates Required
1	1	1	0	1
2	2	4	1	3
3	3	9	3	6
4	4	16	6	10
n	n	n^2	$(n-1)n/2$	$(n+1)n/2$

In the last line we used a recursive sum rule: adding a point to $(n-1)$ points introduces $(n-1)$ *additional* distances. So if S_{n-1} is the number of d_{ij} for $n-1$ points, then $S_n = S_{n-1} + (n-1)$. The number of coordinates required to specify the position and orientation is given by $n^2 - S_n$.

E8-5 (a)   Integrate.

$$\omega_z = \omega_0 + \int_0^t \left(4at^3 - 3bt^2\right) dt = \omega_0 + at^4 - bt^3$$

(b)   Integrate, again.

$$\Delta\theta = \int_0^t \omega_z dt = \int_0^t \left(\omega_0 + at^4 - bt^3\right) dt = \omega_0 t + \frac{1}{5}at^5 - \frac{1}{4}bt^4$$

E8-9 (a) For convenience we will choose the arrow to be shot through the point a point vertical above the axle, and we will measure angles on the wheel from that point. The arrow will pass without hitting a spoke if a spoke passes $\theta = 0$ just before the arrow enters the plane of the wheel, and the next spoke doesn't get to $\theta = 0$ until after the arrow has left the wheel. Since there are eight spokes, this means the wheel can make no more than 1/8 of a revolution while the arrow traverses the plane of the wheel. The wheel rotates at 2.5 rev/s; it makes one revolution every $1/2.5 = 0.4$ s; so the arrow must pass through the wheel in less than $0.4/8 = 0.05$ s.

The arrow is 0.24 m long, so it must move at least one arrow length in 0.05 s. The corresponding minimum speed is $(0.24 \text{ m})/(0.05 \text{ s}) = 4.8$ m/s. Not very fast.

(b) It does not matter where you aim, because the wheel is rigid. It is the angle through which the spokes have turned, not the distance, which matters here.

E8-11 The time required for a planet to make a revolution around the sun relative to Earth is the time between seeing the plane appear in the same position in the sky. For Exercise 8-10, the time could be measured between when the planet appears exactly overhead at midnight. This won't work for Mercury or Venus, since they never appear directly overhead at midnight; but it could be measured between the times the planet crosses the face of the sun. In other words, we look for the times when the sun, the Earth, and the other planet are collinear in some specified order.

Since the inner planets revolve around the sun more quickly than Earth, after one year the Earth has returned to the original position, but the inner planet has completed *more* then one revolution. The inner planet must then have "caught-up" with the Earth *before* the Earth has completed a revolution. If θ_E is the angle through which Earth moved and θ_P is the angle through which the planet moved, then $\theta_P = \theta_E + 2\pi$, since the inner planet completed one more revolution than the Earth.

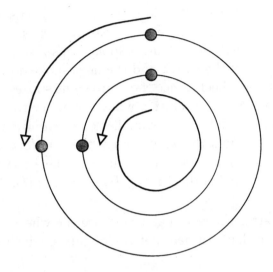

If ω_p is the angular velocity of the planet, then the angle through which it moves during the time T_S (the time for the planet to line up with the Earth). Then

$$\begin{aligned} \theta_P &= \theta_E + 2\pi, \\ \omega_P T_S &= \omega_E T_S + 2\pi, \\ \omega_P &= \omega_E + 2\pi/T_S \end{aligned}$$

The angular velocity of a planet is $\omega = 2\pi/T$, where T is the period of revolution. Substituting this into the last equation above yields

$$1/T_P = 1/T_E + 1/T_S,$$

which is similar to, but different from, the result of Exercise 8-10.

E8-15 (a) This problem is *very* similar to the kinematic problems of Chapter 2, except that now the relevant equations are all Greek to me. $\omega_{0z} = 25.2$ rad/s; $\omega_z = 0$; $t = 19.7$ s; and α_z and ϕ are unknown. From Eq. 8-6,

$$\begin{aligned} \omega_z &= \omega_{0z} + \alpha_z t, \\ (0) &= (25.2 \text{ rad/s}) + \alpha_z(19.7 \text{ s}), \\ \alpha_z &= -1.28 \text{ rad/s}^2 \end{aligned}$$

The negative sign indicates that the wheel is "slowing down", but use this expression carefully, because if the wheel were allowed to continue this negative acceleration it would begin to "speed up" with the opposite rotation.

(b) We use Eq. 8-7 to find the angle through which the wheel rotates.

$$\phi = \phi_0 + \omega_{0z}t + \frac{1}{2}\alpha_z t^2 = (0) + (25.2 \text{ rad/s})(19.7 \text{ s}) + \frac{1}{2}(-1.28 \text{ rad/s}^2)(19.7 \text{ s})^2 = 248 \text{ rad}.$$

(c) Most people can't easily visualize radians, so we often give information in revolutions. Then

$$\phi = 248 \text{ rad} \frac{1 \text{ rev}}{2\pi \text{ rad}} = 39.5 \text{ rev}$$

E8-19 (a) We are given $\phi = 42.3$ rev= 266 rad, $\omega_{0z} = 1.44$ rad/s, and $\omega_z = 0$. Assuming a uniform deceleration, the average angular velocity during the interval is

$$\omega_{\text{av},z} = \frac{1}{2}(\omega_{0z} + \omega_z) = 0.72 \text{ rad/s}.$$

Then the time taken for deceleration is given by $\phi = \omega_{\text{av},z}t$, so $t = 369$ s.

(b) Now that we know the time, the angular acceleration can be found from Eq. 8-6,

$$
\begin{aligned}
\omega_z &= \omega_{0z} + \alpha_z t, \\
(0) &= (1.44 \text{ rad/s}) + \alpha_z(369 \text{ s}), \\
\alpha_z &= -3.9 \times 10^{-3} \text{ rad/s}^2.
\end{aligned}
$$

(c) Since time is squared in Eq. 8-7 we expect that the final answer to this part will somehow involve $\sqrt{2}$. We'll solve Eq. 8-7 for t, except now $\phi = 21.2$ rev= 133 rad:

$$
\begin{aligned}
\phi &= \phi_0 + \omega_{0z}t + \frac{1}{2}\alpha_z t^2, \\
(133 \text{ rad}) &= (0) + (1.44 \text{ rad/s})t + \frac{1}{2}(-3.9 \times 10^{-3} \text{ rad/s}^2)t^2, \\
0 &= -133 + (1.44 \text{ s}^{-1})t - (-1.95 \times 10^{-3}\text{s}^{-2})t^2.
\end{aligned}
$$

Solving this quadratic expression yields two answers: $t = 108$ s and $t = 630$ s. Which is right? Both, in a sense. You can also write these two answers as $t = 369 \pm 261$ s. Notice the 369 s? That's when the flywheel has come to a rest. The 108 s corresponds to when the wheel was still rotating forward, but slowing down. Then it stops and rotates in the opposite direction; it passes the same angle after 630 s from start.

But where is the $\sqrt{2}$ I promised you? Well, $261 = 369/\sqrt{2}$. Maybe a *wee* bit hidden, but it was mildly entertaining to seek it out.

E8-23 (a) The angular speed is given by Eq. 8-9, $v_T = \omega r$. So $\omega = v_T/r = (28,700$ km/hr)$/(3220$ km$) = 8.91$ rad/hr. That's the same thing as 2.48×10^{-3} rad/s.

(b) We find the radial acceleration from Eq. 8-11 $a_R = \omega^2 r = (8.91 \text{ rad/h})^2 (3220$ km$) = 256000 \text{ km/h}^2$. How does this acceleration compare to something more well known, like the acceleration of free fall?

$$a_R = 256000 \text{ km/h}^2 (1/3600 \text{ h/s})^2 (1000 \text{ m/km}) = 19.8 \text{ m/s}^2,$$

or about twice g.

(c) If the speed is constant then the tangential acceleration is constant, regardless of the shape of the trajectory!

E8-27 (a) The pilot sees the propeller rotate, no more. So the tip of the propeller is moving with a tangential velocity of $v_T = \omega r = (2000 \text{ rev/min})(2\pi \text{ rad/rev})(1.5$ m$) = 18900 \text{ m/min}$. This is the same thing as 315 m/s.

(b) The observer on the ground sees this tangential motion and sees the forward motion of the plane. These two velocity components are perpendicular, so the magnitude of the sum can be found from Pythagoras' Theorem.

$$\sqrt{(315 \text{ m/s})^2 + (133 \text{ m/s})^2} = 342 \text{ m/s}.$$

E8-33 (a) The object is "slowing down", so $\vec{\alpha} = (-2.66 \text{ rad/s}^2)\hat{k}$. We know the direction because it is rotating about the z axis and we are given the direction of $\vec{\omega}$. Then from Eq. 8-19, $\vec{v} = \vec{\omega} \times \vec{R} = (14.3 \text{ rad/s})\hat{k} \times [(1.83 \text{ m})\hat{j} + (1.26 \text{ m})\hat{k}]$. But only the cross term $\hat{k} \times \hat{j}$ survives, so $\vec{v} = (-26.2 \text{ m/s})\hat{i}$.

(b) We find the acceleration from Eq. 8-21,

$$
\begin{aligned}
\vec{a} &= \vec{\alpha} \times \vec{R} + \vec{\omega} \times \vec{v}, \\
&= (-2.66 \text{ rad/s}^2)\hat{k} \times [(1.83 \text{ m})\hat{j} + (1.26 \text{ m})\hat{k}] + (14.3 \text{ rad/s})\hat{k} \times (-26.2 \text{ m/s})\hat{i}, \\
&= (4.87 \text{ m/s}^2)\hat{i} + (-375 \text{ m/s}^2)\hat{j}.
\end{aligned}
$$

P8-5 The first problem is to find out through which angle the disk really did turn. There are two parts: the acceleration and the deceleration. They are connected in a way that the final angular velocity of the first part is the initial angular velocity of the second. We can be somewhat sneaky, however, because the *average* angular velocity is really what we want to know.

The final angular velocity during the acceleration phase is $\omega_z = \alpha_z t = (3.0$ rad/s$)(4.0 \text{ s}) = 12.0 \text{ rad/s}$. Since both the acceleration and deceleration phases are uniform with endpoints $\omega_z = 0$, the average angular velocity for both phases is the same, and given by half of the maximum: $\omega_{\text{av},z} = 6.0 \text{ rad/s}$.

The angle through which the wheel turns is then

$$\phi = \omega_{\mathrm{av},z} t = (6.0 \text{ rad/s})(4.1 \text{ s}) = 24.6 \text{ rad}.$$

The time is the total for *both* phases.

(a) The first student sees the wheel rotate through the smallest angle less than one revolution; this student would have no idea that the disk had rotated more than once. Since the disk moved through 3.92 revolutions, the first student will either assume the disk moved forward through 0.92 revolutions or backward through 0.08 revolutions.

(b) According to whom? We've already answered from the perspective of the second student.

P8-9 (a) We don't need to know the radius of the wheel to answer the first part. There are 500 teeth (and 500 spaces between these teeth); so disk rotates $2\pi/500$ rad between the outgoing light pulse and the incoming light pulse. The light traveled 1000 m (it went both ways), so the elapsed time is $t = (1000 \text{ m})/(3\times10^8 \text{ m/s}) = 3.33\times10^{-6}$s.

Then the angular speed of the disk is $\omega_z = \phi/t = 1.26\times10^{-2}$ rad$)/(3.33\times10^{-6}$s$) = 3800$ rad/s.

(b) The linear speed of a point on the edge of the would be

$$v_T = \omega R = (3800 \text{ rad/s})(0.05 \text{ m}) = 190 \text{ m/s}.$$

P8-13 (a) Let the rocket sled move along the line $x = b$. The observer is at the origin and sees the rocket move with a constant angular speed, so the angle made with the x axis increases according to $\theta = \omega t$. The observer, rocket, and starting point form a right triangle; the position y of the rocket is the opposite side of this triangle, so

$$\tan\theta = y/b \text{ implies } y = b/\tan\omega t.$$

We want to take the derivative of this with respect to time and get

$$v(t) = \omega b/\cos^2(\omega t).$$

(b) The speed becomes infinite (which is clearly unphysical) when $t = \pi/2\omega$.

Chapter 9

Rotational Dynamics

E9-1 Use Eq. 9-3, $\vec{\tau} = \vec{r} \times \vec{F}$. In terms of components, we use Eq. 9-4,

$$\vec{\tau} = (yF_z - zF_y)\hat{\mathbf{i}} + (zF_x - xF_z)\hat{\mathbf{j}} + (xF_y - yF_x)\hat{\mathbf{k}}$$

or Eq. 9-2, $\tau = rF\sin\theta$.

It is easiest to use Eq. 9-4. In this exercise $\vec{r} = (2.0\text{ m})\hat{\mathbf{i}} + (3.0\text{ m})\hat{\mathbf{j}}$.

(a) First, $\vec{F} = (5.0\text{ N})\hat{\mathbf{i}}$. With experience you can quickly ignore all of the zero terms in the Eq. 9-4; here we will purposefully do it the long way,

$$
\begin{aligned}
\vec{\tau} &= [yF_z - zF_y]\hat{\mathbf{i}} + [zF_x - xF_z]\hat{\mathbf{j}} + [xF_y - yF_x]\hat{\mathbf{k}}, \\
&= [y(0) - (0)(0)]\hat{\mathbf{i}} + [(0)F_x - x(0)]\hat{\mathbf{j}} + [x(0) - yF_x]\hat{\mathbf{k}}, \\
&= [-yF_x]\hat{\mathbf{k}} = -(3.0\text{ m})(5.0\text{ N})\hat{\mathbf{k}} = -(15.0\text{ N}\cdot\text{m})\hat{\mathbf{k}}.
\end{aligned}
$$

(b) Now $\vec{F} = (5.0\text{ N})\hat{\mathbf{j}}$. We won't repeat finding all of the zero terms, we'll just write the last line:

$$\vec{\tau} = [xF_y]\hat{\mathbf{k}} = (2.0\text{ m})(5.0\text{ N})\hat{\mathbf{k}} = (10\text{ N}\cdot\text{m})\hat{\mathbf{k}}.$$

(c) Finally, $\vec{F} = (-5.0\text{ N})\hat{\mathbf{i}}$. This will look like part (a), except for an extra negative sign:

$$\vec{\tau} = [-yF_x]\hat{\mathbf{k}} = -(3.0\text{ m})(-5.0\text{ N})\hat{\mathbf{k}} = (15.0\text{ N}\cdot\text{m})\hat{\mathbf{k}}.$$

E9-5 The cross product of two vectors will always be perpendicular to the two vectors (*remember this, it is a question in the E & M part of volume 2!*). Since $\vec{r}$ and $\vec{s}$ lie in the xy plane, then $\vec{t} = \vec{r} \times \vec{s}$ must be perpendicular to that plane, and can only point along the z axis. I wrote $\vec{t}$ instead of $\vec{\text{tau}}$ because we are *not* talking about torque here.

The magnitude is found from the equivalent of Eq. 9-2, $\tau = rs\sin\theta$. The angle between $\vec{r}$ and $\vec{s}$ is $320° - 85° = 235°$. So $|\vec{t}| = rs\sin\theta = (4.5)(7.3)\sin(235°) = 27$. You might wonder about the negative sign. Since we are concerned with magnitudes we'll ignore it, but it provides information about the direction of $\vec{t}$.

Now for the direction of $\vec{t}$. The smaller rotation to bring $\vec{r}$ into $\vec{s}$ is through a counterclockwise rotation; the right hand rule would then show that the cross product points along the *positive z* direction.

E9-9   This exercise is a three dimensional generalization of Ex. 9-1, except nothing is zero.

$$\begin{aligned}
\vec{\tau} &= [yF_z - zF_y]\hat{\mathbf{i}} + [zF_x - xF_z]\hat{\mathbf{j}} + [xF_y - yF_x]\hat{\mathbf{k}}, \\
&= [(-2.0\ \text{m})(4.3\ \text{N}) - (1.6\ \text{m})(-2.4\ \text{N})]\hat{\mathbf{i}} + [(1.6\ \text{m})(3.5\ \text{N}) - (1.5\ \text{m})(4.3\ \text{N})]\hat{\mathbf{j}} \\
&\quad + [(1.5\ \text{m})(-2.4\ \text{N}) - (-2.0\ \text{m})(3.5\ \text{N})]\hat{\mathbf{k}}, \\
&= [-4.8\ \text{N}\cdot\text{m}]\hat{\mathbf{i}} + [-0.85\ \text{N}\cdot\text{m}]\hat{\mathbf{j}} + [3.4\ \text{N}\cdot\text{m}]\hat{\mathbf{k}}.
\end{aligned}$$

E9-13 (a)   Negligible mass rods might not be realistic, but they are wonderful in physics! Rotational inertia is additive so long as we consider the inertia about the same axis. We can use Eq. 9-10:

$$I = \sum m_n r_n^2 = (0.075\ \text{kg})(0.42\ \text{m})^2 + (0.030\ \text{kg})(0.65\ \text{m})^2 = 0.026\ \text{kg}\cdot\text{m}^2.$$

(b)   We never needed to know the angular separation of the two objects; putting the two objects on one rod won't change the inertia so long as the distance from the origin are the same.

E9-17   There are two reasonable approaches to this problem; both have merit, so we'll do both.

Method I   Start from Eq. 9-17, which is the rotational inertia for a slab about an axis through the center, and apply the parallel axis theorem. The diagonal distance from the axis through the center of mass and the axis through the edge is $h = \sqrt{(a/2)^2 + (b/2)^2}$, so

$$I = I_{cm} + Mh^2 = \frac{1}{12}M\left(a^2 + b^2\right) + M\left((a/2)^2 + (b/2)^2\right) = \left(\frac{1}{12} + \frac{1}{4}\right)M\left(a^2 + b^2\right).$$

Simplifying, $I = \frac{1}{3}M\left(a^2 + b^2\right)$.

Method II We could also use brute force and integrate Eq. 9-15. This is a *three* dimensional integration, and we will need to take advantage of the fact that

$$\frac{dm}{M} = \frac{dx \, dy \, dz}{abc} \quad \text{and} \quad r^2 = x^2 + y^2.$$

The first expression is possible if we assume the body is uniform; the second is because we will measure distances from the axis of rotation. Then

$$
\begin{aligned}
I &= \int r^2 dm = \int_0^a \int_0^b \int_0^c \left(a^2 + b^2\right) \frac{M}{abc} dx \, dy \, dz, \\
&= \frac{M}{abc} \int_0^a \int_0^b \left(a^2 c + b^2 c\right) dx \, dy \text{ integration over } z, \\
&= \frac{M}{abc} \int_0^a \left(a^2 bc + \frac{1}{3} b^3 c\right) dx \text{ integration over } y, \\
&= \frac{M}{abc} \left(\frac{1}{3} a^3 bc + \frac{1}{3} ab^3 c\right) \text{ integration over } x, \\
&= \frac{1}{3} M \left(a^2 + b^2\right).
\end{aligned}
$$

E9-19 This exercise is similar to Ex. 9-13, except now the rods aren't massless. The rotational inertia is the sum of the rotational inertia of the three parts— that of particle one, that of particle two, and that of the *one* rod of length $2L$ and mass $2M$. We treat the rod as a single entity because this approach cuts in half the amount of work we need to do.

For particle one $I_1 = mr^2 = mL^2$; for particle two $I_2 = mr^2 = m(2L)^2 = 4mL^2$, since the second particle is farther from the axis of rotation. The rotational inertia of the rod can be found from Fig 9-15: $I = \frac{1}{3} ML^2$. Be careful not to get confused here, because the mass of the rod is really $2M$ and the length is really $2L$; these expressions should be substituted into the equation from Fig. 9-15, then $I_{\text{rod}} = \frac{1}{3}(2M)(2L)^2 = \frac{8}{3} ML^2$. Add the three inertias:

$$I = \left(5m + \frac{8}{3} M\right) L^2.$$

E9-25 The ladder slips if the force of static friction required to keep the ladder up exceeds $\mu_s N$. Equations 9-31 give us the normal force in terms of the masses of the ladder and the firefighter, $N = (m + M)g$, and is independent of the location of the firefighter on the ladder. Also from Eq. 9-31 is the relationship between the force from the wall and the force of friction; the condition at which slipping occurs is $F_w \geq \mu_s(m + M)g$.

Now go straight to Eq. 9-32. The $a/2$ in the second term is the location of the firefighter, who in the example was halfway between the base of the ladder and the top of the ladder. In the exercise we don't know where the firefighter is, so we'll replace $a/2$ with x. Then

$$-F_w h + Mgx + \frac{mga}{3} = 0$$

is an expression for rotational equilibrium. Substitute in the condition of F_w when slipping just starts, and we get

$$-\left(\mu_s(m+M)g\right)h + Mgx + \frac{mga}{3} = 0.$$

Solve this for x,

$$x = \mu_s\left(\frac{m}{M}+1\right)h - \frac{ma}{3M} = (0.54)\left(\frac{45 \text{ kg}}{72 \text{ kg}}+1\right)(9.3 \text{ m}) - \frac{(45 \text{ kg})(7.6 \text{ m})}{3(72 \text{ kg})} = 6.6 \text{ m}$$

This is the horizontal distance; the fraction of the total length along the ladder is then given by $x/a = (6.6 \text{ m})/(7.6 \text{ m}) = 0.87$. The firefighter can climb $(0.87)(12\,\text{m}) = 10.4\,\text{m}$ up the ladder.

E9-29 The wheel will make it up the step if the torque about the point of contact with the top of the step from gravity is *slightly over-balanced* by the torque from the applied force $\vec{\mathbf{F}}$.

Before proceeding, however, there are two other forces we at least need to address: the normal force up on the wheel from the point of contact with the floor, and the normal force on the wheel from the point of contact with the top of the step. The normal force from the ground will be zero when the wheel climbs up the step, so we can ignore it. The normal force from the top of the step is at the point where we calculate the torque, so the contribution to the torque is zero.

We can assume that both the force $\vec{\mathbf{F}}$ and the force of gravity $\vec{\mathbf{W}}$ act on the center of the wheel. Then the wheel will just start to lift when

$$\vec{\mathbf{W}} \times \vec{\mathbf{r}} + \vec{\mathbf{F}} \times \vec{\mathbf{r}} = 0,$$

or

$$W \sin\theta = F \cos\theta,$$

where θ is the angle between the vertical (pointing down) and the line between the center of the wheel and the point of contact with the step. The use of the sine on the left is a straightforward application of Eq. 9-2. Why the cosine on the right? Because

$$\sin(90° - \theta) = \cos\theta.$$

Then $F = W \tan\theta$. We can express the angle θ in terms of trig functions, h, and r. $r\cos\theta$ is the vertical distance from the center of the wheel to the top of the step, or $r - h$. Then

$$\cos\theta = 1 - \frac{h}{r} \quad \text{and} \quad \sin\theta = \sqrt{1 - \left(1 - \frac{h}{r}\right)^2}.$$

Finally by combining the above we get

$$F = W \frac{\sqrt{\frac{2h}{r} - \frac{h^2}{r^2}}}{1 - \frac{h}{r}} = W \frac{\sqrt{2hr - h^2}}{r - h}.$$

E9-35 (a) The angular acceleration is

$$\alpha = \frac{\Delta \omega}{\Delta t} = \frac{6.20 \text{ rad/s}}{0.22 \text{ s}} = 28.2 \text{ rad/s}^2$$

(b) From Eq. 9-11, $\tau = I\alpha = (12.0 \text{kg} \cdot \text{m}^2)(28.2 \text{ rad/s}^2) = 338 \text{ N·m}$.

E9-39 The Atwood's machine, again. The heavier block is observed to fall from rest 76.5 cm in 5.11 seconds; we can apply a kinematic equation from chapter 2 to find the acceleration:

$$y = v_{0y}t + \frac{1}{2}a_y t^2,$$
$$a_y = \frac{2y}{t^2} = \frac{2(0.765 \text{ m})}{(5.11 \text{ s})^2} = 0.0586 \text{ m/s}^2$$

We *cannot* simply use the Atwood's machine equations from Chapter 5, because we can no longer ignore the rotational effects of the pulley. We can, however, closely follow the approach in Sample Problem 9-10.

For the heavier block, $m_1 = 0.512$ kg, and Newton's second law gives

$$m_1 g - T_1 = m_1 a_y,$$

where a_y is positive and *down*.

For the lighter block, $m_2 = 0.463$ kg, and Newton's second law gives

$$T_2 - m_2 g = m_2 a_y,$$

where a_y is positive and *up*. T_1 and T_2 are *not* equal, because there is a torque on the pulley. But we do know that $T_1 > T_2$; the net force on the pulley creates a torque which results in the pulley rotating toward the heavier mass. That net force is $T_1 - T_2$; so the rotational form of Newton's second law gives

$$(T_1 - T_2) R = I\alpha_z = Ia_T/R,$$

where $R = 0.049$ m is the radius of the pulley and a_T is the tangential acceleration. But this acceleration is equal to a_y, because everything— both blocks and the pulley— are moving together.

We then have *three* equations and *three* unknowns. We'll add the first two together,

$$m_1 g - T_1 + T_2 - m_2 g = m_1 a_y + m_2 a_y,$$
$$T_1 - T_2 = (g - a_y)m_1 - (g + a_y)m_2,$$

and then combine this with the third equation by substituting for $T_1 - T_2$,

$$(g - a_y)m_1 - (g + a_y)m_2 = I a_y / R^2,$$
$$\left[\left(\frac{g}{a_y} - 1 \right) m_1 - \left(\frac{g}{a_y} + 1 \right) m_2 \right] R^2 = I.$$

Now for the numbers:

$$\left(\frac{(9.81 \,\mathrm{m/s^2})}{(0.0586 \,\mathrm{m/s^2})} - 1 \right) (0.512 \,\mathrm{kg}) - \left(\frac{(9.81 \,\mathrm{m/s^2})}{(0.0586 \,\mathrm{m/s^2})} + 1 \right) (0.463 \,\mathrm{kg}) = 7.23 \,\mathrm{kg},$$
$$(7.23 \,\mathrm{kg})(0.049 \,\mathrm{m})^2 = 0.0174 \,\mathrm{kg \cdot m^2}.$$

E9-43 (a)   There are a number of parts which much be glued together to solve this problem. Frictional force with the conveyor belt is what causes a torque to make the tire rotate, so we need to know the normal force on the tire from the belt. Assuming a perfect hinge at B, the only two vertical forces on the tire will be the normal force from the belt and the force of gravity. Then $N = W = mg$, or $N = (15.0 \,\mathrm{kg})(9.8 \,\mathrm{m/s^2}) = 147$ N.

While the tire skids we have kinetic friction, so $f = \mu_k N = (0.600)(147 \,\mathrm{N}) = 88.2$ N. There are *four* forces on the tire. The force of gravity and and the pull from the holding rod AB both act at the axis of rotation, so can't contribute to the net torque. The normal force acts at a point which is parallel to the displacement from the axis of rotation, so it doesn't contribute to the torque either (because the cross product would vanish); so the only contribution to the torque is from the frictional force.

The frictional force is perpendicular to the radial vector, so the magnitude of the torque is just $\tau = rf = (0.300 \,\mathrm{m})(88.2 \,\mathrm{N}) = 26.5$ N·m. This means the angular acceleration will be $\alpha = \tau/I = (26.5 \,\mathrm{N\cdot m})/(0.750 \,\mathrm{kg\cdot m^2}) = 35.3 \,\mathrm{rad/s^2}$.

Are we done yet? No. Now we need to know when the tire is rotating so that it doesn't slip; this occurs when the tangential velocity of the tire is the same as the belt. For what angular velocity does this occur? When $\omega R = v_T = 12.0$ m/s. We solve for ω and get $\omega = 40$ rad/s.

Now we solve $\omega = \omega_0 + \alpha t$ for the time. The wheel started from rest, so $t = 1.13$ s.

(b)   This part is much easier! The conveyor belt moves at constant speed, so the length of the skid is $x = vt = (12.0 \,\mathrm{m/s})(1.13 \,\mathrm{s}) = 13.6$ m long.

P9-1   The problem of sliding down the ramp has been solved (see Sample Problem 5-8); the critical angle θ_s is given by $\tan\theta_s = \mu_s$.

The problem of tipping is actually not that much harder: an object tips when the center of gravity is no longer over the base. What does that mean, and how did I get that answer so quickly? When the box is just starting to tip there is only one area in contact with the ground— the lower-most edge. This becomes the point of rotation; it is also the point at which the normal force acts, because it is the only point of contact. But if the normal force acts on the point of rotation it can't cause a torque about that point; the only source of torque will be from the force of gravity, which acts on the center of mass. If the center of gravity is over the base, it will cause a torque to rotate the crate back down; if the center of gravity is *not* over the base, it will cause a torque to rotate the box over, to "tip" it.

The important angle for tipping is shown in the figure below; we can find that by trigonometry to be

$$\tan\theta_t = \frac{O}{A} = \frac{(0.56 \text{ m})}{(0.56 \text{ m}) + (0.28 \text{ m})} = 0.67,$$

so $\theta_t = 34°$.

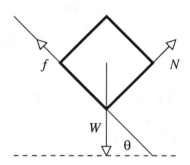

(a)   If $\mu_s = 0.60$ then $\theta_s = 31°$ and the crate slides.

(b)   If $\mu_s = 0.70$ then $\theta_s = 35°$ and the crate tips before sliding; it tips at $34°$. You might find it mildly entertaining to calculate the acceleration of the tipping crate down the ramp...

P9-5 (a)   This is yet another version of the firefighter problem (see Sample Problem 9-7 and the solution to Ex. 9-25). We can solve this problem with Eq. 9-32 after a few modifications. We'll assume the center of mass of the ladder is at the center, then the third term of Eq. 9-32 is $mga/2$. The cleaner didn't climb half-way, he climbed $3.10/5.12 = 60.5\%$ of the way, so the second term of Eq. 9-32 becomes $Mga(0.605)$.

h, L, and a are related by $L^2 = a^2 + h^2$, so $h = \sqrt{(5.12 \text{ m})^2 - (2.45 \text{ m})^2} = 4.5 \text{ m}$. Then, putting the correction into Eq. 9-32,

$$
\begin{aligned}
F_w &= \frac{1}{h}\left[Mga(0.605) + \frac{mga}{2}\right], \\
&= \frac{1}{(4.5 \text{ m})}\left[(74.6 \text{ kg})(9.81 \text{ m/s}^2)(2.45 \text{ m})(0.605)\,, \right. \\
&\quad \left. + (10.3 \text{ kg})(9.81 \text{ m/s}^2)(2.45 \text{ m})/2\right], \\
&= 269 \text{ N}
\end{aligned}
$$

(b)   The vertical component of the force of the ground on the ground is the sum of the weight of the window cleaner and the weight of the ladder, or 833 N. Note, I said *weight* not *mass!*

The horizontal component is equal in magnitude to the force of the ladder on the window, since all horizontal forces on the ladder must balance. Then the net force of the ground on the ladder has magnitude

$$\sqrt{(269 \text{ N})^2 + (833 \text{ N})^2} = 875 \text{ N}$$

and direction

$$\theta = \arctan(833/269) = 72° \text{ above the horizontal.}$$

P9-11   Problem 9-10 says that $I_x + I_y = I_z$ for any thin, flat object which lies only in the $x-y$ plane, this is the perpendicular axis theorem (which is important, but not near as important as the parallel axis theorem.) It doesn't matter in which direction the x and y axes are chosen, so long as they are perpendicular. We can then orient our square as in either of the pictures below:

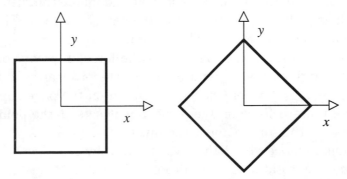

By symmetry $I_x = I_y$ in either picture. Consequently, $I_x = I_y = I_z/2$ for either picture. It is the same square, so I_z is the same for both picture. Then I_x is also the same for both orientations.

P9-13 (a) From Eq. 9-15, $I = \int r^2 dm$ about some axis of rotation when r is measured from that axis. If we consider the x axis as our axis of rotation, then $r = \sqrt{y^2 + z^2}$, since the distance to the x axis depends only on the y and z coordinates. We have similar equations for the y and z axes, so

$$I_x = \int \left(y^2 + z^2\right) dm,$$

$$I_y = \int \left(x^2 + z^2\right) dm,$$

$$I_z = \int \left(x^2 + y^2\right) dm.$$

These three equations can be added together to give

$$I_x + I_y + I_z = 2 \int \left(x^2 + y^2 + z^2\right) dm,$$

so if we *now* define r to be measured from the origin (which is not the definition used above), then we end up with the answer in the text.

(b) The right hand side of the equation is integrated over the entire body, regardless of how the axes are defined. So the integral should be the same, no matter how the coordinate system is rotated. If you've taken linear algebra you'll recognize this as the invariance of the trace under the standard orthogonal transformation.

But why do we really care? In the event that the problem has significant symmetry, like with a sphere, it makes it easier to find the individual I. Try Problem 9-14!

P9-17 I suppose this is somewhat like a glorified Atwood's machine. Since we are only interested in the linear acceleration of the blocks at the very beginning of the motion we can simplify the problem somewhat and ignore the suspended strings; we'll treat the block as if they were attached to the end of the rods. Then we can consider the *rotational* acceleration, which will be given by $\alpha_z = \sum \tau / I$.

The torque about the pivot comes from the force of gravity on each block. This forces will both originally be at right angles to the pivot arm, so the net torque will be $\sum \tau = mgL_2 - mgL_1$, where clockwise as seen on the page is positive. There is a third force, the upward force on the pivot, but it acts at the point of rotation, so it can't contribute to the torque about the pivot.

The rotational inertia about the pivot is given by $I = \sum m_n r_n^2 = m(L_2^2 + L_1^2)$. So we can now find the rotational acceleration,

$$\alpha = \frac{\sum \tau}{I} = \frac{mgL_2 - mgL_1}{m(L_2^2 + L_1^2)} = g\frac{L_2 - L_1}{L_2^2 + L_1^2} = 8.66 \text{ rad/s}^2.$$

The linear acceleration is the tangential acceleration, $a_T = \alpha R$. For the left block, $a_T = 1.73 \text{ m/s}^2$; for the right block $a_T = 6.93 \text{ m/s}^2$.

P9-21 This problem is equivalent to Sample Problem 9-11, except that we have a sphere instead of a cylinder. We'll have the same two equations for Newton's second law,

$$Mg\sin\theta - f = Ma_{cm} \text{ and } N - Mg\cos\theta = 0.$$

Newton's second law for rotation will still look like

$$-fR = I_{cm}\alpha,$$

except that now I_{cm} refers to a sphere. The conditions for accelerating without slipping are $a_{cm} = \alpha R$, then we can rearrange the rotational equation (like was done in the example) to get

$$f = -\frac{I_{cm}\alpha}{R} = -\frac{I_{cm}(-a_{cm})}{R^2},$$

and then substitute this into the x component linear second law equation,

$$Mg\sin\theta - \frac{I_{cm}(a_{cm})}{R^2} = Ma_{cm},$$

and solve for a_{cm}. For fun, let's write the rotational inertia as $I = \beta MR^2$, where $\beta = 2/5$ for the sphere. Then, upon some mild rearranging, we get

$$a_{cm} = g\frac{\sin\theta}{1+\beta}$$

Did we do it right? For the disk, $\beta = 1/2$, and a_{cm} reduces to the expression in the text. For the sphere, $a_{cm} = 5/7g\sin\theta$.

(a) If $a_{cm} = 0.133g$, then $\sin\theta = 7/5(0.133) = 0.186$, and $\theta = 10.7°$.

(b) A frictionless block has no rotational properties; in this case $\beta = 0$! Then $a_{cm} = g\sin\theta = 0.186g$. Is it that easy? Yes.

P9-23 This problem is worked out almost completely in Sample Problem 9-11. The force of friction required to keep the cylinder rolling is given by

$$f = \frac{1}{3}Mg\sin\theta;$$

the normal force is given to be $N = Mg\cos\theta$; so the coefficient of static friction is given by

$$\mu_s \geq \frac{f}{N} = \frac{1}{3}\tan\theta.$$

Chapter 10

Angular Momentum

E10-3 If the angular momentum $\vec{l}$ is constant in time, then $d\vec{l}/dt = 0$. Trying this on Eq. 10-1,

$$
\begin{aligned}
\frac{d\vec{l}}{dt} &= \frac{d}{dt}\left(\vec{r} \times \vec{p}\right), \\
&= \frac{d}{dt}\left(\vec{r} \times m\vec{v}\right), \\
&= m\frac{d\vec{r}}{dt} \times \vec{v} + m\vec{r} \times \frac{d\vec{v}}{dt}, \\
&= m\vec{v} \times \vec{v} + m\vec{r} \times \vec{a}.
\end{aligned}
$$

Now the cross product of a vector with itself is zero, so the first term vanishes. But in the exercise we are told the particle has constant velocity, so $\vec{a} = 0$, and consequently the second term vanishes. Hence, $\vec{l}$ is constant for a single particle if $\vec{v}$ is constant.

E10-9 (a) For constant torque Eq. 10-6 can be written as

$$
\sum \vec{\tau}\Delta t = \Delta \vec{l}.
$$

There is only one torque on the disk and the disk starts from rest, so $\Delta \vec{l} = \vec{l} - \vec{l}_0 = \vec{l}$. We need only concern ourselves with the magnitudes, so

$$
l = \Delta l = \tau \Delta t = (15.8\,\text{N·m})(0.033\,\text{s}) = 0.521\,\text{kg·m}^2/\text{s}.
$$

(b) The angular speed can be found from Eq. 10-11,

$$
\omega = l/I = (0.521\,\text{kg·m}^2/\text{s})/(1.22 \times 10^{-3}\text{kg·m}^2) = 427\,\text{rad}/s.
$$

E10-13 The impulse does two things to the stick: it introduces a linear momentum and it introduces an angular momentum. We'll treat the two cases individually.

An impulse of 12.8 N·s will change the linear momentum by 12.8 N·s; the stick starts from rest, so the final momentum *must* be 12.8 N·s. Since $p = mv$, we then can find $v = p/m = (12.8 \text{ N·s})/(4.42 \text{ kg}) = 2.90$ m/s.

The force did not act on the center of mass, so it imparted an angular momentum to the stick about the center of mass. Let's look at this more closely. Impulse is a vector, given by $\int \vec{F}dt$. We can take the cross product of the impulse with the displacement vector $\vec{r}$ (measured from the axis of rotation to the point where the force is applied) and get

$$\vec{r} \times \int \vec{F}dt \approx \int \vec{r} \times \vec{F}dt,$$

The two sides of the above expression are only equal if $\vec{r}$ has a constant magnitude *and* direction. This won't be true, but if the force is of sufficiently short duration then it hopefully won't change much. The right hand side is an integral over a torque, and will equal the change in angular momentum of the stick.

The exercise states that the force is perpendicular to the stick, then $|\vec{r} \times \vec{F}| = rF$, and the "torque impulse" is then $(0.464 \text{ m})(12.8 \text{ N·s}) = 5.94$ kg·m/s. This "torque impulse" is equal to the change in the angular momentum, but the stick started from rest, so the final angular momentum of the stick is 5.94 kg·m/s.

But how fast is it rotating? We can use Fig. 9-15 to find the rotational inertia about the center of the stick: $I = \frac{1}{12}ML^2 = \frac{1}{12}(4.42 \text{ kg})(1.23 \text{ m})^2 = 0.557$ kg·m². The angular velocity of the stick is $\omega = l/I = (5.94 \text{ kg·m/s})/(0.557 \text{ kg·m}^2) = 10.7$ rad/s.

This problem, fun that it was, has several variations. One very common one is to have the puck "stick" to the stick after the collision. That adds considerable extra work, because we will need to find the center of mass and rotational inertia about this center of mass of this now non-symmetric system. It makes for a nice final exam question.

E10-17 In this exercise we have conservation of angular momentum. The rotational inertia of a solid sphere is $I = \frac{2}{5}MR^2$; so as the sun collapses

$$
\begin{aligned}
\vec{L}_i &= \vec{L}_f, \\
I_i\vec{\omega}_i &= I_f\vec{\omega}_f, \\
\frac{2}{5}MR_i^2\vec{\omega}_i &= \frac{2}{5}MR_f^2\vec{\omega}_f, \\
R_i^2\vec{\omega}_i &= R_f^2\vec{\omega}_f.
\end{aligned}
$$

The angular frequency is inversely proportional to the period of rotation, so

$$T_f = T_i \frac{R_f^2}{R_i^2} = (3.6 \times 10^4 \text{ min}) \left(\frac{(6.37 \times 10^6 \text{m})}{(6.96 \times 10^8 \text{m})} \right)^2 = 3.0 \text{ min}.$$

E10-21 This is a conservation of angular momentum problem. We have two disks which are originally not in contact which then come into contact; there are no external torques. We can write

$$\vec{l}_{1,i} + \vec{l}_{2,i} = \vec{l}_{1,f} + \vec{l}_{2,f},$$
$$I_1\vec{\omega}_{1,i} + I_2\vec{\omega}_{2,i} = I_1\vec{\omega}_{1,f} + I_2\vec{\omega}_{2,f}.$$

We don't need to distinguish between the initial and final values of the rotational inertia in the exercise, but we might need to distinguish them in others.

The final angular velocities of the two disks will be equal, so the above equation can be simplified and rearranged to yield

$$\omega_{\text{f}} = \frac{I_1}{I_1 + I_2}\omega_{1,i} = \frac{(1.27 \text{ kg} \cdot \text{m}^2)}{(1.27 \text{ kg} \cdot \text{m}^2) + (4.85 \text{ kg} \cdot \text{m}^2)}(824 \text{ rev/min}) = 171 \text{ rev/min}$$

We ignored the shaft throughout this problem since we were told the rotational inertia is negligible.

We'll address this question again in Problem 11-25.

E10-27 The relevant precession expression is Eq. 10-22. We know the mass of the gyroscope— $M = (1.14 + 0.130)$ kg. The force of gravity acts at the center of mass, which we will assume is at the center of the axle, so $r = (0.122/2)$ m. The only other variable is the magnitude of the angular momentum.

Since $L = I\omega$, we will need to find the rotational inertia. This will be a sum of the contributions from both the disk and the axle, but the radius of the axle is probably very small compared to the disk, probably as small as 0.5 cm. Since I is proportional to the radius squared, we expect contributions from the axle to be less than $(1/100)^2$ of the value for the disk. For the disk only we use

$$I = \frac{1}{2}MR^2 = \frac{1}{2}(1.14 \text{ kg})(0.487 \text{ m})^2 = 0.135 \text{ kg} \cdot \text{m}^2.$$

One of my proof readers asked why I didn't use 1.27 kg in the above expression. This is because we are only concerned with the contribution to the rotational inertia of the gyroscope from the disk, since by the above argument we expect the contribution from the axle to be much smaller (by a factor of 10^{-4} or so).

Now for ω. We are given the value in rev/min, we want rad/s in order to make our units easier to work with. As such,

$$\omega = 975 \text{ rev/min} \left(\frac{2\pi \text{ rad}}{1 \text{ rev}}\right)\left(\frac{1 \text{ min}}{60 \text{ s}}\right) = 102 \text{ rad/s}.$$

Then $L = I\omega = 13.8 \text{ kg} \cdot \text{m}^2/\text{s}$.

Back to Eq. 10-22,

$$\omega_p = \frac{Mgr}{L} = \frac{(1.27 \text{ kg})(9.81 \text{ m/s}^2)(0.0610 \text{ m})}{13.8 \text{ kg} \cdot \text{m}^2/\text{s}} = 0.0551 \text{ rad/s}.$$

This is consistent with our assumption that precessional angular momentum is small compared to the rotational angular momentum. It would be nice to be finished, but the exercise actually wanted us to find the *time* for one precession. This time is

$$t = \frac{1 \text{rev}}{\omega_p} = \frac{2\pi \text{ rad}}{(0.0551 \text{ rad/s})} = 114 \text{ s}.$$

P10-3 This problem is *closely* related to Ex. 10-13, so go back and look at that solution. We want to impart a linear velocity to the ball at the same time as a rotational velocity so that the tangential velocity of the ball matches the forward velocity.

Assume that the cue stick strikes the ball horizontally with a force of constant magnitude F for a time Δt. Then the magnitude of the change in linear momentum of the ball is given by

$$F\Delta t = \Delta p = p,$$

since the initial momentum is zero.

If the force is applied a distance y above the center of the ball, then the magnitude of the torque about a horizontal axis through the center of the ball is $\tau = xF$. The change in angular momentum of the ball is given by

$$\tau\Delta t = \Delta l = l,$$

since initially the ball is not rotating.

For the ball to roll without slipping we need $v = \omega R$. We can start with this:

$$
\begin{aligned}
v &= \omega R, \\
\frac{m}{m}v &= \frac{I}{I}\omega R, \\
\frac{p}{m} &= \frac{lR}{I}, \\
\frac{F\Delta t}{m} &= \frac{\tau\Delta t R}{I}, \\
\frac{F}{m} &= \frac{xFR}{I}.
\end{aligned}
$$

What did we do? In the second line we began to look at momenta instead of velocities; in the fourth line we substituted expressions for impulses that we had worked out previously; and in the last line we notice that not only does Δt cancel out, but so does F!

81

Then $x = I/mR$ is the condition for rolling without sliding from the start. For a solid sphere, $I = \frac{2}{5}mR^2$, so $x = \frac{2}{5}R$. There is one really neat thing about the way we did this; we could have substituted a disk, a hollow ball, or even a hoop, and we could still easily find x.

P10-7 This problem is basically the same as Ex. 9-43, except that the bowling lane is stationary and the ball is moving, while in the previous exercise the tire was stationary while the conveyor belt beneath it was moving. We can solve the problem in the same manner, or we can solve the problem in terms of angular momentum. we will take the latter approach.

We assume the bowling ball is solid, so the rotational inertia will be $I = (2/5)MR^2$ (see Figure 9-15).

The normal force on the bowling ball will be $N = Mg$, where M is the mass of the bowling ball. We expect M to cancel out of any final expression. The kinetic friction on the bowling ball is $F_f = \mu_k N = \mu_k Mg$. The magnitude of the net torque on the bowling ball (about an axis perpendicular to the direction of motion but parallel to the surface of the bowling lane, just to be specific) while skidding is then $\tau = \mu_k MgR$.

Originally the angular momentum of the ball is zero; the final angular momentum will have magnitude $l = I\omega = Iv/R$, where v is the final translational speed of the ball. Now that we have the ingredients, we can solve the problem.

The easiest order to solve this problem is part (a), then (d), then (b), and finally (c).

(a) The time requires for the ball to stop skidding is the time required to change the angular momentum to l, so

$$\Delta t = \frac{\Delta l}{\tau} = \frac{(2/5)MR^2 v/R}{\mu_k MgR} = \frac{2v}{5\mu_k g}.$$

Not only did M cancel out of this expression, but so did R!

Since we don't know v, we can't solve this for Δt. But the same time through which the angular momentum of the ball is increasing the linear momentum of the ball is decreasing, so we also have

$$\Delta t = \frac{\Delta p}{-F_f} = \frac{Mv - Mv_0}{-\mu_k Mg} = \frac{v_0 - v}{\mu_k g}.$$

The *all* important negative sign is because the friction is slowing the ball down! Combining,

$$\Delta t = \frac{v_0 - v}{\mu_k g},$$

$$= \frac{v_0 - 5\mu_k g\Delta t/2}{\mu_k g},$$

$$2\mu_k g\Delta t = 2v_0 - 5\mu_k g\Delta t,$$
$$\Delta t = \frac{2v_0}{7\mu_k g},$$
$$= \frac{2(8.50\,\mathrm{m/s})}{7(0.210)(9.81\,\mathrm{m/s^2})} = 1.18\,\mathrm{s}.$$

(b) Use the expression for angular momentum and torque,

$$v = 5\mu_k g\Delta t/2 = 5(0.210)(9.81\,\mathrm{m/s^2})(1.18\,\mathrm{s})/2 = 6.08\,\mathrm{m/s}.$$

(c) Since we still don't know any energy conservation principles (see the next chapter!) we need to go back to chapters 2 and 3. The acceleration of the ball is $F/M = -\mu g$. The distance traveled is then given by

$$x = \frac{1}{2}at^2 + v_0 t,$$
$$= -\frac{1}{2}(0.210)(9.81\,\mathrm{m/s^2})(1.18\,\mathrm{s})^2 + (8.50\,\mathrm{m/s})(1.18\,\mathrm{s}) = 8.6\,\mathrm{m},$$

(d) Sole this the same way as part (c), except use the angular equivalent equations. The angular acceleration is $\tau/I = 5\mu_k g/(2R)$. Notice that this is the first time that we find an R which *doesn't* cancel; until this point it didn't matter how big the bowling ball was! Then

$$\theta = \frac{1}{2}\alpha t^2 + \omega_0 t,$$
$$= \frac{5(0.210)(9.81\,\mathrm{m/s^2})}{4(0.11\,\mathrm{m})}(1.18\,\mathrm{s})^2 = 32.6\,\mathrm{rad} = 5.19\,\mathrm{revolutions}.$$

P10-11 Remember that the direction of the angular velocity of an object is perpendicular to the plane in which it rotates. A clockwise rotation (as viewed from above) on a table would correspond to a *negative* value for ω_z. Since *all* relevant angular quantities will be in the z direction, we will drop the subscript. The cockroach initially has an angular speed of $\omega_{c,i} = -v/r$. The rotational inertia of the cockroach about the axis of the turntable is $I_c = mR^2$. Then conservation of angular momentum gives

$$l_{c,i} + l_{s,i} = l_{c,f} + l_{s,f},$$
$$I_c\omega_{c,i} + I_s\omega_{s,i} = I_c\omega_{c,f} + I_s\omega_{s,f},$$
$$-mR^2 v/r + I\omega = (mR^2 + I)\omega_f,$$
$$\omega_f = \frac{I\omega - mvR}{I + mR^2}.$$

Why is there an answer to part (b)? It is now in Chapter 11 as Problem 11-26.

Chapter 11

Energy 1: Work and Kinetic Energy

E11-1 (a) Apply Eq. 11-2,

$$W = Fs\cos\phi = (190\,\text{N})(3.3\,\text{m})\cos(22°) = 580\,\text{J}.$$

(b) The force of gravity is perpendicular to the displacement of the crate, so there is no work done by the force of gravity.

(c) The normal force is perpendicular to the displacement of the crate, so there is no work done by the normal force.

E11-7 Equation 11-5 describes how to find the dot product of two vectors from components

$$
\begin{aligned}
\vec{a} \cdot \vec{b} &= a_x b_x + a_y b_y + a_z b_z, \\
&= (3)(2) + (3)(1) + (3)(3) = 18.
\end{aligned}
$$

Equation 11-3 can be used to find the angle between the vectors, but we must first find the magnitude of $\vec{a}$ and $\vec{b}$. That's easy enough: $A = |\vec{A}| = \sqrt{\vec{A} \cdot \vec{A}}$. Then

$$
\begin{aligned}
a &= \sqrt{(3)^2 + (3)^2 + (3)^2} = 5.19, \\
b &= \sqrt{(2)^2 + (1)^2 + (3)^2} = 3.74.
\end{aligned}
$$

Now use Eq. 11-3,

$$\cos\phi = \frac{\vec{a} \cdot \vec{b}}{ab} = \frac{(18)}{(5.19)(3.74)} = 0.927,$$

and then $\phi = 22.0°$.

E11-11 If the woman is to move *up* the stairs the net force on the woman needs to be zero or slightly upward. There are two forces on the woman, the force of gravity directed down and the normal force of the floor directed up. These will be effectively equal, so $N = W = mg$. One must be careful, though, because the normal force does *not* do work on the woman. It can't, since the point of contact with the woman doesn't move.

The woman, however, pushes down as she moves up. The woman pushing down and the stairs forms an action/reaction pair with the normal force of the stairs on the woman. Consequently, the 57 kg woman must exert a force of $F = (57\text{kg})(9.8\text{m/s}^2) = 560\text{N}$ to propel herself up the stairs.

From the reference frame of the woman the stairs are moving down, and she is exerting a force down, so the work done by the woman is given by

$$W = Fs = (560 \text{ N})(4.5 \text{ m}) = 2500 \text{ J},$$

this work is positive because the force is in the same direction as the displacement; according to the woman the stairs move *down* when she pushes *down* on them.

The average power supplied by the woman is given by Eq. 11-7,

$$P = W/t = (2500 \text{ J})/(3.5 \text{ s}) = 710 \text{ W},$$

about the same as one horsepower.

E11-15 Since the frictional force is parallel to the velocity we can use Eq. 11-11 to find the power.
$$P = Fv = (720 \text{ N})(26 \text{ m/s}) = 19000 \text{ W}$$

In horsepower,
$$P = 19000 \text{ W}(1/745.7 \text{ hp/W}) = 25 \text{ hp}.$$

E11-17 (a) Start by converting kilowatt-hours to Joules:

$$1 \text{ kW} \cdot \text{h} = (1000 \text{ W})(3600 \text{ s}) = 3.6 \times 10^6 \text{ J}.$$

The car gets 30 mi/gal, and one gallon of gas produces 140 MJ of energy. The gas required to produce 3.6×10^6 J is

$$3.6 \times 10^6 \text{ J} \left(\frac{1 \text{ gal}}{140 \times 10^6 \text{ J}} \right) = 0.026 \text{ gal}.$$

The distance traveled on this much gasoline is

$$0.026 \text{ gal} \left(\frac{30 \text{ mi}}{1 \text{ gal}} \right) = 0.78 \text{ mi}.$$

(b) How long did it take to travel 0.78 miles? At 55 mi/h, it will take

$$0.78 \text{ mi} \left(\frac{1 \text{ hr}}{55 \text{ mi}} \right) = 0.014 \text{ h} = 51 \text{ s}.$$

The rate of energy expenditure is then $(3.6 \times 10^6 \text{ J})/(51 \text{ s}) = 71000 \text{ W}$.

E11-21 The acceleration on the object as a function of position is given by

$$a = \frac{20 \text{ m/s}^2}{8 \text{ m}} x,$$

This is found from the slope of the line on the graph and the recognition that the a intercept is zero. From this we know the net force on the object, since $F = ma$. The work done on the object is given by Eq. 11-14,

$$W = \int_0^8 F_x dx = \int_0^8 (10 \text{ kg}) \frac{20 \text{ m/s}^2}{8 \text{ m}} x \, dx = 800 \text{ J}.$$

E11-23 (a) For a perfect spring, $F = -kx$, and $\Delta F = -k\Delta x$. We'll use this to find the spring constant

$$k = -\frac{\Delta F}{\Delta x} = -\frac{(-240 \text{ N}) - (-110 \text{ N})}{(0.060 \text{ m}) - (0.040 \text{ m})} = 6500 \text{ N/m}.$$

The extra negative sign in front of the forces reflects Newton's third law; we are interested in the force exerted by the spring, not the force exerted on the spring.

With no force on the spring,

$$\Delta x = -\frac{\Delta F}{k} = -\frac{(0) - (-110 \text{ N})}{(6500 \text{ N/m})} = -0.017 \text{ m}.$$

This is the amount *less* than the 40 mm mark, so the position of the spring with no force on it is 23 mm.

(b) $\Delta x = -10$ mm compared to the 100 N picture, so

$$\Delta F = -k\Delta x = -(6500 \text{ N/m})(-0.010 \text{ m}) = 65 \text{ N}.$$

The answer is positive, indicating that the force exerted by the spring in the last picture is *less negative* than in the first. Hence the weight of the least object is $110 \text{ N} - 65 \text{ N} = 45 \text{ N}$.

E11-27 The kinetic energy of the electron is

$$4.2 \text{ eV} \left(\frac{1.60 \times 10^{-19} \text{ J}}{1 \text{ eV}} \right) = 6.7 \times 10^{-19} \text{ J}.$$

The expression for kinetic energy is $K = \frac{1}{2}mv^2$, an expression we can rearrange, then

$$v = \sqrt{\frac{2K}{m}} = \sqrt{\frac{2(6.7 \times 10^{-19} \text{J})}{(9.1 \times 10^{-31} \text{kg})}} = 1.2 \times 10^6 \text{m/s}.$$

E11-31 (a) This is a sneaky problem. Find the velocity of the particle by taking the time derivative of the position:

$$v = \frac{dx}{dt} = (3.0 \,\text{m/s}) - (8.0 \,\text{m/s}^2)t + (3.0 \,\text{m/s}^3)t^2.$$

we can square this and multiply by the mass to find the kinetic energy. But it creates a horrible mess, and we are only interested in the kinetic energy at two times: $t = 0$ and $t = 4$ s. So let's save space and just find the velocity at those two times.

$$v(0) = (3.0 \,\text{m/s}) - (8.0 \,\text{m/s}^2)(0) + (3.0 \,\text{m/s}^3)(0)^2 = 3.0 \,\text{m/s},$$
$$v(4) = (3.0 \,\text{m/s}) - (8.0 \,\text{m/s}^2)(4.0 \,\text{s}) + (3.0 \,\text{m/s}^3)(4.0 \,\text{s})^2 = 19.0 \,\text{m/s}$$

The initial kinetic energy is $K_i = \frac{1}{2}(2.80 \,\text{kg})(3.0 \,\text{m/s})^2 = 13 \,\text{J}$, while the final kinetic energy is $K_f = \frac{1}{2}(2.80 \,\text{kg})(19.0 \,\text{m/s})^2 = 505 \,\text{J}$.

The work done by the force is given by Eq. 11-24,

$$W = K_f - K_i = 505 \,\text{J} - 13 \,\text{J} = 492 \,\text{J}.$$

(b) This question is really asking for the instantaneous power when $t = 3.0$ s. We know from Eq. 11-11 that $P = Fv$, so we need to find an expression for F. That means finding a;

$$a = \frac{dv}{dt} = -(8.0 \,\text{m/s}^2) + (6.0 \,\text{m/s}^3)t.$$

Then the power is given by $P = mav$, and when $t = 3$ s this gives

$$P = mav = (2.80 \,\text{kg})(10 \,\text{m/s}^2)(6 \,\text{m/s}) = 168 \,\text{W}.$$

E11-37 We want to find the rotational kinetic energy of the wheel when it is spinning and when it is stopped. I'll do the stopped wheel: $K_f = 0$. Now it's your turn...

From Eq. 11-29, $K_i = \frac{1}{2}I\omega_i^2$. The object is a hoop, so $I = MR^2$. For convenience, we'll change the units of ω to something more familiar,

$$\omega = 283 \text{ rev/min} \left(\frac{2\pi \text{ rad}}{1 \text{ rev}} \right) \left(\frac{1 \text{ min}}{60 \text{ s}} \right) = 29.6 \,\text{rad/s}.$$

Then

$$K_i = \frac{1}{2}MR^2\omega^2 = \frac{1}{2}(31.4\,\text{kg})(1.21\,\text{m})^2(29.6\,\text{rad/s})^2 = 2.01 \times 10^4\,\text{J}.$$

Finally, the average power required to stop the wheel is

$$P = \frac{W}{t} = \frac{K_f - K_i}{t} = \frac{(0) - (2.01 \times 10^4\,\text{J})}{(14.8\,\text{s})} = -1360\,\text{W}.$$

Why negative? because you are taking energy *from* the wheel.

E11-41 First we need to conserve momentum. Let the mass of the freight car be M and the initial speed be v_i. Let the mass of the caboose be m and the final speed of the coupled cars be v_f. The caboose is originally at rest, so the expression of momentum conservation is

$$Mv_i = Mv_f + mv_f = (M + m)v_f$$

The decrease in kinetic energy is given by

$$
\begin{aligned}
K_i - K_f &= \frac{1}{2}Mv_i^2 - \left(\frac{1}{2}Mv_f^2 + \frac{1}{2}mv_f^2\right), \\
&= \frac{1}{2}\left(Mv_i^2 - (M + m)v_f^2\right)
\end{aligned}
$$

What we really want is $(K_i - K_f)/K_i$, so

$$
\begin{aligned}
\frac{K_i - K_f}{K_i} &= \frac{Mv_i^2 - (M + m)v_f^2}{Mv_i^2}, \\
&= 1 - \frac{M + m}{M}\left(\frac{v_f}{v_i}\right)^2, \\
&= 1 - \frac{M + m}{M}\left(\frac{M}{M + m}\right)^2,
\end{aligned}
$$

where in the last line we substituted from the momentum conservation expression.

What are we left with?

$$\frac{K_i - K_f}{K_i} = 1 - \frac{M}{M + m} = 1 - \frac{Mg}{Mg + mg}.$$

The left hand side is 27%. We want to solve this for mg, the weight of the caboose. Upon rearranging,

$$mg = \frac{Mg}{1 - 0.27} - Mg = \frac{(35.0\,\text{ton})}{(0.73)} - (35.0\,\text{ton}) = 12.9\,\text{ton}.$$

P11-1 Change your units! Then

$$F = \frac{W}{s} = \frac{(4.5 \text{ eV})(1.6 \times 10^{-19} \text{ J/eV})}{(3.4 \times 10^{-9} \text{ m})} = 2.1 \times 10^{-10} \text{ N}.$$

P11-7 If the power is constant then the force on the car is given by $F = P/v$. But the force is related to the acceleration by $F = ma$ and to the speed by $F = m\frac{dv}{dt}$ for motion in one dimension. Then

$$
\begin{aligned}
F &= \frac{P}{v}, \\
m\frac{dv}{dt} &= \frac{P}{v}, \\
m\frac{dv}{dt}\frac{dx}{dx} &= \frac{P}{v}, \\
m\frac{dx}{dt}\frac{dv}{dx} &= \frac{P}{v}.
\end{aligned}
$$

What we did in the third line is sneaky! Although we could have rearranged line 2 above and then integrated, it would have given our answer in terms of t, and we want the answer in terms of x. We instead multiplied by $1 = dx/dx$ and then rearranged, because dx/dt is just v. *Learn this trick!* Continuing,

$$
\begin{aligned}
mv\frac{dv}{dx} &= \frac{P}{v}, \\
\int_0^v mv^2 dv &= \int_0^x P dx, \\
\frac{1}{3}mv^3 &= Px.
\end{aligned}
$$

We can rearrange this final expression to get v as a function of x,

$$v = (3xP/m)^{1/3}.$$

P11-11 (b) Integrate,

$$W = \int_0^{3x_0} \vec{F} \cdot d\vec{s} = \frac{F_0}{x_0} \int_0^{3x_0} (x - x_0)dx = F_0 x_0 \left(\frac{9}{2} - 3\right),$$

or $W = 3F_0 x_0/2$.

P11-13 The work required to stretch the spring from x_i to x_f is given by

$$W = \int_{x_i}^{x_f} kx^3 dx = \frac{k}{4}x_f{}^4 - \frac{k}{4}x_i{}^4.$$

What about the negative sign in $-kx^3$? Remember, the force law gives the force exerted by the spring, but to find the work to stretch the spring we want to know the force exerted on the spring.

The problem gives

$$W_0 = \frac{k}{4}(l)^4 - \frac{k}{4}(0)^4 = \frac{k}{4}l^4.$$

We then want to find the work required to stretch from $x = l$ to $x = 2l$, so

$$
\begin{aligned}
W_{l \to 2l} &= \frac{k}{4}(2l)^4 - \frac{k}{4}(l)^4, \\
&= 16\frac{k}{4}l^4 - \frac{k}{4}l^4, \\
&= 15\frac{k}{4}l^4 = 15W_0.
\end{aligned}
$$

P11-17 This reminds me of an elementary school word problem: "John is half as old as Peter now, but five years ago John was one third as old as Peter..."

Let M be the mass of the man and m be the mass of the boy. Let v_M be the original speed of the man and v_m be the original speed of the boy. Then

$$\frac{1}{2}Mv_M^2 = \frac{1}{2}\left(\frac{1}{2}mv_m^2\right)$$

and

$$\frac{1}{2}M(v_M + 1.0\,\mathrm{m/s})^2 = \frac{1}{2}mv_m^2.$$

Combine these two expressions and solve for v_M,

$$
\begin{aligned}
\frac{1}{2}Mv_M^2 &= \frac{1}{2}\left(\frac{1}{2}M(v_M + 1.0\,\mathrm{m/s})^2\right), \\
v_M^2 &= \frac{1}{2}(v_M + 1.0\,\mathrm{m/s})^2, \\
0 &= -v_M^2 + (2.0\,\mathrm{m/s})v_M + (1.0\,\mathrm{m/s})^2.
\end{aligned}
$$

The last line can be solved as a quadratic, and $v_M = (1.0\,\mathrm{m/s}) \pm (1.41\,\mathrm{m/s})$. Both the positive and the negative answers should be legitimate solutions; so what if the man is running backwards? This is because velocity is a vector, and writing "speeding-up" as $v + 1$ means increasing in the positive direction. But the text probably only meant for us to use $v_M = 2.41\,\mathrm{m/s}$. Now we use the very first equation to find the speed of the boy,

$$
\begin{aligned}
\frac{1}{2}Mv_M^2 &= \frac{1}{2}\left(\frac{1}{2}mv_m^2\right), \\
v_M^2 &= \frac{1}{4}v_m^2, \\
2v_M &= v_m.
\end{aligned}
$$

The boy, then, is running with twice the speed of the man.

P11-21 (a) We can solve this with a trick of integration.

$$W = \int_0^x F\,dx,$$

$$= \int_0^x ma_x \frac{dt}{dt}\,dx = ma_x \int_0^t \frac{dx}{dt}\,dt$$

$$= ma_x \int_0^t v_x\,dt = ma_x \int_0^t at\,dt,$$

$$= \frac{1}{2}ma_x^2 t^2.$$

Basically, we changed the variable of integration from x to t, and then used the fact the the acceleration was constant so $v_x = v_{0x} + a_x t$. The object started at rest so $v_{0x} = 0$, and we are given in the problem that $v_f = at_f$. Combining,

$$W = \frac{1}{2}ma_x^2 t^2 = \frac{1}{2}m\left(\frac{v_f}{t_f}\right)^2 t^2.$$

(b) Instantaneous power will be the derivative of this, so

$$P = \frac{dW}{dt} = m\left(\frac{v_f}{t_f}\right)^2 t$$

Does this make sense? A constant force is applied to an object, yet the power *increases* with time? Yes, it does. Power depends on the force and the speed; as the object moves faster, it will take more power to keep a constant force on the object.

P11-25 We did the first part of the solution in Ex. 10-21. The initial kinetic energy is (again, ignoring the shaft),

$$K_i = \frac{1}{2}I_1\vec{\omega}_{1,i}{}^2,$$

since the second wheel was originally at rest. The final kinetic energy is

$$K_f = \frac{1}{2}(I_1 + I_2)\vec{\omega}_f{}^2,$$

since the two wheel moved as one. Then

$$\frac{K_i - K_f}{K_i} = \frac{\frac{1}{2}I_1\vec{\omega}_{1,i}{}^2 - \frac{1}{2}(I_1 + I_2)\vec{\omega}_f{}^2}{\frac{1}{2}I_1\vec{\omega}_{1,i}{}^2},$$

$$= 1 - \frac{(I_1 + I_2)\vec{\omega}_f{}^2}{I_1\vec{\omega}_{1,i}{}^2},$$

$$= 1 - \frac{I_1}{I_1 + I_2},$$

where in the last line we substituted from the results of Ex. 10-21.
Using the numbers from Ex. 10-21,

$$\frac{K_i - K_f}{K_i} = 1 - \frac{(1.27\,\text{kg·m}^2)}{(1.27\,\text{kg·m}^2) + (4.85\,\text{kg·m}^2)} = 79.2\%.$$

P11-29 There's nothing to integrate here! Start with the work-energy theorem

$$
\begin{aligned}
W &= K_{\text{f}} - K_{\text{i}} = \frac{1}{2}mv_{\text{f}}^2 - \frac{1}{2}mv_{\text{i}}^2, \\
&= \frac{1}{2}m\left(v_{\text{f}}^2 - v_{\text{i}}^2\right), \\
&= \frac{1}{2}m\left(v_{\text{f}} - v_{\text{i}}\right)\left(v_{\text{f}} + v_{\text{i}}\right),
\end{aligned}
$$

where in the last line we factored the difference of two squares. If you look, you might see the hint of a change in momentum in that last line as well, so

$$
\begin{aligned}
W &= \frac{1}{2}\left(mv_{\text{f}} - mv_{\text{i}}\right)\left(v_{\text{f}} + v_{\text{i}}\right), \\
&= \frac{1}{2}\left(\Delta p\right)\left(v_{\text{f}} + v_{\text{i}}\right),
\end{aligned}
$$

but $\Delta p = J$, the impulse. That finishes this problem.

Chapter 12

Energy 2: Potential Energy

E12-3 Start with Eq. 12-6.

$$
\begin{aligned}
U(x) - U(x_0) &= -\int_{x_0}^{x} F_x(x)dx, \\
&= -\int_{x_0}^{x} \left(-\alpha x e^{-\beta x^2}\right) dx
\end{aligned}
$$

We can solve this by hand if we substitute $u = x^2$, then $du = 2x\,dx$, and the integral becomes

$$
\begin{aligned}
U(x) - U(x_0) &= \int_{x_0}^{x} \left(\alpha e^{-\beta u}\right) \frac{1}{2} du, \\
&= \left. \frac{-\alpha}{2\beta} e^{-\beta u} \right|_{x_0}^{x}, \\
&= \left. \frac{-\alpha}{2\beta} e^{-\beta x^2} \right|_{x_0}^{x}.
\end{aligned}
$$

A purist would have insisted that I change my limits of integration to match the variable of integration, but I knew I was going to change back, so I was lazy. Finishing the integration,

$$
U(x) = U(x_0) + \frac{\alpha}{2\beta} \left(e^{-\beta x_0^2} - e^{-\beta x^2}\right).
$$

Can we simplify this? Not really, but if we *choose $x_0 = \infty$ and $U(x_0) = 0$* we would be left with

$$
U(x) = -\frac{\alpha}{2\beta} e^{-\beta x^2},
$$

which looks much neater. Note that our choice of x_0 doesn't affect $\vec{F}$ at all, we could have picked anywhere!

E12-7 This is a classical conservation of energy problem. We apply Eq. 12-15,

$$K_f + U_f = K_i + U_i,$$
$$\frac{1}{2}mv_f{}^2 + mgy_f = \frac{1}{2}mv_i{}^2 + mgy_i,$$
$$\frac{1}{2}v_f{}^2 + g(-r) = \frac{1}{2}(0)^2 + g(0).$$

In the second line we used the relevant expressions for kinetic and potential energy; in the third we canceled out the mass, and then choose y so that the top of the bowl was $y = 0$. The particle started from rest. Rearranging,

$$v_f = \sqrt{-2g(-r)} = \sqrt{-2(9.81\,\text{m/s}^2)(-0.236\,\text{m})} = 2.15\,\text{m/s}.$$

What ever you do, don't fall into the *common* trap of saying that the kinetic energy equaled the potential energy!

E12-11 (a) The force constant of the spring is

$$k = F/x = mg/x = (7.94\,\text{kg})(9.81\,\text{m/s}^2)/(0.102\,\text{m}) = 764\,\text{N/m}.$$

(b) The potential energy stored in the spring is given by Eq. 12-8,

$$U = \frac{1}{2}kx^2 = \frac{1}{2}(764\,\text{N/m})(0.286\,\text{m} + 0.102\,\text{m})^2 = 57.5\,\text{J}.$$

(c) Conservation of energy,

$$K_f + U_f = K_i + U_i,$$
$$\frac{1}{2}mv_f{}^2 + mgy_f + \frac{1}{2}kx_f{}^2 = \frac{1}{2}mv_i{}^2 + mgy_i + \frac{1}{2}kx_i{}^2,$$
$$\frac{1}{2}(0)^2 + mgh + \frac{1}{2}k(0)^2 = \frac{1}{2}(0)^2 + mg(0) + \frac{1}{2}kx_i{}^2.$$

In the second line we showed *all* contributions to the potential energy; in the third we have chosen $y = 0$ for the bottom starting position and $y = h$ for the final highest point. Rearranging,

$$h = \frac{k}{2mg}x_i{}^2 = \frac{(764\,\text{N/m})}{2(7.94\,\text{kg})(9.81\,\text{m/s}^2)}(0.388\,\text{m})^2 = 0.738\,\text{m}.$$

E12-15 The angle makes no difference as to the *vertical* height to which the block rises. The working is identical to Ex. 12-11,

$$K_f + U_f = K_i + U_i,$$

$$\frac{1}{2}mv_f{}^2 + mgy_f + \frac{1}{2}kx_f{}^2 = \frac{1}{2}mv_i{}^2 + mgy_i + \frac{1}{2}kx_i{}^2,$$

$$\frac{1}{2}(0)^2 + mgh + \frac{1}{2}k(0)^2 = \frac{1}{2}(0)^2 + mg(0) + \frac{1}{2}kx_i{}^2,$$

so

$$h = \frac{k}{2mg}x_i{}^2 = \frac{(2080\,\text{N/m})}{2(1.93\,\text{kg})(9.81\,\text{m/s}^2)}(0.187\,\text{m})^2 = 1.92\,\text{m}.$$

The distance up the incline is given by a trig relation,

$$d = h/\sin\theta = (1.92\,\text{m})/\sin(27^\circ) = 4.23\,\text{m}.$$

E12-19 Regardless of how fast the marble is moving when it leaves the gun, the marble will fall for the same length of time, because it falls through the same vertical distance. We'll call this time t_f, and we neither know this time nor can we ever find it from the information given. But that's okay.

The horizontal distance traveled by the marble is $R = vt_f$, where v is the speed of the marble when it leaves the gun. I used R (range) here because we're going to have an x associated with the spring later. We find *that* speed using energy conservation principles applied to the spring just before it is released and just after the marble leaves the gun.

$$K_i + U_i = K_f + U_f,$$

$$0 + \frac{1}{2}kx^2 = \frac{1}{2}mv^2 + 0.$$

$K_i = 0$ because the marble isn't moving originally, and $U_f = 0$ because the spring is no longer compressed. Substituting R into this,

$$\frac{1}{2}kx^2 = \frac{1}{2}m\left(\frac{R}{t_f}\right)^2.$$

We have two values for the compression, x_1 and x_2, and two ranges, R_1 and R_2. We can put both pairs into the above equation and get two expressions; if we divide one expression by the other we get

$$\left(\frac{x_2}{x_1}\right)^2 = \left(\frac{R_2}{R_1}\right)^2.$$

Notice that everything else has canceled out– no k, no m, and no t_f to clutter our efforts. We can easily take the square root of both sides, then

$$\frac{x_2}{x_1} = \frac{R_2}{R_1},$$

which is much simpler than I expected. R_1 was Bobby's try, and was equal to $2.20 - 0.27 = 1.93\,\mathrm{m}$. $x_1 = 1.1$ cm was his compression. If Rhoda wants to score, she wants $R_2 = 2.2\,\mathrm{m}$, then

$$x_2 = \frac{2.2\,\mathrm{m}}{1.93\,\mathrm{m}} 1.1\ \mathrm{cm} = 1.25\ \mathrm{cm}.$$

E12-23 Another energy conservation problem, except now there are *three* contributions to the kinetic energy: rotational kinetic energy of the shell (K_s), rotational kinetic energy of the pulley (K_p), and translational kinetic energy of the block (K_b). There is only one object with a changing potential energy— the block— so that's all we'll consider.

The conservation of energy statement is then

$$
\begin{aligned}
K_{\mathrm{s,i}} + K_{\mathrm{p,i}} + K_{\mathrm{b,i}} + U_\mathrm{i} &= K_{\mathrm{s,f}} + K_{\mathrm{p,f}} + K_{\mathrm{b,f}} + U_\mathrm{f}, \\
(0) + (0) + (0) + (0) &= \frac{1}{2} I_\mathrm{s} \omega_\mathrm{s}^2 + \frac{1}{2} I_\mathrm{p} \omega_\mathrm{p}^2 + \frac{1}{2} m v_\mathrm{b}^2 + mgy.
\end{aligned}
$$

y is the vertical distance from the start, and is negative because the block is moving down. Then $y = -h$.

Don't assume that $\omega_\mathrm{s} = \omega_\mathrm{p}$, instead we have that the tangential velocity of both the pulley and the sphere are equal to the linear velocity of the block. This is because they are all tied to the same string. Then

$$\omega_\mathrm{s} R = \omega_\mathrm{p} r = v_\mathrm{b}.$$

Combine all of this together, and our energy conservation statement will look like this:

$$0 = \frac{1}{2}\left(\frac{2}{3} M R^2\right)\left(\frac{v_\mathrm{b}}{R}\right)^2 + \frac{1}{2} I_\mathrm{p}\left(\frac{v_\mathrm{b}}{r}\right)^2 + \frac{1}{2} m v_\mathrm{b}^2 - mgh$$

which can be fairly easily rearranged into

$$v_\mathrm{b}^2 = \frac{2mgh}{2M/3 + I_\mathrm{P}/r^2 + m}.$$

The answer is found by taking the square root of both sides. You might consider it assymetrical that the radius of the shell doesn't matter, while the radius of the pulley does. However, if we knew the mass and shape of the pulley (disk, hoop, or otherwise), we would find that the radius cancels out as well. In other words, the radius is hidden within I_p.

E12-27 Didn't we do this one already? Check back to the solution to Problem 10-7. In that exercise we first solved for Δt, here we want to first solve for v. As written, this exercise here has nothing to do with conservation of energy. In future editions we will probably add a part (b): find an expression for the fractional loss in kinetic energy.

E12-31 (a)   We can find F_x and F_y from the appropriate derivatives of the potential,

$$F_x = -\frac{\partial U}{\partial x} = -kx,$$
$$F_y = -\frac{\partial U}{\partial y} = -ky.$$

The force at point (x, y) is then

$$\vec{F} = F_x\hat{\mathbf{i}} + F_y\hat{\mathbf{j}} = -kx\hat{\mathbf{i}} - ky\hat{\mathbf{j}}.$$

But what does it mean? It is a vector which points *directly* toward the origin and has magnitude $k\sqrt{x^2 + y^2}$, which is proportional to the distance from the origin.

(b)   Since the force vector points directly toward the origin there is *no* angular component, and $F_\theta = 0$. Then $F_r = -kr$ where r is the distance from the origin.

(c)   A spring which is attached to a point; the spring is free to rotate, perhaps?

P12-1 (a)   We need to integrate

$$U(z) = -\int_\infty^z F_z \, dz.$$

We could look it up in a table, but we'll do it the hard way. Make the substitution $u = z + l$ and the first part of the integral becomes

$$-\int_\infty^z \frac{k}{(z+l)^2} \, dz = -\int_\infty^z \frac{k}{u^2} \, du,$$
$$= \left. \frac{k}{u} \right|_\infty^z,$$
$$= \frac{k}{z+l}$$

The second half is dealt with in a similar manner, yielding

$$U(z) = \frac{k}{z+l} - \frac{k}{z-l}.$$

(b)   If $z \gg l$ then we can expand the denominators according to

$$\frac{1}{z \pm l} = \frac{1}{z} \mp \frac{l}{z^2} + \frac{l^2}{z^3} \cdots$$

which is just the Taylor expansion of the fraction about $l = 0$. We need only keep the first two terms, then

$$
\begin{aligned}
U(z) &= \frac{k}{z+l} - \frac{k}{z-l}, \\
&\approx \left(\frac{k}{z} - \frac{kl}{z^2} \right) - \left(\frac{k}{z} + \frac{kl}{z^2} \right), \\
&= -\frac{2kl}{z^2}.
\end{aligned}
$$

P12-5 (a) This is an energy conservation problem, although it is somewhat disguised. In order to find the net force at Q we need to know how fast it is moving, and we find that from energy conservation principles. Considering points P and Q we have

$$
\begin{aligned}
K_P + U_P &= K_Q + U_Q, \\
(0) + mg(5R) &= \frac{1}{2}mv^2 + mg(R), \\
4mgR &= \frac{1}{2}mv^2, \\
\sqrt{8gR} &= v.
\end{aligned}
$$

That wasn't so hard.

The net force on an object in uniform circular motion is of magnitude mv^2/R. But the block isn't in uniform circular motion, it is slowing down! However, at Q there are two forces on the block: the normal force from the wall pushing it toward the center of the circle and the force of gravity. Since the force of gravity is perpendicular to the radial direction at Q, we can fairly assume that mv^2/R is the normal force.

So there are two forces on the block, the normal force from the track,

$$
N = \frac{mv^2}{R} = \frac{m(8gR)}{R} = 8mg,
$$

and the force of gravity $W = mg$. They are orthogonal (at right angles), so we can find the magnitude of the net force by

$$
F_{\text{net}} = \sqrt{(8mg)^2 + (mg)^2} = \sqrt{65}\, mg
$$

and the angle from the horizontal by

$$
\tan\theta = \frac{-mg}{8mg} = -\frac{1}{8},
$$

or $\theta = 7.13°$ below the horizontal.

(b)   If the block *barely* makes it over the top of the track then the speed at the top of the loop (point S, perhaps?) is just fast enough so that the centripetal force is equal in magnitude to the weight,

$$mv_S{}^2/R = mg.$$

Assume the block was released from point T. The energy conservation problem is then

$$
\begin{aligned}
K_T + U_T &= K_S + U_S, \\
(0) + mgy_T &= \frac{1}{2}mv_S^2 + mgy_S, \\
y_T &= \frac{1}{2}(R) + m(2R), \\
&= 5R/2.
\end{aligned}
$$

P12-9   This is same as Problem 12-5, except we now have $L \geq 5r/2$, or $r \leq 2L/5$, which means

$$d = L - r > L - 2L/5 = 3L/5.$$

P12-11   Let the angle θ be measured from the horizontal to the point on the hemisphere where the boy is located. θ starts at $90°$, but as the boy slides down it decreases.

There are then two components to the force of gravity— a component tangent to the hemisphere, $W_\| = mg\cos\theta$, and a component directed radially toward the center of the hemisphere, $W_\perp = mg\sin\theta$.

While the boy is in contact with the hemisphere the motion is circular; although not uniform, we can write

$$mv^2/R = W_\perp - N.$$

When the boy leaves the surface we have $mv^2/R = W_\perp$, or $mv^2 = mgR\sin\theta$. Note that $\sin\theta = y/R$.

Now for energy conservation,

$$
\begin{aligned}
K + U &= K_0 + U_0, \\
\frac{1}{2}mv^2 + mgy &= \frac{1}{2}m(0)^2 + mgR, \\
\frac{1}{2}gR\sin\theta + mgy &= mgR, \\
\frac{1}{2}y + y &= R, \\
y &= 2R/3.
\end{aligned}
$$

P12-17 The function needs to fall off at infinity in both directions; an exponential envelope would work, but it will need to have an $-x^2$ term to force the potential to zero on *both* sides. So we propose something of the form

$$U(x) = P(x)e^{-\beta x^2}$$

where $P(x)$ is a polynomial in x and β is a positive constant.

We proposed the *polynomial* because we need a symmetric function which has **two** zeroes. A quadratic of the form $\alpha x^2 - U_0$ would work, it has two zeroes, a minimum at $x = 0$, and is symmetric.

So our *trial* function is

$$U(x) = \left(\alpha x^2 - U_0\right) e^{-\beta x^2}.$$

This function should have *three* extrema. Take the derivative, and then we'll set it equal to zero,

$$\frac{dU}{dx} = 2\alpha x e^{-\beta x^2} - 2\left(\alpha x^2 - U_0\right)\beta x e^{-\beta x^2}.$$

Setting this equal to zero leaves two possibilities,

$$x = 0,$$
$$2\alpha - 2\left(\alpha x^2 - U_0\right)\beta = 0.$$

The first equation is trivial, the second is easily rearranged to give

$$x = \pm\sqrt{\frac{\alpha + \beta U_0}{\beta\alpha}}$$

These are the points $\pm x_1$. We can, if we wanted, try to find α and β from the picture, but you might notice we have one equation, $U(x_1) = U_1$ and two unknowns. It really isn't very illuminating to take this problem much farther, but we could.

(b) The force is the derivative of the potential; this expression was found above.

(c) As long as the energy is *less* than the two peaks, then the motion would be oscillatory, trapped in the well.

Chapter 13

Energy 3: Conservation of Energy

E13-1 If the projectile had *not* experienced air drag it would have risen to a height y_2, but because of air drag 68 kJ of mechanical energy was dissipated so it only rose to a height y_1. In either case the initial velocity, and hence initial kinetic energy, was the same; and the velocity at the highest point was zero. Then $W = \Delta U$, so the potential energy would have been 68 kJ greater, and

$$\Delta y = \Delta U / mg = (68 \times 10^3 \, \text{J})/(9.4 \, \text{kg})(9.81 \, \text{m/s}^2) = 740 \, \text{m}$$

is how much higher it would have gone without air friction.

E13-5 (a) Applying conservation of energy to the points where the ball was dropped and where it entered the oil,

$$\frac{1}{2}mv_\text{f}^2 + mgy_\text{f} = \frac{1}{2}mv_\text{i}^2 + mgy_\text{i},$$
$$\frac{1}{2}v_\text{f}^2 + g(0) = \frac{1}{2}(0)^2 + gy_\text{i},$$
$$v_\text{f} = \sqrt{2gy_\text{i}},$$
$$= \sqrt{2(9.81 \, \text{m/s}^2)(0.76 \, \text{m})} = 3.9 \, \text{m/s}.$$

(b) The change in internal energy of the ball + oil can be found by considering the points where th ball was released and where the ball reached the bottom of the container.

$$\Delta E = K_\text{f} + U_\text{f} - K_\text{i} - U_\text{i},$$
$$= \frac{1}{2}mv_\text{f}^2 + mgy_\text{f} - \frac{1}{2}m(0)^2 - mgy_\text{i},$$
$$= \frac{1}{2}(12.2 \times 10^{-3} \text{kg})(1.48 \text{m/s})^2 - (12.2 \times 10^{-3} \text{kg})(9.81 \text{m/s}^2)(-0.55 \text{m} - 0.76 \text{m}),$$
$$= -0.143 \, \text{J}$$

E13-9 Let m be the mass of the water under consideration. Then the percentage of the potential energy "lost" which appears as kinetic energy is

$$\frac{K_f - K_i}{U_i - U_f}.$$

Then

$$\begin{aligned}
\frac{K_f - K_i}{U_i - U_f} &= \frac{1}{2}m\left(v_f^2 - v_i^2\right)/(mgy_i - mgy_f), \\
&= \frac{v_f^2 - v_i^2}{-2g\Delta y}, \\
&= \frac{(13\,\text{m/s})^2 - (3.2\,\text{m/s})^2}{-2(9.81\,\text{m/s}^2)(-15\,\text{m})}, \\
&= 54\,\%.
\end{aligned}$$

The rest of the energy would have been converted to sound and thermal energy.

E13-13 When the ball bounces off of the ground all of the energy is in the form of kinetic energy, and none is (gravitational) potential energy. When the ball is at the highest point after the return bounce all of the energy is potential. But this final potential energy is 15 % less than the initial kinetic plus potential energy of the ball, so

$$0.85(K_i + U_i) = U_f.$$

But the ball started and ended at the same height, so $U_i = U_f$, then

$$K_i = 0.15U_f/0.85,$$

or

$$v_i = \sqrt{\frac{0.15}{0.85}2gh} = \sqrt{2(0.176)(9.81\,\text{m/s}^2)(12.4\,\text{m})} = 6.54\,\text{m/s}.$$

It is interesting (to me) that the mass of the ball doesn't matter!

E13-17 We can find the kinetic energy of the center of mass of the woman when her feet leave the ground by considering energy conservation and her highest point. Then

$$\begin{aligned}
\frac{1}{2}mv_i^2 + mgy_i &= \frac{1}{2}mv_f^2 + mgy_f, \\
\frac{1}{2}mv_i &= mg\Delta y, \\
&= (55.0\,\text{kg})(9.81\,\text{m/s}^2)(1.20\,\text{m} - 0.90\,\text{m}) = 162\,\text{J}.
\end{aligned}$$

(a) This kinetic energy is the final kinetic energy when she leaves the ground, and is then the change in kinetic energy while the ground is exerting an upward force on her. During the jumping phase her potential energy also changed by

$$\Delta U = mg\Delta y = (55.0\,\text{kg})(9.81\,\text{m/s}^2)(0.50\,\text{m}) = 270\,\text{J}$$

while she was moving up. Then

$$F_{\text{ext}} = \frac{\Delta K + \Delta U}{\Delta s} = \frac{(162\,\text{J}) + (270\,\text{J})}{(0.5\,\text{m})} = 864\,\text{N}.$$

(b) Her fastest speed was when her feet left the ground,

$$v = \frac{2K/m}{=} \frac{2(162\,\text{J})/(55.0\,\text{kg})}{=} 2.42\,\text{m/s}.$$

E13-21 Momentum conservation requires

$$mv_0 = mv + MV,$$

where the sign indicates the direction. We are assuming one dimensional collisions. Energy conservation requires

$$\frac{1}{2}mv_0^2 = \frac{1}{2}mv^2 + \frac{1}{2}MV^2 + E.$$

Now for a little math,

$$\frac{1}{2}mv_0^2 = \frac{1}{2}mv^2 + \frac{1}{2}M\left(\frac{m}{M}v_0 - \frac{m}{M}v\right)^2 + E,$$
$$Mv_0^2 = Mv^2 + m\left(v_0 - v\right)^2 + 2(M/m)E.$$

We can rearrange this as a quadratic in v,

$$(M + m)\,v^2 - (2mv_0)\,v + \left(2(M/m)E + mv_0^2 - Mv_0^2\right) = 0.$$

This will only have real solutions if the discriminant $(b^2 - 4ac)$ is greater than or equal to zero. Then

$$(2mv_0)^2 \geq 4\left(M + m\right)\left(2(M/m)E + mv_0^2 - Mv_0^2\right)$$

is the condition for the minimum v_0. Solving the equality condition,

$$4m^2v_0^2 = 4(M + m)\left(2(M/m)E + (m - M)v_0^2\right),$$

or

$$M^2v_0^2 = 2(M + m)(M/m)E.$$

One last rearrangement, and

$$v_0 = \sqrt{2(M + m)E/(mM)}.$$

P13-1 (a) An energy expression can be used to find out how high the stone rises. The initial kinetic energy will equal the potential energy at the highest point *plus* the amount of energy which is dissipated because of air drag.

$$mgh + fh = \frac{1}{2}mv_0^2,$$

$$h = \frac{v_0^2}{2(g + f/m)} = \frac{v_0^2}{2g(1 + f/w)}.$$

(b) The final kinetic energy when the stone lands will be equal to the initial kinetic energy *minus* twice the energy dissipated on the way up, so

$$\frac{1}{2}mv^2 = \frac{1}{2}mv_0^2 - 2fh,$$

$$= \frac{1}{2}mv_0^2 - 2f\frac{v_0^2}{2g(1 + f/w)},$$

$$= \left(\frac{m}{2} - \frac{f}{g(1 + f/w)}\right)v_0^2,$$

$$v^2 = \left(1 - \frac{2f}{w + f}\right)v_0^2,$$

$$v = \left(\frac{w - f}{w + f}\right)^{1/2}v_0.$$

P13-7 The net force on the top block while it is being pulled is

$$11.0\,\text{N} - F_f = 11.0\,\text{N} - (0.35)(2.5\,\text{kg})(9.81\,\text{m/s}^2) = 2.42\,\text{N}.$$

This means it is accelerating at $(2.42\,\text{N})/(2.5\,\text{kg}) = 0.968\,\text{m/s}^2$. That acceleration will last a time $t = \sqrt{2d/a}$, or $\sqrt{2(0.30\,\text{m})/(0.968\,\text{m/s}^2)} = 0.787\,\text{s}$. The speed of the top block after the force stops pulling is then $(0.968\,\text{m/s}^2)(0.787\,\text{s}) = 0.762\,\text{m/s}$. The force on the bottom block is F_f, so the acceleration of the bottom block is

$$(0.35)(2.5\,\text{kg})(9.81\,\text{m/s}^2)/(10.0\,\text{kg}) = 0.858\,\text{m/s}^2,$$

and the speed after the force stops pulling on the top block is $(0.858\,\text{m/s}^2)(0.787\,\text{s}) = 0.675\,\text{m/s}$.

(a) $W = Fs = (11.0\,\text{N})(0.30\,\text{m}) = 3.3\,\text{J}$ of energy were delivered to the system, but after the force stops pulling only

$$\frac{1}{2}(2.5\,\text{kg})(0.762\,\text{m/s})^2 + \frac{1}{2}(10.0\,\text{kg})(0.675\,\text{m/s})^2 = 3.004\,\text{J}$$

were present as kinetic energy. So $0.296\,\text{J}$ is "missing" and would be now present as internal energy.

(b) The impulse received by the two block system is then $J = (11.0\,\text{N})(0.787\,\text{s}) = 8.66\,\text{N}\cdot\text{s}$. This impulse causes a change in momentum, so the speed of the two block system after the external force stops pulling and both blocks move as one is $(8.66\,\text{N}\cdot\text{s})(12.5\,\text{kg}) = 0.693\,\text{m/s}$. The final kinetic energy is

$$\frac{1}{2}(12.5\,\text{kg})(0.693\,\text{m/s})^2 = 3.002\,\text{J};$$

this means that $0.002\,\text{J}$ are dissipated. This answer is *smaller* than the answer in the back of the book. If you leave your intermediate answers with five or so significant figures, then you will get the back of the book answers. This reflects the loss of accuracy that occurs when you subtract numbers.

Chapter 14

Gravitation

E14-3 We'll assume that the entire mass of the echo satellite is located at the center. This point is proved in Section 14-5, and the application of the shell theorem is fundamental in gravitation, electrostatics, or any inverse square force law. We'll make a similar assumption for the meteor, but a 7.0 kg meteor is likely to have a diameter on the order 10 cm, which is much smaller than the distance between the meteor and the satellite center.

The masses of each object are $m_1 = 20.0 \, \text{kg}$ and $m_2 = 7.0 \, \text{kg}$; the distance between the centers of the two objects is $15 + 3 = 18 \, \text{m}$.

The magnitude of the force from Newton's law of gravitation is then

$$F = \frac{Gm_1m_2}{r^2} = \frac{(6.67 \times 10^{-11} \text{N} \cdot \text{m}^2/\text{kg}^2)(20.0 \, \text{kg})(7.0 \, \text{kg})}{(18 \, \text{m})^2} = 2.9 \times 10^{-11} \text{N}.$$

E14-5 Again, we assume that the entire mass of the Earth is concentrated at the center of the Earth. The force of gravity on an object near the surface of the earth is given by

$$F = \frac{GMm}{(r_e + y)^2},$$

where M is the mass of the Earth, m is the mass of the object, r_e is the radius of the Earth, and y is the height above the surface of the Earth. There are two ways we can proceed. One is the "brute force" method, and simply subtract the force of gravity on the man at the top of the tower from the force on the man at the bottom.

The other method is *considerably* more elegant— expand the expression since $y \ll r_e$. We'll use a Taylor expansion, where $F(r_e + y) \approx F(r_e) + y \partial F / \partial r_e$;

$$F \approx \frac{GMm}{r_e^2} - 2y\frac{GMm}{r_e^3}$$

Since we are interested in the difference between the force at the top and the bottom, we really want

$$\Delta F = 2y\frac{GMm}{r_e^3} = 2\frac{y}{r_e}\frac{GMm}{r_e^2} = 2\frac{y}{r_e}W,$$

where in the last part we substituted for the weight, which is the same as the force of gravity,

$$W = \frac{GMm}{r_e^2}.$$

The true power of this approach is that we don't need to know G, M, or m.

Finally,

$$\Delta F = 2(411\,\text{m})/(6.37 \times 10^6\text{m})(120\text{ lb}) = 0.015\text{ lb}.$$

E14-9   We'll repeat this question in a different form when we consider the pendulum. If an object is dropped from rest, the relevant kinematic expression is

$$y = -\frac{1}{2}gt^2.$$

The object fell through $y = -10.0\,\text{m}$; the time required to fall would then be

$$t = \sqrt{-2y/g} = \sqrt{-2(-10.0\,\text{m})/(9.81\,\text{m/s}^2)} = 1.43\,\text{s}.$$

We are interested in the *error*, however, which is the behavior of the equation under small variations. That means taking the total derivative of

$$y = -\frac{1}{2}gt^2.$$

and getting

$$\delta y = -\frac{1}{2}\delta g\, t^2 - gt\,\delta t,$$

where we use the δ notation because the variations, although small, are not infinitesimal.

It is given that y is measured *perfectly*. This means that $\delta y = 0$. We are left with

$$-\frac{1}{2}\delta g\, t = g\delta\, t,$$

which, since we are interested in the necessary precision for t, can be rearranged as

$$\delta t = -\frac{\delta g\, t}{2g}.$$

Interestingly enough, the tolerable percentage error in t is half the tolerable percentage error in g, regardless of how far the object falls!

The percentage error in t needs to be $\delta t/t = 0.1\,\%/2 = 0.05\,\%$. The absolute error is then $\delta t = (0.05\,\%)(1.43\,\text{s}) = 0.7\text{ ms}$.

E14-11 The force of gravity on an object of mass m located inside a solid sphere is given by

$$F = \frac{GM(r)\,m}{r^2},$$

where r is the distance of the object from the center of the sphere and $M(r)$ is the amount of the mass of the sphere which is located within a distance r from the center of the sphere.

For a sphere of uniform density and radius $R > r$,

$$\frac{M(r)}{\frac{4}{3}\pi r^3} = \frac{M}{\frac{4}{3}\pi R^3},$$

where M is the total mass. This expression is fundamentally the same as the one derived in Sample Problem 14-4.

The force of gravity on the object of mass m is then

$$F = \frac{GMm}{r^2}\frac{r^3}{R^3} = \frac{GMmr}{R^3},$$

as was found in Sample Problem 14-4.

g is the free-fall acceleration of the object, and is the gravitational force divided by the mass, so

$$g = \frac{GMr}{R^3}.$$

Let's massage this expression some.

$$g = \frac{GMr}{R^3} = \frac{GM}{R^2}\frac{r}{R} = \frac{GM}{R^2}\frac{R-D}{R}.$$

Since R is the distance from the center to the surface, and D is the distance of the object beneath the surface, then $r = R - D$ is the distance from the center to the object. The first fraction is the free-fall acceleration on the surface, so

$$g = \frac{GM}{R^2}\frac{R-D}{R} = g_s\frac{R-D}{R} = g_s\left(1 - \frac{D}{R}\right)$$

E14-15 (a) We traditionally set the gravitational potential energy of a system of two particles to be zero when the objects are infinitely far apart. In this case the potential energy will be negative when the objects are closer together. Escaping will only happen if the total energy is greater than zero.

Near the surface of the Earth the total energy is then

$$E = K + U = \frac{1}{2}m\left(2\sqrt{gR_{\mathrm{E}}}\right)^2 - \frac{GM_{\mathrm{E}}m}{R_{\mathrm{E}}}$$

but, according to Eq. 14-5,

$$g = \frac{GM}{R_E^2},$$

so the total energy is

$$
\begin{aligned}
E &= 2mgR_E - \frac{GM_Em}{R_E}, \\
&= 2m\left(\frac{GM}{R_E^2}\right)R_E - \frac{GM_Em}{R_E}, \\
&= \frac{GM_Em}{R_E}
\end{aligned}
$$

This is a positive number, so the rocket will escape, regardless of which direction it is originally pointing (unless it crashes by colliding with the Earth, of course.)

(b) Far from earth there is no gravitational potential energy, so

$$\frac{1}{2}mv^2 = \frac{GM_Em}{R_E} = \frac{GM_E}{R_E^2}mR_E = gmR_E,$$

with solution $v = \sqrt{2gR_E}$.

E14-17 Energy conservation is $K_i + U_i = K_f + U_f$, but at the highest point $K_f = 0$, so

$$
\begin{aligned}
U_f &= K_i + U_i, \\
-\frac{GM_Em}{R} &= \frac{1}{2}mv_0^2 - \frac{GM_Em}{R_E}, \\
\frac{1}{R} &= \frac{1}{R_E} - \frac{1}{2GM_E}v_0^2, \\
\frac{1}{R} &= \frac{1}{(6.37\times10^6\text{m})} - \frac{(9.42\times10^3\text{m/s})^2)}{2(6.67\times10^{-11}\text{N}\cdot\text{m}^2/\text{kg}^2)(5.98\times10^{24}\text{kg})}, \\
R &= 2.19\times10^7 \text{ m}.
\end{aligned}
$$

The distance above the Earth's surface is 2.19×10^7 m $- 6.37\times10^6$ m $= 1.55\times10^6$ m.

E14-23 We can use Eq. 14-23 to find the mass of Mars; all we need to do is rearrange to solve for M—

$$M = \frac{4\pi^2r^3}{GT^2} = \frac{4\pi^2(9.4\times10^6\text{m})^3}{(6.67\times10^{-11}\text{N}\cdot\text{m}^2/\text{kg}^2)(2.75\times10^4\text{s})^2} = 6.5\times10^{23}\text{kg}.$$

E14-27 (b) It is actually easier to find the period of motion first. We'll make the assumption that the altitude of the satellite is so low that the radius of the orbit is effectively the radius of the moon. Then, from Eq. 14-23,

$$T^2 = \left(\frac{4\pi^2}{GM}\right)r^3,$$

$$= \left(\frac{4\pi^2}{(6.67\times10^{-11}\mathrm{N\cdot m^2/kg^2})(7.36\times10^{22}\mathrm{kg})}\right)(1.74\times10^6\mathrm{m})^3 = 4.24\times10^7\mathrm{s^2}.$$

We want the square root of this, so $T = 6.5\times10^3\mathrm{s}$.

(a) The speed of the satellite is the circumference divided by the period, or

$$v = \frac{2\pi r}{T} = \frac{2\pi(1.74\times10^6\mathrm{m})}{(6.5\times10^3\mathrm{s})} = 1.68\times10^3\mathrm{m/s}.$$

E14-31 Kepler's third law states $T^2 \propto r^3$, where r is the mean distance from the *Sun* and T is the period of revolution. Newton was in a position to find the acceleration of the Moon toward the Earth by assuming the Moon moved in a circular orbit, since $a_c = v^2/r = 4\pi^2 r/T^2$. But this means that, because of Kepler's law, $a_c \propto r/T^2 \propto 1/r^2$.

E14-35 (a) The approximate force of gravity on a 2000 kg pickup truck on Eros will be

$$F = \frac{GMm}{r^2} = \frac{(6.67\times10^{-11}\mathrm{N\cdot m^2/kg^2})(5.0\times10^{15}\,\mathrm{kg})(2000\,\mathrm{kg})}{(7000\,\mathrm{m})^2} = 13.6\,\mathrm{N}.$$

Could I lift it? Yes, it is only about 3 pounds!

(b) Orbital velocity can be found by (1) repeating the steps in Ex. 14-27, or (2) equating centripetal force to gravitational force, or (3) applying Eq. 14-27 (which is actually just found by applying method 2.)

However you arrive, we will get to

$$v = \sqrt{\frac{GM}{r}} = \sqrt{\frac{(6.67\times10^{-11}\mathrm{N\cdot m^2/kg^2})(5.0\times10^{15}\,\mathrm{kg})}{(7000\,\mathrm{m})}} = 6.9\,\mathrm{m/s}.$$

Note that you *could not* apply Eq. 14-19, because escape velocity and orbital velocity are not the same thing! Can I put myself in orbit? Definitely. Shortly before typing this we landed a satellite on Eros.

E14-39 (a) The Starshine satellite was approximately 275 km above the surface of the Earth on 1 January 2000. We can find the orbital period from Eq. 14-23,

$$
\begin{aligned}
T^2 &= \left(\frac{4\pi^2}{GM}\right) r^3, \\
&= \left(\frac{4\pi^2}{(6.67\times 10^{-11}\text{N·m}^2/\text{kg}^2)(5.98\times 10^{24}\text{kg})}\right)(6.65\times 10^6\text{m})^3 = 2.91\times 10^7 \text{s}^2,
\end{aligned}
$$

so $T = 5.39\times 10^3$s.

(b) Equation 14-25 gives the total energy of the system of a satellite of mass m in a circular orbit of radius r around a stationary body of mass $M \gg m$,

$$
E = -\frac{GMm}{2r}.
$$

We want the rate of change of this with respect to time, so

$$
\frac{dE}{dt} = \frac{GMm}{2r^2}\frac{dr}{dt}
$$

We can estimate the value of dr/dt from the diagram. I'll choose February 1 and December 1 as my two reference points.

$$
\frac{dr}{dt}\bigg|_{t=t_0} \approx \frac{\Delta r}{\Delta t} = \frac{(240\text{ km}) - (300\text{ km})}{(62\text{ days})} \approx -1\text{ km/day}
$$

Don't worry too much about accuracy; anytime you subtract two numbers the error/uncertainty grows. We can only hope to get accuracy to within some 10 km from the graph, so our rate of altitude loss has a relative uncertainty of more than 10%.

The rate of energy loss is then

$$
\frac{dE}{dt} = \frac{(6.67\times 10^{-11}\text{N·m}^2/\text{kg}^2)(5.98\times 10^{24}\text{kg})(39\text{ kg})}{2(6.65\times 10^6\text{m})^2}\frac{-1000\text{ m}}{8.64\times 10^4\text{ s}} = -2.0\text{ J/s}.
$$

Your answer may differ depending on your approximation of the orbit decay rate.

P14-1 The object on the top experiences a force down from gravity W_1 and a force down from the tension in the rope T. The object on the bottom experiences a force down from gravity W_2 and a force up from the tension in the rope.

In either case, the magnitude of W_i is

$$
W_i = \frac{GMm}{r_i^2}
$$

where r_i is the distance of the ith object from the center of the Earth. While the objects fall they have the same acceleration (they are attached by a rope, eh?), and since they have the same mass we can quickly write

$$\frac{GMm}{r_1^2} + T = \frac{GMm}{r_2^2} - T,$$

or

$$
\begin{aligned}
T &= \frac{GMm}{2r_2^2} - \frac{GMm}{2r_1^2}, \\
&= \frac{GMm}{2}\left(\frac{1}{r_1^2} - \frac{1}{r_2^2}\right), \\
&= \frac{GMm}{2}\frac{r_2^2 - r_1^2}{r_1^2 r_2^2}.
\end{aligned}
$$

Now $r_1 \approx r_2 \approx R$ in the denominator, but $r_2 = r_1 + l$, so $r_2^2 - r_1^2 \approx 2Rl$ in the numerator. Then

$$T \approx \frac{GMml}{R^3}.$$

P14-5 (a) The magnitude of the gravitational force from the Moon on a particle at A is

$$F_A = \frac{GMm}{(r-R)^2},$$

where the denominator is the distance from the center of the moon to point A.

(b) At the center of the Earth the gravitational force of the moon on a particle of mass m is

$$F_C = \frac{GMm}{r^2},$$

which is slightly *smaller* than the previous expression.

(c) Now we want to know the difference between these two expressions:

$$
\begin{aligned}
F_A - F_C &= \frac{GMm}{(r-R)^2} - \frac{GMm}{r^2}, \\
&= GMm\left(\frac{r^2}{r^2(r-R)^2} - \frac{(r-R)^2}{r^2(r-R)^2}\right), \\
&= GMm\left(\frac{r^2 - (r-R)^2}{r^2(r-R)^2}\right), \\
&= GMm\left(\frac{R(2r-R)}{r^2(r-R)^2}\right).
\end{aligned}
$$

This is as simple as we can get without making some approximations. Really we should do a Taylor expansion, but we'll simply assume $R \ll r$ and then substitute $(r - R) \approx r$. The force difference simplifies to

$$F_T = GMm\frac{R(2r)}{r^2(r)^2} = \frac{2GMmR}{r^3}$$

(d) Repeat part (c) except we want $r + R$ instead of $r - R$. Then

$$
\begin{aligned}
F_A - F_C &= \frac{GMm}{(r+R)^2} - \frac{GMm}{r^2}, \\
&= GMm\left(\frac{r^2}{r^2(r+R)^2} - \frac{(r+R)^2}{r^2(r+R)^2}\right), \\
&= GMm\left(\frac{r^2 - (r+R)^2}{r^2(r+R)^2}\right), \\
&= GMm\left(\frac{-R(2r+R)}{r^2(r+R)^2}\right).
\end{aligned}
$$

This is as simple as we can get without making some approximations. Really we should do a Taylor expansion, but we'll simply assume $R \ll r$ and then substitute $(r + R) \approx r$. The force difference simplifies to

$$F_T = GMm\frac{-R(2r)}{r^2(r)^2} = -\frac{2GMmR}{r^3}$$

The negative sign indicates that this "apparent" force points *away* from the moon, not toward it.

(e) Consider the directions: the water is effectively attracted to the moon when closer, but repelled when farther.

P14-11 This problem is *fun*.

The force of gravity on the small sphere of mass m is equal to the force of gravity from a solid lead sphere minus the force which would have been contributed by the smaller lead sphere which would have filled the hole. So we need to know about the size and mass of the lead which was removed to make the hole.

The density of the lead is given by

$$\rho = \frac{M}{\frac{4}{3}\pi R^3}$$

The hole has a radius of $R/2$, so if the density is constant the mass of the hole will be

$$M_h = \rho V = \left(\frac{M}{\frac{4}{3}\pi R^3}\right)\frac{4}{3}\pi\left(\frac{R}{2}\right)^3 = \frac{M}{8}$$

The "hole" is closer to the small sphere; the center of the hole is $d - R/2$ away. The force of the whole lead sphere minus the force of the "hole" lead sphere is

$$\frac{GMm}{d^2} - \frac{G(M/8)m}{(d - R/2)^2}$$

P14-15 (a) We will use part of the hint, but we will integrate instead of assuming the bit about g_{av}; doing it this way will become important for later chapters. Consider a small horizontal slice of the column of thickness dr. The weight of the material above the slice exerts a force $F(r)$ on the top of the slice; there is a force of gravity on the slice given by

$$dF = \frac{GM(r)\,dm}{r^2},$$

where $M(r)$ is the mass contained in the sphere of radius r,

$$M(r) = \frac{4}{3}\pi r^3 \rho.$$

Lastly, the mass of the slice dm is related to the thickness and cross sectional area by $dm = \rho A\,dr$. Then

$$dF = \frac{4\pi G A \rho^2}{3} r\,dr.$$

Integrate both sides of this expression. On the left the limits are 0 to F_{center}, on the right the limits are R to 0; we need to throw in an extra negative sign because the force increases as r decreases. Then

$$F = \frac{2}{3}\pi G A \rho^2 R^2.$$

Divide both sides by A to get the compressive stress.

(b) Put in the numbers!

$$S = \frac{2}{3}\pi(6.67\times10^{-11}\text{N}\cdot\text{m}^2/\text{kg}^2)(4000\,\text{kg/m}^3)^2(3.0\times10^5\text{m})^2 = 2.0\times10^8\text{N/m}^2.$$

(c) Rearrange, and then put in numbers;

$$R = \sqrt{\frac{3(4.0\times10^7\text{N/m}^2)}{2\pi(6.67\times10^{-11}\text{N}\cdot\text{m}^2/\text{kg}^2)(3000\,\text{kg/m}^3)^2}} = 1.8\times10^5\text{ m}.$$

P14-21 (a) The force of one star on the other is given by $F = Gm^2/d^2$, where d is the distance between the stars. The stars revolve around the center of mass, which is half way between the stars (which have equal mass!), so $r = d/2$ is the radius of the orbit of the stars. If a is the centripetal acceleration of the stars, the period of revolution is then

$$T = \sqrt{\frac{4\pi^2 r}{a}} = \sqrt{\frac{4m\pi^2 r}{F}} = \sqrt{\frac{16\pi^2 r^3}{Gm}}.$$

The numerical value is

$$T = \sqrt{\frac{16\pi^2 (1.12 \times 10^{11}\text{m})^3}{(6.67 \times 10^{-11}\text{N} \cdot \text{m}^2/\text{kg}^2)(3.22 \times 10^{30}\text{kg})}} = 3.21 \times 10^7 \text{s} = 1.02 \text{ y}.$$

(b) The gravitational potential energy per kilogram midway between the stars is

$$-2\frac{Gm}{r} = -2\frac{(6.67 \times 10^{-11}\text{N} \cdot \text{m}^2/\text{kg}^2)(3.22 \times 10^{30}\text{kg})}{(1.12 \times 10^{11}\text{m})} = -3.84 \times 10^9 \text{J/kg}.$$

An object of mass M at the center between the stars would need $(3.84 \times 10^9 \text{J/kg})M$ kinetic energy to escape, this corresponds to a speed of

$$v = \sqrt{2K/M} = \sqrt{2(3.84 \times 10^9 \text{J/kg})} = 8.76 \times 10^4 \text{m/s}.$$

P14-23 (a) Consider the following diagram.

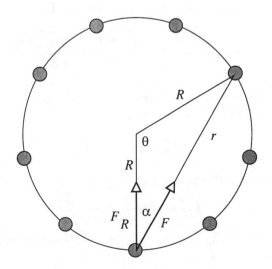

The distance r is given by the cosine law to be

$$r^2 = R^2 + R^2 - 2R^2 \cos\theta = 2R^2(1 - \cos\theta).$$

The force between two particles is then

$$F = Gm^2/r^2.$$

Each particle has a symmetric partner, so only the force component directed toward the center contributes. If we call this the R component we have

$$F_R = F\cos\alpha = F\cos(90° - \theta/2) = F\sin(\theta/2).$$

Combining,

$$F_R = \frac{Gm^2}{2R^2}\frac{\sin(\theta/2)}{1 - \cos\theta}.$$

But *each* of the other particles contributes to this force, so

$$F_{\text{net}} = \frac{Gm^2}{2R^2}\sum_i \frac{\sin(\theta_i/2)}{1 - \cos\theta_i}.$$

When there are only 9 particles the angles are in steps of 40°; the θ_i are then 40°, 80°, 120°, 160°, 200°, 240°, 280°, and 320°. With a little patience you will find

$$\sum_i \frac{\sin(\theta_i/2)}{1 - \cos\theta_i} = 6.649655,$$

using these angles. Then

$$F_{\text{net}} = 3.32 Gm^2/R^2.$$

(b) The rotational period of the ring would need to be

$$T = \sqrt{\frac{4\pi^2 R}{a}} = \sqrt{\frac{4m\pi^2 R}{F}} = \sqrt{\frac{16\pi^2 R^3}{3.32 Gm}}.$$

P14-33 There are three forces on loose matter (of mass m_0) sitting on the moon: the force of gravity toward the moon,

$$F_m = Gmm_0/a^2,$$

the force of gravity toward the planet,

$$F_M = GMm_0/(r - a)^2,$$

and the normal force N of the moon pushing the loose matter away from the center of the moon.

The net force on this loose matter is $F_M + N - F_m$, this value is *exactly* equal to the centripetal force necessary to keep the loose matter moving in a uniform circle. The period of revolution of the loose matter is identical to that of the moon,

$$T = 2\pi\sqrt{r^3/GM},$$

but since the loose matter is actually revolving at a radial distance $r - a$ the centripetal force is

$$F_{\text{c}} = \frac{4\pi^2 m_0 (r - a)}{T^2} = \frac{GMm_0(r - a)}{r^3}.$$

Only if the normal force is zero can the loose matter can lift off, and this will happen when $F_{\text{c}} = F_M - F_m$, or

$$
\begin{aligned}
\frac{M(r - a)}{r^3} &= \frac{M}{(r - a)^2} - \frac{m}{a^2}, \\
&= \frac{Ma^2 - m(r - a)^2}{a^2(r - a)^2}, \\
Ma^2(r - a)^3 &= Mr^3a^2 - mr^3(r - a)^2, \\
-3r^2a^3 + 3ra^4 - a^4 &= \frac{m}{M}\left(-r^5 + 2r^4a - r^3a^2\right)
\end{aligned}
$$

Gosh, that's ugly. Let $r = ax$, then x is dimensionless; let $\beta = m/M$, then β is dimensionless. The expression then simplifies to

$$-3x^2 + 3x - 1 = \beta(-x^5 + 2x^4 - x^3).$$

If we assume than x is very large ($r \gg a$) then only the largest term on each side survives. This means

$$3x^2 \approx \beta x^5,$$

or

$$x = (3/\beta)^{1/3}.$$

In that case, $r = a(3M/m)^{1/3}$. For the Earth's moon $r_{\text{c}} = 1.1 \times 10^7$m, which is only 4,500 km away from the surface of the Earth. It is somewhat interesting to note that the radius r is actually independent of both a and m if the moon has a uniform density!

Chapter 15

Fluid Statics

E15-1 From Eq. 15-2, $p = \Delta F / \Delta A$. The area in question is the surface area of the piston, so $\Delta A = \pi r^2$. The force was exerted by the nurse, so the pressure in the syringe is

$$p = \frac{(42.3\,\text{N})}{\pi (1.12 \times 10^{-2}\text{m}/s)^2} = 4.29 \times 10^5 \text{Pa}.$$

E15-5 There is an inward force F_1 pushing the lid closed from the pressure of the air outside the box; there is an outward force F_2 pushing the lid open from the pressure of the air inside the box. To lift the lid we need to exert an additional outward force F_3 to get a net force of zero. Then $F_1 = F_2 + F_3$ is the condition for the lid to be opened.

The magnitude of the inward force is $F_1 = P_{\text{out}} A$, where A is the area of the lid and P_{out} is the pressure outside the box. The magnitude of the outward force F_2 is $F_2 = P_{\text{in}} A$. We are told $F_3 = 108$ lb. Combining,

$$
\begin{aligned}
F_2 &= F_1 - F_3, \\
P_{\text{in}} A &= P_{\text{out}} A - F_3, \\
P_{\text{in}} &= P_{\text{out}} - F_3/A,
\end{aligned}
$$

so $P_{\text{in}} = (15\ \text{lb/in}^2 - (108\ \text{lb})/(12\ \text{in}^2) = 6.0\ \text{lb/in}^2$.

E15-9 Using Eq. 15-8 we can find the pressure differential assuming we don't have a sewage pump.

$$p_2 - p_1 = -\rho g\,(y_2 - y_1) = (926\,\text{kg/m}^3)(9.81\,\text{m/s}^2)(8.16\,\text{m} - 2.08\,\text{m}) = 5.52 \times 10^4 \text{Pa}.$$

We need to overcome this pressure difference with the pump.

E15-13 (a) Equation 15-8 can be used to find the height y_2 of the atmosphere if the density is constant. The pressure at the top of the atmosphere would be $p_2 = 0$, and the height of the bottom y_1 would be zero. Then

$$p_1 = \rho g y_2 \text{ or } y_2 = p_1/(\rho g).$$

With numbers,

$$y_2 = (1.01 \times 10^5 \text{Pa})/\left[(1.21 \text{ kg/m}^3)(9.81 \text{ m/s}^2)\right] = 8.51 \times 10^3 \text{m}.$$

Airplanes fly *much* higher than this; the assumption that the atmosphere is of uniform density is *bad*.

(b) We have to go back to Eq. 15-7 for an atmosphere which has a density which varies linearly with altitude. Linear variation of density means

$$\rho = \rho_0 \left(1 - \frac{y}{y_{\text{max}}}\right)$$

How did we do this? We want a linear function of the form

$$\rho = my + b,$$

where $\rho = \rho_0$ when $y = 0$, so $b = \rho_0$. But we also require that when $y = y_{\text{max}}$ $\rho = 0$. This means $m = -\rho_0/y_{\text{max}}$.

We substitute this into Eq. 15-7,

$$
\begin{aligned}
p_2 - p_1 &= -\int_0^{y_{\text{max}}} \rho g \, dy, \\
&= -\int_0^{y_{\text{max}}} \rho_0 g \left(1 - \frac{y}{y_{\text{max}}}\right) dy, \\
&= -\rho_0 g \left(y - \frac{y^2}{2y_{\text{max}}}\right)\bigg|_0^{y_{\text{max}}}, \\
&= -\rho g y_{\text{max}}/2.
\end{aligned}
$$

In this case we have $y_{\text{max}} = 2p_1/(\rho g)$, so the answer is twice that in part (a), or **17 km**. It is still wrong.

E15-17 There are *three* force on the block: gravity ($W = mg$), a buoyant force $B_0 = m_w g$, and a tension T_0. When the container is at rest all three forces balance, so $B_0 - W - T_0 = 0$. The tension in this case is $T_0 = (m_w - m)g$.

When the container accelerates upward we now have $B - W - T = ma$. Note that neither the tension *nor* the buoyant force stay the same; the buoyant force increases according to $B = m_w(g + a)$. The new tension is then

$$T = m_w(g + a) - mg - ma = (m_w - m)(g + a) = T_0(1 + a/g).$$

119

E15-21 The can will float if the buoyant force is equal to the *total* weight of the can. Interestingly enough, we don't care about the density of lead. The can has a volume of 1200 cm^3, so it can displace that much water. This would provide a buoyant force of

$$B = \rho V g = (998 \text{kg/m}^3)(1200 \times 10^{-6} \text{m}^3)(9.81 \text{ m/s}^2) = 11.7 \text{ N}.$$

This force can then support a total mass of $(11.7 \text{ N})/(9.81 \text{ m/s}^2) = 1.20 \text{ kg}$. If 130 g belong to the can, then the can will be able to carry 1.07 kg of lead.

E15-27 There are *three* force on the dirigible: gravity $(W = m_\text{g} g)$, a buoyant force $B = m_\text{a} g$, and a tension T. Since these forces must balance we have $T = B - W$. The masses are related to the densities, so we can write

$$T = (\rho_\text{a} - \rho_\text{g}) V g = (1.21 \text{ kg/m}^3 - 0.796 \text{ kg/m}^3)(1.17 \times 10^6 \text{m}^3)(9.81 \text{m/s}^2) = 4.75 \times 10^6 \text{ N}.$$

E15-31 If there were *no* water vapor pressure above the barometer then the height of the water would be $y_1 = P/(\rho g)$, where $P = P_0$ is the atmospheric pressure. If there is water vapor where there should be a vacuum, then P is the difference, and we would have $y_2 = (P_0 - P_\text{v})/(\rho g)$. The relative error is

$$
\begin{aligned}
(y_1 - y_2)/y_1 &= [P_0/(\rho g) - (P_0 - P_\text{v})/(\rho g)] / [P_0/(\rho g)], \\
&= P_\text{v}/P_0 = (3169 \text{ Pa})/(1.01 \times 10^5 \text{Pa}) = 3.14 \%.
\end{aligned}
$$

I found the vapor pressure searching the web with google; the sixth entry was

http://dbhs.wvusd.k12.ca.us/GasLaw/Vapor-Pressure-Data.html,

but it might not be there when you check.

E15-35 The force required is just the surface tension times the circumference of the circular patch. Then

$$F = (0./072 \text{ N/m}) 2\pi (0.12 \text{ m}) = 5.43 \times 10^{-2} \text{N}.$$

P15-3 (a) The gauge pressure will be given by $P = -\rho g y$, where y is the (negative) distance beneath the surface. The resultant force on the wall will then be

$$
\begin{aligned}
F &= \int \int P \, dx \, dy, \\
&= \int (-\rho g y) W \, dy, \\
&= \rho g D^2 W / 2.
\end{aligned}
$$

(b) The torque will is given by $\tau = F(D - y)$ (the distance is from the bottom) so if we generalize,

$$\begin{aligned}
\tau &= \int \int P y \, dx \, dy, \\
&= \int (-\rho g(D - y)) \, yW \, dy, \\
&= \rho g D^3 W / 6.
\end{aligned}$$

(c) Dividing to find the location of the equivalent resultant force,

$$d = \tau / F = (\rho g D^3 W / 6) / (\rho g D^2 W / 2) = D/3,$$

this distance being measured from the bottom.

P15-7 (a) Use Eq. 15-10, $p = (p_0/\rho_0)\rho$, then Eq. 15-13 will look like

$$(p_0/\rho_0)\rho = (p_0/\rho_0)\rho_0 e^{-h/a}.$$

(b) The upward velocity of the rocket as a function of time is given by $v = a_r t$. The height of the rocket above the ground is given by $y = \frac{1}{2}a_r t^2$. Combining,

$$v = a_r \sqrt{\frac{2y}{a_r}} = \sqrt{2y a_r}.$$

Put this into the expression for drag, along with the equation for density variation with altitude;

$$D = CA\rho v^2 = CA\rho_0 e^{-y/a} 2y a_r.$$

Now take the derivative with respect to y,

$$dD/dy = (-1/a)CA\rho_0 e^{-y/a}(2y a_r) + CA\rho_0 e^{-y/a}(2a_r).$$

This will vanish when $y = a$, regardless of the acceleration a_r.

P15-11 We can start with Eq. 15-11, except that we'll write our distance in terms of r instead if y. Into this we can substitute our expression for g,

$$g = g_0 \frac{R^2}{r^2},$$

which provides the $1/r^2$ behavior from Chapter 14. The value of g at the surface is g_0. Substituting, then integrating,

$$\frac{dp}{p} = -\frac{g\rho_0}{p_0}dr,$$

$$\frac{dp}{p} = -\frac{g_0 \rho_0 R^2}{p_0} \frac{dr}{r^2},$$

$$\int_{p_0}^{p} \frac{dp}{p} = -\int_{R}^{r} \frac{g_0 \rho_0 R^2}{p_0} \frac{dr}{r^2},$$

$$\ln \frac{p}{p_0} = \frac{g_0 \rho_0 R^2}{p_0} \left(\frac{1}{r} - \frac{1}{R} \right)$$

If $k = g_0 \rho_0 R^2 / p_0$, then

$$p = p_0 e^{k(1/r - 1/R)}.$$

P15-15 Sample Problem 15-3 outlines a very important fact for floating bodies: the fraction of the volume submerged for a floating object is equal to the ratio of the densities:

$$\frac{V_{\text{submerged}}}{V_o} = \frac{\rho_o}{\rho_{\text{fluid}}}$$

where ρ_{fluid} is the density of the supporting fluid, V_o is the volume of the object, and $V_{\text{submerged}}$ is the volume of the object beneath the surface of the fluid.

Then we initially have

$$\frac{1}{4} = \frac{\rho_o}{\rho_{\text{mercury}}}.$$

When water is poured over the object the simple relation derived in Sample Problem 15-3 no longer works. A good question to ask at this point (and surely a multiple choice question in the next edition) is "will the object float higher or lower in the mercury when water is poured over it?" Since the water will provide an additional buoyant force upward, the object will float higher in the mercury.

Once the water is over the object there are two buoyant forces: one from mercury, F_1, and one from the water, F_2. Following a derivation which is similar to Sample Problem 15-3, we have

$$F_1 = \rho_1 V_1 g \text{ and } F_2 = \rho_2 V_2 g$$

where ρ_1 is the density of mercury, V_1 the volume of the object which is in the mercury, ρ_2 is the density of water, and V_2 is the volume of the object which is in the water. We also have

$$F_1 + F_2 = \rho_o V_o g \text{ and } V_1 + V_2 = V_o$$

as expressions for the net force on the object (zero) and the total volume of the object. Combining these four expressions,

$$\rho_1 V_1 + \rho_2 V_2 = \rho_o V_o,$$

or

$$\rho_1 V_1 + \rho_2 (V_o - V_1) = \rho_o V_o,$$
$$(\rho_1 - \rho_2) V_1 = (\rho_o - \rho_2) V_o,$$
$$\frac{V_1}{V_o} = \frac{\rho_o - \rho_2}{\rho_1 - \rho_2}.$$

This last expression is quite similar to the one derived in the Sample Problem, but corrects for the fluid density above the supporting fluid. The left hand side is the fraction that is submerged in the mercury, so we just need to substitute our result for the density of the material from the beginning to solve the problem. The fraction submerged after adding water is then

$$
\begin{aligned}
\frac{V_1}{V_o} &= \frac{\rho_o - \rho_2}{\rho_1 - \rho_2}, \\
&= \frac{\rho_1/4 - \rho_2}{\rho_1 - \rho_2}, \\
&= \frac{(13600\,\mathrm{kg/m^3})/4 - (998\,\mathrm{kg/m^3})}{(13600\,\mathrm{kg/m^3}) - (998\,\mathrm{kg/m^3})} = 0.191.
\end{aligned}
$$

P15-19 Pretend the bubble consists of two hemispheres. The force from surface tension holding the hemispheres together is $F = 2\gamma L = 4\pi r\gamma$. The "extra" factor of two occurs because *each* hemisphere has a circumference which "touches" the boundary that is held together by the surface tension of the liquid. The pressure difference between the inside and outside is $\Delta p = F/A$, where A is the area of the flat side of one of the hemispheres, so $\Delta p = (4\pi r\gamma)/(\pi r^2) = 4\gamma/r$.

Chapter 16

Fluid Dynamics

E16-3 We'll assume that each river has a rectangular cross section, despite what the picture implies. The cross section area of the two streams is then

$$A_1 = (8.2\,\text{m})(3.4\,\text{m}) = 28\,\text{m}^2 \text{ and } A_2 = (6.8\,\text{m})(3.2\,\text{m}) = 22\,\text{m}^2.$$

The volume flow rate in the first stream is

$$R_1 = A_1 v_1 = (28\,\text{m}^2)(2.3\,\text{m/s}) = 64\,\text{m}^3/\text{s},$$

while the volume flow rate in the second stream is

$$R_2 = A_2 v_2 = (22\,\text{m}^2)(2.6\,\text{m/s}) = 57\,\text{m}^3/\text{s}.$$

The amount of fluid in the stream/river system is conserved, so

$$R_3 = R_1 + R_2 = (64\,\text{m}^3/\text{s}) + (57\,\text{m}^3/\text{s}) = 121\,\text{m}^3/\text{s}.$$

where R_3 is the volume flow rate in the river. Then

$$D_3 = R_3/(v_3 W_3) = (121\,\text{m}^3/\text{s})/[(10.7\,\text{m})(2.9\,\text{m/s})] = 3.9\,\text{m}.$$

E16-5 To find the speed of the water in the river we need to know the volume flow rate; this can be found by considering how much water needs to be drained from the watershed each year.

There are 8500 km^2 which collects an average of $(0.75)(0.48\text{ m/y})$, where the 0.75 reflects the fact that 1/4 of the water evaporates, so

$$R = \left[8500(10^3\,\text{m})^2\right](0.75)(0.48\,\text{m/y})\left(\frac{1\,\text{y}}{365 \times 24 \times 60 \times 60\,\text{s}}\right) = 97\,\text{m}^3/\text{s}.$$

Then the speed of the water in the river is

$$v = R/A = (97\,\text{m}^3/\text{s})/[(21\,\text{m})(4.3\,\text{m})] = 1.1\,\text{m/s}.$$

E16-9 (b) We will do part (b) first. This exercise starts off the same way as Ex. 16-5: we need to find the necessary volume flow rate.

$$R = (100\,\text{m}^2)(1.6\,\text{m/y}) \left(\frac{1\,\text{y}}{365 \times 24 \times 60 \times 60\,\text{s}} \right) = 5.1 \times 10^{-6}\,\text{m}^3/\text{s}.$$

This may seem absurdly small, but it is 5.1×10^{-3} liters/second, or 440 liters/day.

(b) The speed of the flow R through a hole of cross sectional area a will be $v = R/a$. Pascal's equation can be used to find the pressure of the liquid at the level of the hole,

$$p = p_0 + \rho g h,$$

where $h = 2.0\,\text{m}$ is the depth of the hole. Bernoulli's equation can be applied to find the speed of the water as it travels a horizontal stream line out the hole,

$$p_0 + \frac{1}{2}\rho v^2 = p,$$

where we drop any terms which are either zero or the same on both sides. Then

$$v = \sqrt{2(p - p_0)/\rho} = \sqrt{2gh} = \sqrt{2(9.81\,\text{m/s}^2)(2.0\,\text{m})} = 6.3\,\text{m/s}.$$

Finally, $a = (5.1 \times 10^{-6}\,\text{m}^3/\text{s})/(6.3\,\text{m/s}) = 8.1 \times 10^{-7}\text{m}^2$, or about 0.81 mm^2. Not a big hole, but if it isn't plugged, the water will leak out!

E16-13 We have two conditions that must be satisfied: the equation of continuity and Bernoulli's equation. First we'll set up the equation of continuity.

The lower pipe has a radius $r_1 = 2.52$ cm, and a cross sectional area of $A_1 = \pi r_1^2$. The speed of the fluid flow at this point is v_1. The higher pipe has a radius of $r_2 = 6.14$ cm, a cross sectional area of $A_2 = \pi r_2^2$, and a fluid speed of v_2. Then

$$A_1 v_1 = A_2 v_2 \text{ or } r_1^2 v_1 = r_2^2 v_2.$$

Now Bernoulli's equation. We'll set $y_1 = 0$ for the lower pipe. The problem specifies that the pressures in the two pipes are the same, so

$$p_0 + \frac{1}{2}\rho v_1^2 + \rho g y_1 = p_0 + \frac{1}{2}\rho v_2^2 + \rho g y_2,$$
$$\frac{1}{2}v_1^2 = \frac{1}{2}v_2^2 + g y_2,$$

We can combine the results of the equation of continuity with this and get

$$v_1^2 = v_2^2 + 2g y_2,$$
$$v_1^2 = \left(v_1 r_1^2/r_2^2 \right)^2 + 2g y_2,$$
$$v_1^2 \left(1 - r_1^4/r_2^4 \right) = 2g y_2,$$
$$v_1^2 = 2g y_2/ \left(1 - r_1^4/r_2^4 \right).$$

125

To avoid the plethora of algebraic symbols we will go ahead and find this quantity,

$$v_1^2 = 2(9.81\,\text{m/s}^2)(11.5\,\text{m}) / \left(1 - (0.0252\,\text{m})^4/(0.0614\,\text{m})^4\right) = 232\,\text{m}^2/\text{s}^2$$

The volume flow rate in the bottom (and top) pipe is

$$R = \pi r_1^2 v_1 = \pi(0.0252\,\text{m})^2(15.2\,\text{m/s}) = 0.0303\,\text{m}^3/\text{s}.$$

That is the same as 30.3 liters/second.

E16-15 A quick application of Bernoulli's equation will solve this problem. We assume that we can construct a streamline from the surface through the hole into the submarine. We can't, but the approximation works. The pressure is the same at the surface and inside the submarine. Sea level will be defined as $y = 0$, and at that point the fluid is assumed to be at rest. Then

$$p_0 + \frac{1}{2}\rho v_1^2 + \rho g y_1 = p_0 + \frac{1}{2}\rho v_2^2 + \rho g y_2,$$

$$0 = \frac{1}{2}v_2^2 + g y_2,$$

where $y_2 = -200$ m. Then

$$v_2 = \sqrt{-2g y_2} = \sqrt{-2(9.81\,\text{m/s}^2)(-200\,\text{m})} = 63\,\text{m/s}.$$

E16-19 (a) The pressure of the water 6.15 m beneath the surface can be found from Pascal's principle.

$$\Delta P = -\rho g y$$

There are three forces on the plug. The force from the pressure of the water, $F_1 = P_1 A$, the force from the pressure of the air, $F_2 = P_2 A$, and the force of friction, F_3. These three forces must balance, so

$$F_3 = F_1 - F_2,$$

or

$$F_3 = P_1 A - P_2 A.$$

But $P_1 - P_2$ is the pressure difference between the surface and the water 6.15 m below the surface, so

$$
\begin{aligned}
F_3 &= \Delta P A = -\rho g y A, \\
&= -(998\,\text{kg/m}^3)(9.81\,\text{m/s}^2)(-6.15\,\text{m})\pi(0.0215\,\text{m})^2, \\
&= 87.4\,\text{N}
\end{aligned}
$$

(b) To find the volume of water which flows out in three hours we need to know the volume flow rate, and for that we need both the cross section area of the hole and the speed of the flow. The speed of the flow can be found by an application of Bernoulli's equation. We'll consider the horizontal motion only— a point just inside the hole, and a point just outside the hole. These points are at the same level, so

$$p_1 + \frac{1}{2}\rho v_1^2 + \rho g y_1 = p_2 + \frac{1}{2}\rho v_2^2 + \rho g y_2,$$

$$p_1 = p_2 + \frac{1}{2}\rho v_2^2.$$

Combine this with the results of Pascal's principle above, and

$$v_2 = \sqrt{2(p_1 - p_2)/\rho} = \sqrt{-2gy} = \sqrt{-2(9.81\,\mathrm{m/s^2})(-6.15\,\mathrm{m})} = 11.0\,\mathrm{m/s}.$$

The volume of water which flows out in three hours is

$$V = Rt = (11.0\,\mathrm{m/s})\pi(0.0215\,\mathrm{m})^2(3 \times 3600\,\mathrm{s}) = 173\,\mathrm{m^3}.$$

E16-21 This is a fairly simple application of Bernoulli's equation. We'll assume that the central column of air down the pipe exerts minimal force on the card when it is deflected to the sides. Then

$$p_1 + \frac{1}{2}\rho v_1^2 + \rho g y_1 = p_2 + \frac{1}{2}\rho v_2^2 + \rho g y_2,$$

$$p_1 = p_2 + \frac{1}{2}\rho v_2^2.$$

The resultant upward force on the card is the area of the card times the pressure difference, or

$$F = (p_1 - p_2)A = \frac{1}{2}\rho A v^2.$$

E16-25 Despite what you might be tempted to try, this problem is not an application of the Venturi tube Eq. 16-11. It is instead an application of Bernoulli's equation and the equation of continuity, much like Ex. 16-13. First we'll set up the equation of continuity.

The larger pipe has a radius $r_1 = 12.7$ cm, and a cross sectional area of $A_1 = \pi r_1^2$. The speed of the fluid flow at this point is v_1. The smaller pipe has a radius of $r_2 = 5.65$ cm, a cross sectional area of $A_2 = \pi r_2^2$, and a fluid speed of v_2. Then

$$A_1 v_1 = A_2 v_2 \text{ or } r_1^2 v_1 = r_2^2 v_2.$$

Now Bernoulli's equation. The two pipes are at the same level, so $y_1 = y_2$. Then

$$p_1 + \frac{1}{2}\rho v_1^2 + \rho g y_1 = p_2 + \frac{1}{2}\rho v_2^2 + \rho g y_2,$$

$$p_1 + \frac{1}{2}\rho v_1^2 = p_2 + \frac{1}{2}\rho v_2^2.$$

Combining this with the results from the equation of continuity,

$$p_1 + \frac{1}{2}\rho v_1^2 = p_2 + \frac{1}{2}\rho v_2^2,$$

$$v_1^2 = v_2^2 + \frac{2}{\rho}(p_2 - p_1),$$

$$v_1^2 = \left(v_1 \frac{r_1^2}{r_2^2}\right)^2 + \frac{2}{\rho}(p_2 - p_1),$$

$$v_1^2\left(1 - \frac{r_1^4}{r_2^4}\right) = \frac{2}{\rho}(p_2 - p_1),$$

$$v_1^2 = \frac{2(p_2 - p_1)}{\rho(1 - r_1^4/r_2^4)}.$$

It may look a mess, but we can solve it to find v_1,

$$v_1 = \sqrt{\frac{2(32.6 \times 10^3 \text{Pa} - 57.1 \times 10^3 \text{Pa})}{(998 \text{ kg/m}^3)(1 - (0.127 \text{ m})^4/(0.0565 \text{ m})^4)}} = 1.41 \text{ m/s}.$$

The volume flow rate is then

$$R = Av = \pi(0.127 \text{ m})^2(1.41 \text{ m/s}) = 7.14 \times 10^{-3} \text{m}^3/\text{s}.$$

That's about 71 liters/second.

In retrospect we could have started from the Venturi tube equation and then applied Pascal's principle, but it makes more sense to start from the fundamentals like we did above, and you can use the steps above to help in your solution to Problem 8!

E16-29 (a) We need to find Reynold's number for the flow to determine if it is laminar. The volume flux is given; from that we can find the average speed of the fluid in the pipe.

$$v = \frac{5.35 \times 10^{-2} \text{ L/min}}{\pi(1.88 \text{ cm})^2} = 4.81 \times 10^{-3} \text{ L/cm}^2 \cdot \text{min}.$$

But 1 L is the same as 1000 cm^3 and 1 min is equal to 60 seconds, so $v = 8.03 \times 10^{-4}$ m/s.

Reynold's number from Eq. 16-22 is then

$$R = \frac{\rho D v}{\eta} = \frac{(13600 \text{ kg/m}^3)(0.0376 \text{ m})(8.03 \times 10^{-4} \text{ m/s})}{(1.55 \times 10^{-3} \text{N} \cdot \text{s/m}^2)} = 265.$$

This is well below the critical value of 2000.

(b) Poiseuille's Law, Eq. 16-20, can be used to find the pressure difference between the ends of the pipe. But first, note that the mass flux dm/dt is equal to the volume rate times the density when the density is constant. Then $\rho\, dV/dt = dm/dt$, and Poiseuille's Law can be written as

$$\delta p = \frac{8\eta L}{\pi R^4}\frac{dV}{dt} = \frac{8(1.55\times 10^{-3}\mathrm{N}\cdot\mathrm{s/m^2})(1.26\,\mathrm{m})}{\pi(1.88\times 10^{-2}\ \mathrm{m})^4}(8.92\times 10^{-7}\mathrm{m^3/s}) = 0.0355\,\mathrm{Pa}.$$

Note that we converted the volume flow rate from the awkward units of L/min to $\mathrm{m^3/s}$.

P16-3 (a) If we can find the horizontal speed of the water as it leaves the tank then the remainder of the problem is one of kinematics. We'll start by using Bernoulli's equation in the form of Eq. 16-9, and consider the top of the fluid in the tank as point 1 and the exit point for the fluid as point 2. Both points are at atmospheric pressure, so

$$p_0 + \frac{1}{2}\rho v_1^2 + \rho g y_1 = p_0 + \frac{1}{2}\rho v_2^2 + \rho g y_2,$$

$$\frac{1}{2}\rho v_1^2 + \rho g y_1 = \frac{1}{2}\rho v_2^2 + \rho g y_2,$$

$$\frac{1}{2} v_1^2 + g y_1 = \frac{1}{2} v_2^2 + g y_2.$$

At the top the fluid isn't moving, so $v_1 = 0$. As such, we drop the subscript on v_2 and just write v. For convenience we'll make our vertical distance measurements from the top of the fluid. Then

$$0 = \frac{1}{2}v^2 - gh$$

is the expression for the exit speed of the fluid. This expression can be rearranged as $v = \sqrt{2gh}$, and is known as Torricelli's law. You needed to do a similar derivation to solve Exercise 16-14.

The speed v is a horizontal velocity, and serves as the initial horizontal velocity of the fluid "projectile" after it leaves the tank. There is no initial vertical velocity.

This fluid "projectile" falls through a vertical distance $H - h$ before splashing against the ground. The equation governing the time t for it to fall is

$$-(H - h) = -\frac{1}{2}gt^2,$$

Solve this for the time, and $t = \sqrt{2(H - h)/g}$. The equation which governs the horizontal distance traveled during the fall is $x = v_x t$, but $v_x = v$ and we just found t, so

$$x = v_x t = \sqrt{2gh}\,\sqrt{2(H - h)/g} = 2\sqrt{h(H - h)}.$$

(b) How many values of h will lead to a distance of x? We need to invert the expression, and we'll start by squaring both sides

$$x^2 = 4h(H - h) = 4hH - 4h^2,$$

and then solving the resulting quadratic expression for h,

$$h = \frac{4H \pm \sqrt{16H^2 - 16x^2}}{8} = \frac{1}{2}\left(H \pm \sqrt{H^2 - x^2}\right).$$

For values of x between 0 and H there are two real solutions, if $x = H$ there is one real solution, and if $x > H$ there are no real solutions.

If h_1 is a solution, then we can write $h_1 = (H + \Delta)/2$, where $\Delta = 2h_1 - H$ could be positive or negative. Then $h_2 = (H + \Delta)/2$ is also a solution, and

$$h_2 = (H + 2h_1 - 2H)/2 = h_1 - H$$

is also a solution.

(c) The farthest distance is $x = H$, and this happens when $h = H/2$, as we can see from the previous section.

P16-7 The greatest possible value for v will be the value over the wing which results in an air pressure of zero. If the air at the leading edge is stagnant (not moving) and has a pressure of p_0, then Bernoulli's equation gives

$$p_0 = \frac{1}{2}\rho v^2,$$

or $v = \sqrt{2p_0/\rho} = \sqrt{2(1.01 \times 10^5 \,\text{Pa})/(1.2\,\text{kg/m}^3)} = 410\,\text{m/s}$. This value is only slightly larger than the speed of sound; they are related because sound waves involve the movement of air particles which "shove" other air particles out of the way.

P16-13 A flow will be irrotational if and only if $\oint \vec{v} \cdot d\vec{s} = 0$ for all possible paths. It is fairly easy to construct a rectangular path which is parallel to the flow on the top and bottom sides, but perpendicular on the left and right sides. Then only the top and bottom paths contribute to the integral. $\vec{v}$ is constant for either path (but not the same), so the magnitude v will come out of the integral sign. Since the lengths of the two paths are the same but v is different the two terms *don't* cancel, so the flow is not irrotational.

P16-15 The volume flux (called R_f to distinguish it from the radius R) through an annular ring of radius r and width δr is

$$\delta R_f = \delta A\, v = 2\pi r\,\delta r\, v,$$

where v is a function of r given by Eq. 16-18. The mass flux is the volume flux times the density, so the total mass flux is

$$
\begin{aligned}
\frac{dm}{dt} &= \rho \int_0^R \frac{\delta R_f}{\delta r} dr, \\
&= \rho \int_0^R 2\pi r \left(\frac{\Delta p}{4\eta L}(R^2 - r^2) \right) dr, \\
&= \frac{\pi\rho\Delta p}{2\eta L} \int_0^R (rR^2 - r^3)dr, \\
&= \frac{\pi\rho\Delta p}{2\eta L}(R^4/2 - R^4/4), \\
&= \frac{\pi\rho\Delta p R^4}{8\eta L}.
\end{aligned}
$$

That wasn't so bad, was it?

Chapter 17

Oscillations

E17-1 For a perfect spring $|F| = k|x|$. We are told that the extension of the spring is $x = 0.157$ m when a 3.94 kg is suspended from it. The net force on the mass will be zero when it is suspended from the spring. There would be two forces on the mass—the force of gravity, $W = mg$, and the force of the spring, F. These two force must balance, so $mg = kx$ or

$$k = \frac{mg}{x} = \frac{(3.94\,\mathrm{kg})(9.81\,\mathrm{m/s^2})}{(0.157\,\mathrm{m})} = 0.246\,\mathrm{N/m}.$$

It would have certainly been less work to just write this last expression down and solve it. The result would be the same, but the physics would be wrong; it really is necessary to step through the problem, because small changes in the problem can fundamentally alter the final equations!

Now that we know k, the spring constant, we can find the period of oscillations from Eq. 17-8,

$$T = 2\pi\sqrt{\frac{m}{k}} = 2\pi\sqrt{\frac{(0.520\,\mathrm{kg})}{(0.246\,\mathrm{N/m})}} = 0.289\,\mathrm{s}.$$

E17-5 (a) The amplitude is half of the distance between the extremes of the motion, so $A = (2.00\ \mathrm{mm})/2 = 1.00$ mm.

(b) The maximum blade speed is given by Eq. 17-11 (read the paragraph which follows it), or $v_\mathrm{m} = \omega x_\mathrm{m}$. The blade oscillates with a frequency of 120 Hz, so

$$\omega = 2\pi f = 2\pi(120\,\mathrm{s^{-1}}) = 754\ \mathrm{rad/s},$$

and then $v_\mathrm{m} = (754\ \mathrm{rad/s})(0.001\,\mathrm{m}) = 0.754\,\mathrm{m/s}.$

(c) The same set of equations will allow us to find the maximum acceleration from $a_\mathrm{m} = \omega^2 x_\mathrm{m}$,

$$a_\mathrm{m} = (754\ \mathrm{rad/s})^2(0.001\,\mathrm{m}) = 568\,\mathrm{m/s^2}.$$

E17-9 If the drive wheel rotates at 193 rev/min then

$$\omega = (193 \text{ rev/min})(2\pi \text{ rad/rev})(1/60 \text{ s/min}) = 20.2 \text{ rad/s},$$

then $v_{\mathrm{m}} = \omega x_{\mathrm{m}} = (20.2 \text{ rad/s})(0.3825 \text{ m}) = 7.73 \text{ m/s}.$

E17-15 (a) To find the frequency of oscillation of the two carts we need to first know the elastic constant of the cable. The net force on the three cars is zero when before the cable breaks. There are three forces on the cars: the weight, W, a normal force, N, and the upward force from the cable, F. What interests us most is the component of the weight which is parallel to the surface, $W \sin \theta$, since this component is what balances F. Then

$$F = W \sin \theta = 3mg \sin \theta.$$

The factor of "3" is because there are three cars of equal mass. This force is from the elastic properties of the cable, so

$$k = \frac{F}{x} = \frac{3mg \sin \theta}{x}$$

The frequency of oscillation of the remaining two cars after the bottom car is released is

$$f = \frac{1}{2\pi}\sqrt{\frac{k}{2m}} = \frac{1}{2\pi}\sqrt{\frac{3mg\sin\theta}{2mx}} = \frac{1}{2\pi}\sqrt{\frac{3g\sin\theta}{2x}}.$$

The factor of "2" is because there are still two carts connected to the cable. Note that although we need to take into account the angle of the incline when finding the elastic constant, we didn't need to worry about the angle of incline when worrying about the mass $2m$. In fact, the mass *appears* to not have any effect on the frequency. It does, but can you tell where?

Numerically, the frequency is

$$f = \frac{1}{2\pi}\sqrt{\frac{3g\sin\theta}{2x}} = \frac{1}{2\pi}\sqrt{\frac{3(9.81 \text{ m/s}^2)\sin(26°)}{2(0.142 \text{ m})}} = 1.07 \text{ Hz}.$$

(b) The amplitude is measured from the equilibrium position to the maximum extension. We know the maximum extension: the oscillations start when the third car decouples, this happens 14.2 cm away from the unstretched position of the cable. But where is the new equilibrium position of the two car + cable system?

Let's do this the easiest possible way. Each car contributes equally to the stretching of the cable, so one car causes the cable to stretch $14.2/3 = 4.73$ cm. The equilibrium position of the two car system would then be $14.2 - 4.73 = 9.5$ cm. How far is the new equilibrium position from the maximum displacement? Well, 4.73 cm.

E17-17 (a) There are two forces on the log. The weight, $W = mg$, is fixed; the buoyant force B, depends on the amount of displaced water and changes as the log bobs up and down.

We'll assume the log is cylindrical. The person who wrote the problem certainly did. If x is the length of the log beneath the surface and A the cross sectional area of the log, then $V = Ax$ is the volume of the displaced water. Furthermore, $m_w = \rho_w V$ is the mass of the displaced water and $B = m_w g$ is then the buoyant force on the log. Combining,

$$B = \rho_w Agx,$$

where ρ_w is the density of water. This certainly looks similar to an elastic spring force law, with $k = \rho_w Ag$. We would then expect the motion to be simple harmonic.

(b) The period of the oscillation would be

$$T = 2\pi\sqrt{\frac{m}{k}} = 2\pi\sqrt{\frac{m}{\rho_w Ag}},$$

where m is the total mass of the log and lead. It might look like we can't go any further, but we are told the log is in equilibrium when $x = L = 2.56$ m. This would give us the *weight* of the log, since $W = B$ is the condition for the log to float. Then

$$mg = B = \rho_w AgL,$$

so

$$m = \frac{\rho_w AgL}{g} = \rho AL.$$

From this we can write the period of the motion as

$$T = 2\pi\sqrt{\frac{\rho AL}{\rho_w Ag}} = 2\pi\sqrt{L/g} = 2\pi\sqrt{\frac{(2.56\,\mathrm{m})}{(9.81\,\mathrm{m/s^2})}} = 3.21\,\mathrm{s},$$

an expression which looks quite similar to the pendulum! Note that we never really cared about the density of the water when finding the period of oscillation. Or did we?

E17-21 (a) We'll apply the information in the paragraph which follows Eq. 17-11 to start this problem,

$$a_m = \omega^2 x_m.$$

so

$$\omega = \sqrt{\frac{a_m}{x_m}} = \sqrt{\frac{(7.93 \times 10^3\,\mathrm{m/s^2})}{(1.86 \times 10^{-3}\,\mathrm{m})}} = 2.06 \times 10^3\,\mathrm{rad/s}$$

The period of the motion is then

$$T = \frac{2\pi}{\omega} = 3.05 \times 10^{-3}\mathrm{s}.$$

(b) The maximum speed of the particle is found by

$$v_{\mathrm{m}} = \omega x_{\mathrm{m}} = (2.06 \times 10^3 \ \mathrm{rad/s})(1.86 \times 10^{-3} \ \mathrm{m}) = 3.83 \ \mathrm{m/s}.$$

(c) The mechanical energy is given by Eq. 17-15, except that we will focus on when $v_x = v_{\mathrm{m}}$, because then $x = 0$ and

$$E = \frac{1}{2} m v_{\mathrm{m}}^2 = \frac{1}{2}(12.3 \ \mathrm{kg})(3.83 \ \mathrm{m/s})^2 = 90.2 \ \mathrm{J}.$$

E17-27 This is a straightforward, but very important, exercise. In physics we regularly make approximations to a physical problem. It is crucial that we always consider how good the approximation is, and considering higher order terms is one way to do this.

We are interested in the value of θ_{m} which will make the second term 2% of the first term. We want to solve

$$0.02 = \frac{1}{2^2} \sin^2 \frac{\theta_{\mathrm{m}}}{2},$$

which has solution

$$\sin \frac{\theta_{\mathrm{m}}}{2} = \sqrt{0.08}$$

or $\theta_m = 33°$. Note that this is a fairly large angle, so Eq. 17-24 is an excellent approximation for angles less than 30°.

(b) How large is the third term at this angle?

$$\frac{3^2}{2^2 4^2} \sin^4 \frac{\theta_m}{2} = \frac{3^2}{2^2}\left(\frac{1}{2^2}\sin^2\frac{\theta_m}{2}\right)^2 = \frac{9}{4}(0.02)^2$$

or 0.0009, which is very small.

E17-29 Let the period of the clock in Paris be T_1. In a day of length $D_1 = 24$ hours it will undergo $n = D/T_1$ oscillations. In Cayenne the period is T_2. n oscillations should occur in 24 hours, but since the clock runs slow, D_2 is 24 hours + 2.5 minutes elapse. So

$$T_2 = D_2/n = (D_2/D_1)T_1 = [(1442.5 \ \mathrm{min})/(1440.0 \ \mathrm{min})]T_1 = 1.0017T_1.$$

Since the ratio of the periods is $(T_2/T_1) = \sqrt{(g_1/g_2)}$, the g_2 in Cayenne is

$$g_2 = g_1(T_1/T_2)^2 = (9.81 \ \mathrm{m/s}^2)/(1.0017)^2 = 9.78 \ \mathrm{m/s}^2.$$

E17-33 We'll assume that the hoop is oscillating in the plane of the hoop. It makes a difference!

The frequency of oscillation is the reciprocal of the period, so from Eq. 17-28,

$$f = \frac{1}{2\pi}\sqrt{\frac{Mgd}{I}},$$

where d is the distance from the pivot about which the hoop oscillates and the center of mass of the hoop.

The rotational inertia I is about an axis through the pivot, so we will need to apply the parallel axis theorem. Then

$$I = Md^2 + I_{\text{cm}} = Md^2 + Mr^2.$$

But d is r, since the pivot point is on the rim of the hoop. So $I = 2Md^2$, and the frequency is

$$f = \frac{1}{2\pi}\sqrt{\frac{Mgd}{2Md^2}} = \frac{1}{2\pi}\sqrt{\frac{g}{2d}} = \frac{1}{2\pi}\sqrt{\frac{(9.81\,\text{m/s}^2)}{2(0.653\,\text{m})}} = 0.436\,\text{Hz}.$$

(b) Note the above expression looks like the simple pendulum equation if we replace $2d$ with l. Then the equivalent length of the simple pendulum is $2(0.653\,\text{m}) = 1.31\,\text{m}$.

E17-37 We will eventually apply the same formula that was used for Ex. 17-33. First, however, we need to deal with the rotational inertia and center of mass of the meter stick.

For a stick of length L which can pivot about the end, $I = \frac{1}{3}ML^2$. The center of mass of such a stick is located $d = L/2$ away from the end.

The frequency of oscillation of such a stick is

$$f = \frac{1}{2\pi}\sqrt{\frac{Mgd}{I}},$$

$$f = \frac{1}{2\pi}\sqrt{\frac{Mg\,(L/2)}{\frac{1}{3}ML^2}},$$

$$f = \frac{1}{2\pi}\sqrt{\frac{3g}{2L}}.$$

This means that f is proportional to $\sqrt{1/L}$, regardless of the mass or density of the stick. The ratio of the frequency of two such sticks is then $f_2/f_1 = \sqrt{L_1/L_2}$, which in our case gives

$$f_2 = f_1\sqrt{L_2/L_1} = f_1\sqrt{(L_1)/(2L_1/3)} = 1.22f_1.$$

E17-43 The ω which describes the angular velocity in uniform circular motion is effectively the same ω which describes the angular frequency of the corresponding simple harmonic motion. Since $\omega = \sqrt{k/m}$, we can find the effective force constant k from knowledge of the Moon's mass and the period of revolution.

The moon orbits with a period of $T =$, so

$$\omega = \frac{2\pi}{T} = \frac{2\pi}{(27.3 \times 24 \times 3600\,\text{s})} = 2.66 \times 10^{-6}\text{rad/s}.$$

This can be used to find the value of the effective force constant k from

$$k = m\omega^2 = (7.36 \times 10^{22}\text{kg})(2.66 \times 10^{-6}\text{rad/s})^2 = 5.21 \times 10^{11}\text{N/m}.$$

E17-45 Equation 17-39 is

$$x = x_\text{m}e^{-bt/2m}\cos(\omega't + \phi)$$

The first derivative of this is

$$
\begin{aligned}
\frac{dx}{dt} &= x_\text{m}(-b/2m)e^{-bt/2m}\cos(\omega't + \phi) + x_\text{m}e^{-bt/2m}(-\omega')\sin(\omega't + \phi), \\
&= -x_\text{m}e^{-bt/2m}\left((b/2m)\cos(\omega't + \phi) + \omega'\sin(\omega't + \phi)\right)
\end{aligned}
$$

The second derivative is quite a bit messier;

$$
\begin{aligned}
\frac{d^2}{dx^2} &= -x_\text{m}(-b/2m)e^{-bt/2m}\left((b/2m)\cos(\omega't + \phi) + \omega'\sin(\omega't + \phi)\right) \\
&\quad -x_\text{m}e^{-bt/2m}\left((b/2m)(-\omega')\sin(\omega't + \phi) + (\omega')^2\cos(\omega't + \phi)\right), \\
&= x_\text{m}e^{-bt/2m}\left((\omega'b/m)\sin(\omega't + \phi) + (b^2/4m^2 - \omega'^2)\cos(\omega't + \phi)\right).
\end{aligned}
$$

We need to substitute these three expressions into Eq. 17-38. There are, however, some fairly obvious simplifications that we can make. Every one of the terms above has a factor of x_m, and every term above has a factor of $e^{-bt/2m}$, so simultaneously with the substitution we will cancel out those factors. Then Eq. 17-38 will look like

$$
\begin{aligned}
m\left[(\omega'b/m)\sin(\omega't + \phi) + (b^2/4m^2 - \omega'^2)\cos(\omega't + \phi)\right] \\
-b\left[(b/2m)\cos(\omega't + \phi) + \omega'\sin(\omega't + \phi)\right] + k\cos(\omega't + \phi) = 0
\end{aligned}
$$

Now we collect terms with cosine and terms with sine,

$$(\omega'b - \omega'b)\sin(\omega't + \phi) + \left(mb^2/4m^2 - \omega'^2 - b^2/2m + k\right)\cos(\omega't + \phi) = 0.$$

The coefficient for the sine term is identically zero; furthermore, because the cosine term must then vanish regardless of the value of t, the coefficient for the sine term must also vanish. Then

$$\left(mb^2/4m^2 - m\omega'^2 - b^2/2m + k\right) = 0,$$

or

$$\omega'^2 = \frac{k}{m} - \frac{b^2}{4m^2}.$$

If this condition is met, then Eq. 17-39 is indeed a solution of Eq. 17-38.

E17-49 The steps to follow for the exercise are much the same as Ex. 17-45. We need the first two derivatives of

$$x = \frac{F_{\mathrm{m}}}{G} \cos(\omega''' t - \beta)$$

The derivatives are easy enough to find,

$$\frac{dx}{dt} = \frac{F_{\mathrm{m}}}{G}(-\omega''') \sin(\omega''' t - \beta),$$

and

$$\frac{d^2 x}{dt^2} = -\frac{F_{\mathrm{m}}}{G}(\omega''')^2 \cos(\omega''' t - \beta),$$

We'll substitute this into Eq. 17-42,

$$m\left(-\frac{F_{\mathrm{m}}}{G}(\omega''')^2 \cos(\omega''' t - \beta)\right)$$
$$+ b\left(\frac{F_{\mathrm{m}}}{G}(-\omega''') \sin(\omega''' t - \beta)\right) + k\frac{F_{\mathrm{m}}}{G} \cos(\omega''' t - \beta) = F_{\mathrm{m}} \cos \omega'' t.$$

Then we'll cancel out as much as we can and collect the sine and cosine terms,

$$\left(k - m(\omega''')^2\right) \cos(\omega''' t - \beta) - (b\omega''') \sin(\omega''' t - \beta) = G \cos \omega'' t.$$

We can write the left hand side of this equation in the form

$$A \cos \alpha_1 \cos \alpha_2 - A \sin \alpha_1 \sin \alpha_2,$$

if we let $\alpha_2 = \omega''' t - \beta$ and choose A and α_1 correctly. The best choice is

$$A \cos \alpha_1 = k - m(\omega''')^2,$$
$$A \sin \alpha_1 = b\omega''',$$

and then taking advantage of the fact that $\sin^2 + \cos^2 = 1$,

$$A^2 = \left(k - m(\omega''')^2\right)^2 + (b\omega''')^2,$$

which looks like Eq. 17-44! But then we can apply the cosine angle addition formula, and

$$A \cos(\alpha_1 + \omega''' t - \beta) = G \cos \omega'' t.$$

This expression needs to be true for all time. This means that $A = G$ and $\alpha_1 + \omega''' t - \beta = \omega'' t$ and $\alpha_1 = \beta$ and $\omega''' = \omega''$.

E17-51 We know the speed of the car, we want to know how often the car hits a bump. The time between bumps is the solution to

$$vt = x,$$

$$t = \frac{(13 \text{ ft})}{(10 \text{ mi/hr})}\left(\frac{1 \text{ mi}}{5280 \text{ ft}}\right)\left(\frac{3600 \text{ s}}{1 \text{ hr}}\right) = 0.886 \text{ s}$$

The angular frequency is

$$\omega = \frac{2\pi}{T} = 7.09 \text{ rad/s}$$

This is the driving frequency, and the problem states that at this frequency the up-down bounce oscillation is at a maximum. This occurs when the driving frequency is approximately equal to the natural frequency of oscillation. See the last paragraph of Section 17-8 for more details.

The force constant for the car is k, and this is related to the natural angular frequency by

$$k = m\omega^2 = \frac{W}{g}\omega^2,$$

where $W = (2200 + 4 \times 180) \text{ lb} = 2920 \text{ lb}$ is the weight of the car and occupants and g the acceleration of free fall. Then

$$k = \frac{(2920 \text{ lb})}{(32 \text{ ft/s}^2)}(7.09 \text{ rad/s})^2 = 4590 \text{ lb/ft}$$

When the four people get out of the car there is less downward force on the car springs. The important relationship is

$$\Delta F = k\Delta x.$$

In this case $\Delta F = 720$ lb, the weight of the four people who got out of the car. Δx is the distance the car will rise when the people get out. So

$$\Delta x = \frac{\Delta F}{k} = \frac{(720 \text{ lb})}{4590 \text{ lb/ft}} = 0.157 \text{ ft} \approx 2 \text{ in.}$$

Just think. Next time you go out for a drive with friends over a bumpy road you can predict how much the suspension will sag when you get into the car. Think how impressed your friends will be!

P17-3 The larger the amplitude of the oscillations, the larger the maximum acceleration of the two masses. In order to generate this larger acceleration, we need a larger force. But the nature of static friction between the top and bottom block sets an upper limit to the force.

The maximum static friction is $F_f \leq \mu_s N$. Since the top block isn't moving up or down the normal force on the upper block is equal to the weight of the upper block. Then

$$F_f = \mu_s N = \mu_s W = \mu_s mg$$

is the maximum available force to accelerate the upper block. So the maximum acceleration is

$$a_{\mathrm{m}} = \frac{F_f}{m} = \mu_s g$$

The maximum possible amplitude of the oscillation is then given by

$$x_{\mathrm{m}} = \frac{a_{\mathrm{m}}}{\omega^2} = \frac{\mu_s g}{k/(m+M)},$$

where in the last part we substituted the total mass of the two blocks because both blocks are oscillating. Now we put in numbers, and find

$$x_{\mathrm{m}} = \frac{(0.42)(1.22\,\mathrm{kg} + 8.73\,\mathrm{kg})(9.81\,\mathrm{m/s^2})}{(344\,\mathrm{N/m})} = 0.119\,\mathrm{m}.$$

That wasn't so bad. Using a computer you could try to describe the motion if the initial displacement is larger than this.

P17-7 (a) When a spring is stretched the tension is the same everywhere in the spring. The stretching, however, is distributed over the entire length of the spring, so that the relative amount of stretch is proportional to the length of the spring under consideration. Half a spring, half the extension. But $k = -F/x$, so half the extension means twice the spring constant.

In short, cutting the spring in half will create two stiffer springs with twice the spring constant, so $k = 7.20\,\mathrm{N/cm}$ for each spring.

(b) The two spring halves now support a mass M. We can view this as each spring is holding one-half of the total mass, so in effect

$$f = \frac{1}{2\pi}\sqrt{\frac{k}{M/2}}$$

or, solving for M,

$$M = \frac{2k}{4\pi^2 f^2} = \frac{2(720\,\mathrm{N/m})}{4\pi^2(2.87\,\mathrm{s^{-1}})^2} = 4.43\,\mathrm{kg}.$$

P17-11 This would make for a good exam question. It isn't that hard, but there are a number of parts that need to be glued together. First, we use conservation of momentum to find the speed of the block + bullet immediately after the bullet collides with the block. This speed is maximum speed of the oscillation. We can use the mass of the bullet and block together with the spring constant to find the frequency of oscillations, and this can be combined with the maximum speed to find the maximum amplitude.

Now for the math. Conservation of momentum for the bullet block collision gives

$$mv = (m + M)v_{\mathrm{f}}$$

or

$$v_{\mathrm{f}} = \frac{m}{m + M}v.$$

This v_{f} will be equal to the maximum oscillation speed v_{m}. The angular frequency for the oscillation is given by

$$\omega = \sqrt{\frac{k}{m + M}}.$$

Then the amplitude for the oscillation is

$$x_{\mathrm{m}} = \frac{v_{\mathrm{m}}}{\omega} = v\frac{m}{m + M}\sqrt{\frac{m + M}{k}} = \frac{mv}{\sqrt{k(m + M)}}.$$

P17-17 (a) The rotational inertia of a stick about an axis through a point which is a distance d from the center of mass is given by the parallel axis theorem,

$$I = I_{\mathrm{cm}} + md^2 = \frac{1}{12}mL^2 + md^2.$$

The period of oscillation is given by Eq. 17-28,

$$T = 2\pi\sqrt{\frac{I}{mgd}} = 2\pi\sqrt{\frac{L^2 + 12d^2}{12gd}}$$

(b) We want to find the minimum period, so we need to take the derivative of T with respect to d. It'll look weird, but

$$\frac{dT}{dd} = \pi\frac{12d^2 - L^2}{\sqrt{12gd^3(L^2 + 12d^2)}}.$$

This will vanish when $12d^2 = L^2$, or when $d = L/sqrt12$.

P17-23 (a) Consider an object of mass m at a point P on the axis of the ring. It experiences a gravitational force of attraction to all points on the ring; by symmetry, however, the net force is not directed toward the circumference of the ring, but instead along the axis of the ring. There is then a factor of $\cos\theta$ which will be thrown in to the mix.

The distance from P to *any* point on the ring is $r = \sqrt{R^2 + z^2}$, and θ is the angle between the axis on the line which connects P and *any* point on the circumference. Consequently,

$$\cos\theta = z/r,$$

and then the net force on the star of mass m at P is

$$F = \frac{GMm}{r^2}\cos\theta = \frac{GMmz}{r^3} = \frac{GMmz}{(R^2 + z^2)^{3/2}}.$$

You will have the chance to do this calculation again in the first few chapters of Volume 2.

(b) If $z \ll R$ we can apply the binomial expansion to the denominator, and

$$(R^2 + z^2)^{-3/2} = R^{-3}\left(1 + \left(\frac{z}{R}\right)^2\right)^{-3/2} \approx R^{-3}\left(1 - \frac{3}{2}\left(\frac{z}{R}\right)^2\right).$$

Keeping terms only linear in z we have

$$F = \frac{GMm}{R^3}z,$$

which corresponds to a spring constant $k = GMm/R^3$. The frequency of oscillation is then

$$f = \sqrt{k/m}/(2\pi) = \sqrt{GM/R^3}/(2\pi).$$

(c) Using some numbers from the Milky Way galaxy,

$$f = \sqrt{(7\times 10^{-11}\mathrm{N}\cdot\mathrm{m}^2/\mathrm{kg}^2)(2\times 10^{43}\mathrm{kg})/(6\times 10^{19}\mathrm{m})^3}/(2\pi) = 1\times 10^{-14}\ \mathrm{Hz}.$$

Chapter 18

Wave Motion

E18-3 (a) The time for a particular point to move from maximum displacement to zero displacement is one-quarter of a period; the point must then go to maximum negative displacement, zero displacement, and finally maximum positive displacement to complete a cycle. So the period is 4(178 ms = 712 ms.

(b) The frequency is $f = 1/T = 1/(712 \times 10^{-3}\text{s}) = 1.40$ Hz.

(c) The wave-speed is $v = f\lambda = (1.40 \text{ Hz})(1.38 \text{ m}) = 1.93 \text{ m/s}$.

E18-5 The dimensions for tension are

$$[F] = \frac{[M][L]}{[T]^2}$$

where M stands for mass, L for length, T for time, and F stands for force.
   The dimensions for linear mass density are

$$\frac{[M]}{[L]}.$$

The dimensions for velocity are

$$\frac{[L]}{[T]}.$$

Inserting this into the expression $v = F^a/\mu^b$,

$$\frac{[L]}{[T]} = \left(\frac{[M][L]}{[T]^2}\right)^a / \left(\frac{[M]}{[L]}\right)^b,$$

$$\frac{[L]}{[T]} = \frac{[M]^a[L]^a}{[T]^{2a}}\frac{[L]^b}{[M]^b},$$

$$\frac{[L]}{[T]} = \frac{[M]^{a-b}[L]^{a+b}}{[T]^{2a}}$$

143

There are three equations here. One for time, $-1 = -2a$; one for length, $1 = a+b$; and one for mass, $0 = a - b$. We need to satisfy all three equations. The first is fairly quick; $a = 1/2$. Either of the other equations can be used to show that $b = 1/2$.

Note that if we couldn't satisfy all three equations, then there *must* be additional variables which are relevant to finding the wave velocity in a string.

E18-9 We'll first find the linear mass density by rearranging Eq. 18-19,

$$\mu = \frac{F}{v^2}$$

Since this is the same string, we expect that changing the tension will not significantly change the linear mass density. Then for the two different instances,

$$\frac{F_1}{v_1^2} = \frac{F_2}{v_2^2}$$

We want to know the new tension, so

$$F_2 = F_1 \frac{v_2^2}{v_1^2} = (123\,\text{N})\frac{(180\,\text{m/s})^2}{(172\,\text{m/s})^2} = 135\,\text{N}$$

E18-13 We need to know the wave speed before we do anything else. This is found from Eq. 18-19,

$$v = \sqrt{\frac{F}{\mu}} = \sqrt{\frac{F}{m/L}} = \sqrt{\frac{(248\,\text{N})}{(0.0978\,\text{kg})/(10.3\,\text{m})}} = 162\,\text{m/s}.$$

The two pulses travel in opposite directions on the wire; one travels as distance x_1 in a time t, the other travels a distance x_2 in a time $t + 29.6$ ms, and since the pulses meet, we have $x_1 + x_2 = 10.3$ m.

Our equations are then
$$x_1 = vt = (162\,\text{m/s})t,$$

and

$$x_2 = v(t + 29.6\,\text{ms}) = (162\,\text{m/s})(t + 29.6\,\text{ms}) = (162\,\text{m/s})t + 4.80\,\text{m}.$$

We can add these two expressions together to solve for the time t at which the pulses meet,

$$10.3\,\text{m} = x_1 + x_2 = (162\,\text{m/s})t + (162\,\text{m/s})t + 4.80\,\text{m} = (324\,\text{m/s})t + 4.80\,\text{m}.$$

which has solution $t = 0.0170\,\text{s}$. The two pulses meet at $x_1 = (162\,\text{m/s})(0.0170\,\text{s}) = 2.75\,\text{m}$, or $x_2 = 7.55\,\text{m}$.

E18-17 The intensity is the average power per unit area (Eq. 18-33); as you get farther from the source the intensity falls off because the perpendicular area increases. At some distance r from the source the total possible area is the area of a spherical shell of radius r, so intensity as a function of distance from the source would be

$$I = \frac{P_{av}}{4\pi r^2}$$

We are given two intensities: $I_1 = 1.13\,\text{W/m}^2$ at a distance r_1; $I_2 = 2.41\,\text{W/m}^2$ at a distance $r_2 = r_1 - 5.30\,\text{m}$. Since the average power of the source is the same in both cases we can equate these two values as

$$\begin{aligned} 4\pi r_1^2 I_1 &= 4\pi r_2^2 I_2, \\ 4\pi r_1^2 I_1 &= 4\pi (r_1 - d)^2 I_2, \end{aligned}$$

where $d = 5.30\,\text{m}$, and then solve for r_1. Doing this we find a quadratic expression which is

$$\begin{aligned} r_1^2 I_1 &= (r_1^2 - 2dr_1 + d^2) I_2, \\ 0 &= \left(1 - \frac{I_1}{I_2}\right) r_1^2 - 2dr_1 + d^2, \\ 0 &= \left(1 - \frac{(1.13\,\text{W/m}^2)}{(2.41\,\text{W/m}^2)}\right) r_1^2 - 2(5.30\,\text{m})r_1 + (5.30\,\text{m})^2, \\ 0 &= (0.531) r_1^2 - (10.6\,\text{m})r_1 + (28.1\,\text{m}^2). \end{aligned}$$

The solutions to this are $r_1 = 16.8\,\text{m}$ and $r_1 = 3.15\,\text{m}$; but since the person walked 5.3 m toward the lamp we'll assume they started at least that far away. Then the power output from the light is

$$P = 4\pi r_1^2 I_1 = 4\pi (16.8\,\text{m})^2 (1.13\,\text{W/m}^2) = 4.01 \times 10^3\,\text{W}.$$

E18-19 Refer to Eq. 18-40, where the amplitude of the combined wave is

$$2y_m \cos(\Delta\phi/2),$$

where y_m is the amplitude of the combining waves. Then

$$\cos(\Delta\phi/2) = (1.65 y_m)/(2y_m) = 0.825,$$

which has solution $\Delta\phi = 68.8°$.

E18-25 (a) The wave speed can be found from Eq. 18-19; we need to know the linear mass density, which is $\mu = m/L = (0.122\,\text{kg})/(8.36\,\text{m}) = 0.0146\,\text{kg/m}$. The wave speed is then given by

$$v = \sqrt{\frac{F}{\mu}} = \sqrt{\frac{(96.7\,\text{N})}{(0.0146\,\text{kg/m})}} = 81.4\,\text{m/s}.$$

(b) The longest possible standing wave will be twice the length of the string; so $\lambda = 2L = 16.7$ m.

(c) The frequency of the wave is found from Eq. 18-13, $v = f\lambda$.

$$f = \frac{v}{\lambda} = \frac{(81.4\,\text{m/s})}{(16.7\,\text{m})} = 4.87\ \text{Hz}$$

E18-29 (a) We are given the wave frequency and the wave-speed, the wavelength is found from Eq. 18-13,

$$\lambda = \frac{v}{f} = \frac{(388\,\text{m/s})}{(622\ \text{Hz})} = 0.624\,\text{m}$$

The standing wave has four loops, so from Eq. 18-45

$$L = n\frac{\lambda}{2} = (4)\frac{(0.624\,\text{m})}{2} = 1.25\,\text{m}$$

is the length of the string.

(b) We can just write it down,

$$y = (1.90\ \text{mm})\sin[(2\pi/0.624\,\text{m})x]\cos[(2\pi 622\,\text{s}^{-1})t].$$

E18-33 Although the tied end of the string forces it to be a node, the fact that the other end is loose means that it should be an anti-node. This arrangement acts identically to a pipe which is closed at one end and open at the other, and you'll discuss this in the next chapter.

The discussion of Section 18-10 indicated that the spacing between nodes is always $\lambda/2$. Since anti-nodes occur between nodes, we can expect that the distance between a node and the nearest anti-node is $\lambda/4$.

The longest possible wavelength will have one node at the tied end, an anti-node at the loose end, and no other nodes or anti-nodes. In this case $\lambda/4 = 120$ cm, or $\lambda = 480$ cm.

The next longest wavelength will have a node somewhere in the middle region of the string. But this means that there must be an anti-node between this new node and the node at the tied end of the string. Moving from left to right, we then have an anti-node at the loose end, a node, and anti-node, and finally a node at the tied end. There are four points, each separated by $\lambda/4$, so the wavelength would be given by $3\lambda/4 = 120$ cm, or $\lambda = 160$ cm.

To progress to the next wavelength we will add another node, and another anti-node. This will add another two lengths of $\lambda/4$ that need to be fit onto the string; hence $5\lambda/4 = 120$ cm, or $\lambda = 100$ cm.

This pattern will go on forever, for a string which is tied at one end but loose at the other,

$$L = n\frac{\lambda_n}{4} \text{ with } n = 1, 3, 5, 7...$$

In the figure below we have sketched the first three standing waves.

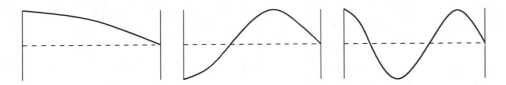

P18-3 (a) This problem really isn't as bad as it might look. The tensile stress S is tension per unit cross sectional area, so

$$S = \frac{F}{A} \text{ or } F = SA.$$

Note that this has dimensions of pressure. We already know that linear mass density is $\mu = m/L$, where L is the length of the wire. Substituting into Eq. 18-19,

$$v = \sqrt{\frac{F}{\mu}} = \sqrt{\frac{SA}{m/L}}\sqrt{\frac{S}{m/(AL)}}.$$

But AL is the volume of the wire, so the denominator is just the mass density ρ.

(b) The maximum speed of the transverse wave will be

$$v = \sqrt{\frac{S}{\rho}} = \sqrt{\frac{(720 \times 10^6 \text{ Pa})}{(7800 \text{ kg/m}^3)}} = 300 \text{ m/s}.$$

This isn't as fast as the speed of sound in air; yet we are usually led to believe that sound travels faster in steel than in air. What's wrong? Sound is a longitudal wave; the derivation above was for a transverse wave.

P18-7 (a) The linear mass density changes as the rubber band is stretched! In this case,

$$\mu = \frac{m}{L + \Delta L}.$$

The tension in the rubber band is given by $F = k\Delta L$. Substituting this into Eq. 18-19,

$$v = \sqrt{\frac{F}{\mu}} = \sqrt{\frac{k\Delta L(L + \Delta L)}{m}}.$$

147

(b) We want to know the time it will take to travel the length of the rubber band, so

$$v = \frac{L + \Delta L}{t} \text{ or } t = \frac{L + \Delta L}{v}.$$

Into this we will substitute our expression for wave speed

$$t = (L + \Delta L)\sqrt{\frac{m}{k\Delta L(L + \Delta L)}} = \sqrt{\frac{m(L + \Delta L)}{k\Delta L}}$$

We have two possibilities to consider: either $\Delta L \ll L$ or $\Delta L \gg L$. In either case we are only interested in the part of the expression with $L + \Delta L$; whichever term is much larger than the other will be the only significant part.

Then if $\Delta L \ll L$ we get $L + \Delta L \approx L$ and

$$t = \sqrt{\frac{m(L + \Delta L)}{k\Delta L}} \approx \sqrt{\frac{mL}{k\Delta L}},$$

so that t is proportional to $1/\sqrt{\Delta L}$.

But if $\Delta L \gg L$ we get $L + \Delta L \approx \Delta L$ and

$$t = \sqrt{\frac{m(L + \Delta L)}{k\Delta L}} \approx \sqrt{\frac{m\Delta L}{k\Delta L}} = \sqrt{\frac{m}{k}},$$

so that t is constant.

You can actually try this with a rubber band. Pluck it, and then stretch it while listening to the pitch.

P18-11 If we assume that Handel wanted his violins to play in tune with the other instruments then all we need to do is find an instrument from Handel's time that will accurately keep pitch over a period of several hundred years. Most instruments won't keep pitch for even a few days because of temperature and humidity changes; some (like the piccolo?) can't even play in tune for more than a few notes! But if someone found a tuning fork...

Since the length of the string doesn't change, and we are using a string with the same mass density, the only choice is to change the tension. But $f \propto v \propto \sqrt{T}$, so the percentage change in the tension of the string is

$$\frac{T_f - T_i}{T_i} = \frac{f_f{}^2 - f_i{}^2}{f_i{}^2} = \frac{(440 \text{ Hz})^2 - (422.5 \text{ Hz})^2}{(422.5 \text{ Hz})^2} = 8.46\,\%.$$

P18-15 Before we can solve this problem we need to make sure that we have the right answer from Problem 18-14. I'm not going to do all the work here, but I will point out some of the more obvious facts.

The wavelength is

$$\lambda = v/f = (3.00 \times 10^8 \text{m/s})/(13.0 \times 10^6 \text{Hz}) = 23.1\,\text{m}.$$

The direct wave travels a distance d from S to D. The wave which reflects off the original layer travels a distance $\sqrt{d^2 + 4H^2}$ between S and D. The wave which reflects off the layer one minute layer travels a distance $\sqrt{d^2 + 4(H+h)^2}$. Waves will interfere constructively if there is a difference of an integer number of wavelengths between the two path lengths. In other words originally we have

$$\sqrt{d^2 + 4H^2} - d = n_1\lambda,$$

and then one minute later we have

$$\sqrt{d^2 + 4(H+h)^2} - d = n_2\lambda.$$

We don't know either n_1 or n_2, but we do know the difference is 6, so we can subtract the top equation from the bottom and get

$$\sqrt{d^2 + 4(H+h)^2} - \sqrt{d^2 + 4H^2} = 6\lambda$$

This isn't the answer to the previous problem, but it is close.

We could use that expression as written, do some really obnoxious algebra, and then get the answer. But we don't want to; we want to take advantage of the fact that h is small compared to d and H. Then the first term can be written as

$$\begin{aligned}
\sqrt{d^2 + 4(H+h)^2} &= \sqrt{d^2 + 4H^2 + 8Hh + 4h^2}, \\
&\approx \sqrt{d^2 + 4H^2 + 8Hh}, \\
&\approx \sqrt{d^2 + 4H^2}\sqrt{1 + \frac{8H}{d^2 + 4H^2}h}, \\
&\approx \sqrt{d^2 + 4H^2}\left(1 + \frac{1}{2}\frac{8H}{d^2 + 4H^2}h\right).
\end{aligned}$$

Between the second and the third lines we factored out $d^2 + 4H^2$; that last line is from the binomial expansion theorem. We put this into the previous expression, and

$$\begin{aligned}
\sqrt{d^2 + 4(H+h)^2} - \sqrt{d^2 + 4H^2} &= 6\lambda, \\
\sqrt{d^2 + 4H^2}\left(1 + \frac{4H}{d^2 + 4H^2}h\right) - \sqrt{d^2 + 4H^2} &= 6\lambda, \\
\frac{4H}{\sqrt{d^2 + 4H^2}}h &= 6\lambda.
\end{aligned}$$

Now what were we doing? We were trying to find the speed at which the layer is moving. We know H, d, and λ; we can then find h,

$$h = \frac{6(23.1\,\text{m})}{4(510 \times 10^3\text{m})}\sqrt{(230 \times 10^3\text{m})^2 + 4(510 \times 10^3\text{m})^2} = 71.0\,\text{m}.$$

The layer is then moving at $v = (71.0\,\text{m})/(60\,\text{s}) = 1.18\,\text{m/s}$.

P18-19 (a) Call the three waves

$$
\begin{aligned}
y_i &= A \sin k_1 (x - v_1 t), \\
y_t &= B \sin k_2 (x - v_2 t), \\
y_r &= C \sin k_1 (x + v_1 t),
\end{aligned}
$$

where the subscripts i, t, and r refer to the incident, transmitted, and reflected waves respectively.

We'll apply the principle of superposition. Just to the left of the knot the wave has amplitude $y_i + y_r$ while just to the right of the knot the wave has amplitude y_t. These two amplitudes must line up at the knot for all times t, or the knot will come undone. Remember the knot is at $x = 0$, so

$$
\begin{aligned}
y_i + y_r &= y_t, \\
A \sin k_1 (-v_1 t) + C \sin k_1 (+v_1 t) &= B \sin k_2 (-v_2 t), \\
-A \sin k_1 v_1 t + C \sin k_1 v_1 t &= -B \sin k_2 v_2 t
\end{aligned}
$$

We know that $k_1 v_1 = k_2 v_2 = \omega$, so the three sin functions are all equivalent, and can be canceled. This leaves $A = B + C$.

(b) We need to match more than the displacement, we need to match the slope just on either side of the knot. In that case we need to take the derivative of

$$
y_i + y_r = y_t
$$

with respect to x, and then set $x = 0$. First we take the derivative,

$$
\begin{aligned}
\frac{d}{dx}(y_i + y_r) &= \frac{d}{dx}(y_t), \\
k_1 A \cos k_1 (x - v_1 t) + k_1 C \cos k_1 (x + v_1 t) &= k_2 B \cos k_2 (x - v_2 t),
\end{aligned}
$$

and then we set $x = 0$ and simplify,

$$
\begin{aligned}
k_1 A \cos k_1 (-v_1 t) + k_1 C \cos k_1 (+v_1 t) &= k_2 B \cos k_2 (-v_2 t), \\
k_1 A \cos k_1 v_1 t + k_1 C \cos k_1 v_1 t &= k_2 B \cos k_2 v_2 t.
\end{aligned}
$$

This last expression simplifies like the one in part (a) to give

$$
k_1 (A + C) = k_2 B
$$

We can combine this with $A = B + C$ to solve for C,

$$
\begin{aligned}
k_1 (A + C) &= k_2 (A - C), \\
C (k_1 + k_2) &= A (k_2 - k_1), \\
C &= A \frac{k_2 - k_1}{k_1 + k_2}.
\end{aligned}
$$

If $k_2 < k_1$ C will be negative; this means the reflected wave will be inverted.

Chapter 19

Sound Waves

E19-3 (a) The wavelength is given by $\lambda = v/f = (343\,\text{m/s})/(4.50 \times 10^6\text{Hz} = 7.62 \times 10^{-5}\text{m}$.

(b) The wavelength is given by $\lambda = v/f = (1500\,\text{m/s})/(4.50 \times 10^6\text{Hz} = 3.33 \times 10^{-4}\text{m}$.

E19-7 Marching at 120 paces per minute means that you move a foot every half a second. The soldiers in the back are moving the wrong foot, which means they are moving the correct foot half a second later than they should. If the speed of sound is $343\,\text{m/s}$, then the column of soldiers must be $(343\,\text{m/s})(0.5\,\text{s}) = 172\,\text{m}$ long.

E19-11 If the source emits equally in all directions the intensity at a distance r is given by the average power divided by the surface area of a sphere of radius r centered on the source.

The power output of the source can then be found from

$$P = IA = I(4\pi r^2) = (197 \times 10^{-6}\text{W/m}^2)4\pi(42.5\,\text{m})^2 = 4.47\,\text{W}.$$

E19-15 (a) Relative sound level is given by Eq. 19-21,

$$SL_1 - SL_2 = 10 \log \frac{I_1}{I_2} \text{ or } \frac{I_1}{I_2} = 10^{(SL_1 - SL_2)/10},$$

so if $\Delta SL = 30$ then $I_1/I_2 = 10^{30/10} = 1000$.

(b) Intensity is proportional to pressure amplitude squared according to Eq. 19-19; so

$$\Delta p_{\text{m},1}/\Delta p_{\text{m},2} = \sqrt{I_1/I_2} = \sqrt{1000} = 32.$$

E19-19 The sound level is given by Eq. 19-20,

$$SL = 10 \log \frac{I}{I_0}$$

where I_0 is the threshold intensity of 10^{-12} W/m^2. Intensity is given by Eq. 19-19,

$$I = \frac{(\Delta p_{\mathrm{m}})^2}{2\rho v}$$

If we assume the maximum possible pressure amplitude is equal to one atmosphere, then

$$I = \frac{(\Delta p_{\mathrm{m}})^2}{2\rho v} = \frac{(1.01 \times 10^5 \mathrm{Pa})^2}{2(1.21\,\mathrm{kg/m^3})(343\,\mathrm{m/s})} = 1.22 \times 10^7 \mathrm{W/m^2}.$$

The sound level would then be

$$SL = 10 \log \frac{I}{I_0} = 10 \log \frac{1.22 \times 10^7 \mathrm{W/m^2}}{(10^{-12}\ \mathrm{W/m^2})} = 191\ \mathrm{dB}$$

E19-23 A minimum will be heard at the detector if the path length difference between the straight path and the path through the curved tube is half of a wavelength. Both paths involve a straight section from the source to the start of the curved tube, and then from the end of the curved tube to the detector. Since it is the path difference that matters, we'll only focus on the part of the path between the start of the curved tube and the end of the curved tube. The length of the straight path is one diameter, or $2r$. The length of the curved tube is half a circumference, or πr. The difference is $(\pi - 2)r$. This difference is equal to half a wavelength, so

$$(\pi - 2)r = \lambda/2,$$

$$r = \frac{\lambda}{2\pi - 4} = \frac{(42.0\ \mathrm{cm})}{2\pi - 4} = 18.4\ \mathrm{cm}.$$

E19-29 The well is a tube open at one end and closed at the other; Eq. 19-28 describes the allowed frequencies of the resonant modes. The lowest frequency is when $n = 1$, so $f_1 = v/4L$. We know f_1; to find the depth of the well, L, we need to know the speed of sound.

We should use the information provided, instead of looking up the speed of sound, because maybe the well is filled with some kind of strange gas.

Then, from Eq. 19-14,

$$v = \sqrt{\frac{B}{\rho}} = \sqrt{\frac{(1.41 \times 10^5\ \mathrm{Pa})}{(1.21\ \mathrm{kg/m^3})}} = 341\ \mathrm{m/s}.$$

The depth of the well is then

$$L = v/(4f_1) = (341\ \mathrm{m/s})/[4(7.20\ \mathrm{Hz})] = 11.8\ \mathrm{m}.$$

E19-31 The maximum reflected frequencies will be the ones that undergo constructive interference, which means the path length difference will be an integer multiple of a wavelength. A wavefront will strike a terrace wall and part will reflect, the other part will travel on to the next terrace, and then reflect. Since part of the wave had to travel to the next terrace and back, the path length difference will be $2 \times 0.914\,\text{m} = 1.83\,\text{m}$.

If the speed of sound is $v = 343\,\text{m/s}$, the lowest frequency wave which undergoes constructive interference will be

$$f = \frac{v}{\lambda} = \frac{(343\,\text{m/s})}{(1.83\,\text{m})} = 187\ \text{Hz}$$

Any integer multiple of this frequency will also undergo constructive interference, and will also be heard. The ear and brain, however, will most likely interpret the complex mix of frequencies as a single tone of frequency 187 Hz.

E19-35 The mass of the string doesn't make a difference for this problem! The speed of a wave on the string is the same, regardless of where you put your finger, so $f\lambda$ is a constant. The string will vibrate (mostly) in the lowest harmonic, so that $\lambda = 2L$, where L is the length of the part of the string that is allowed to vibrate. Then

$$
\begin{aligned}
f_2\lambda_2 &= f_1\lambda_1, \\
2f_2L_2 &= 2f_1L_1, \\
L_2 &= L_1\frac{f_1}{f_2} = (30\ \text{cm})\frac{(440\ \text{Hz})}{(528\ \text{Hz})} = 25\ \text{cm}.
\end{aligned}
$$

So you need to place your finger 5 cm from the end. Small changes in where you place your finger can result in significant pitch differences; a good violin player practices enough to know within a millimeter or so where the finger needs to be positioned; an excellent violin player can do this, but will listen to the pitch when it is first played and within a very small amount of time slightly shift the position of the finger to get the note in tune.

E19-37 The unknown frequency is either 3 Hz higher or lower than the standard fork. A small piece of wax placed on the fork of this unknown frequency tuning fork will result in a lower frequency because $f \propto \sqrt{k/m}$. If the beat frequency decreases then the two tuning forks are getting *closer* in frequency, so the frequency of the first tuning fork must be above the frequency of the standard fork. Hence, 387 Hz.

E19-41 We'll use Eq. 19-44, since both the observer and the source are in motion. Then

$$f' = f\frac{v \pm v_O}{v \mp v_S} = (15.8\ \text{kHz})\frac{(343\,\text{m/s}) + (246\,\text{m/s})}{(343\,\text{m/s}) + (193\,\text{m/s})} = 17.4\ \text{kHz}$$

The sign in the numerator is positive because the observer is moving in a direction toward the source; the sign in the denominator is positive because the source is moving in a direction away from the observer.

E19-49 First, the trumpet player is out of tune, so he should push his tuning slide in about 3 mm.

(a) Although the observer is at rest, we'll use Eq. 19-44 instead of Eq. 19-43, because we only want to remember *one* equation, not three. The frequency "heard" by the wall is

$$f' = f\frac{v + v_O}{v - v_S} = (438\text{ Hz})\frac{(343\text{ m/s}) + (0)}{(343\text{ m/s}) - (19.3\text{ m/s})} = 464\text{ Hz}$$

(b) The wall then reflects a frequency of 464 Hz back to the trumpet player. Sticking with Eq. 19-44, the source is now at rest while the observer moving,

$$f' = f\frac{v + v_O}{v - v_S} = (464\text{ Hz})\frac{(343\text{ m/s}) + (19.3\text{ m/s})}{(343\text{ m/s}) - (0)} = 490\text{ Hz}$$

P19-3 (a) The intensity at 28.5 m is found from the $1/r^2$ dependence;

$$I_2 = I_1(r_1/r_2)^2 = (962\,\mu\text{W/m}^2)(6.11\text{ m}/28.5\text{ m})^2 = 44.2\,\mu\text{W/m}^2.$$

(c) We'll do this part first. The pressure amplitude is found from Eq. 19-19,

$$\Delta p_\text{m} = \sqrt{2\rho v I} = \sqrt{2(1.21\text{ kg/m}^3)(343\text{ m/s})(962\times 10^{-6}\text{W/m}^2)} = 0.894\,\text{Pa}.$$

(b) The displacement amplitude is found from Eq. 19-8,

$$s_\text{m} = \Delta p_\text{m}/(kB),$$

where $k = 2\pi f/v$ is the wave number. From Eq. 19-14 w know that $B = \rho v^2$, so

$$s_\text{m} = \frac{\Delta p_\text{m}}{2\pi f \rho v} = \frac{(0.894\text{ Pa})}{2\pi(2090\text{ Hz})(1.21\text{ kg/m}^3)(343\text{ m/s})} = 1.64\times 10^{-7}\text{m}.$$

P19-9 What the device is doing is taking all of the energy which strikes a large surface area and concentrating it into a small surface area. It doesn't succeed; only 12% of the energy is concentrated. We can think, however, in terms of power: 12% of the average power which strikes the parabolic reflector is transmitted into the tube.

If the sound intensity on the reflector is I_1, then the average power is $P_1 = I_1 A_1 = I_1 \pi r_1^2$, where r_1 is the radius of the reflector. The average power in the tube will be $P_2 = 0.12 P_1$, so the intensity in the tube will be

$$I_2 = \frac{P_2}{A_2} = \frac{0.12 I_1 \pi r_1^2}{\pi r_2^2} = 0.12 I_1 \frac{r_1^2}{r_2^2}$$

Since the lowest audible sound has an intensity of $I_0 = 10^{-12}$ W/m^2, we can set $I_2 = I_0$ as the condition for "hearing" the whisperer through the apparatus. The minimum sound intensity at the parabolic reflector is

$$I_1 = \frac{I_0}{0.12} \frac{r_2^2}{r_1^2}.$$

Now for the whisperers. Intensity falls off as $1/d^2$, where d is the distance from the source. We are told that when $d = 1.0$ m the sound level is 20 dB; this sound level has an intensity of

$$I = I_0 10^{20/10} = 100 I_0$$

Then at a distance d from the source the intensity must be

$$I_1 = 100 I_0 \frac{(1\,\text{m})^2}{d^2}.$$

This would be the intensity "picked-up" by the parabolic reflector. Combining this with the condition for being able to hear the whisperers through the apparatus, we have

$$\frac{I_0}{0.12} \frac{r_2^2}{r_1^2} = 100 I_0 \frac{(1\,\text{m})^2}{d^2}$$

or, upon some very mild mannered rearranging,

$$d = (\sqrt{12}\,\text{m}) \frac{r_1}{r_2} = (\sqrt{12}\,\text{m}) \frac{(0.50\,\text{m})}{(0.005\,\text{m})} = 346\,\text{m}.$$

This answer might seem somewhat puzzling at first, because it appears to be independent of the threshold of hearing, as if the distance from the whisperers isn't affected by how well you can hear! However, since the problem was expressed in terms of sound levels (20 dB at 1.0 m), this information on threshold levels was included in the beginning, and is actually part of the factor of 100 which appears in the equation previous to the last one. If I_0 were to be defined differently, then the 20 dB at 1.0 m would also change.

P19-13 A 30.0 cm string fixed at both ends will resonate if the frequency is given by $f = nv/2L$, where n is an integer, L the length of the string, and v the wave-speed on the string. In this problem the string is observed to resonate at 880 Hz and then

155

again at 1320 Hz, so the two corresponding values of n must differ by 1. We can then write two equations

$$(880 \text{ Hz}) = \frac{nv}{2L} \text{ and } (1320 \text{ Hz}) = \frac{(n+1)v}{2L}$$

and solve these for v. It is somewhat easier to first solve for n, and since this is a common midterm exam question, we'll do it this way. Rearranging both equations, we get

$$\frac{(880 \text{ Hz})}{n} = \frac{v}{2L} \text{ and } \frac{(1320 \text{ Hz})}{n+1} = \frac{v}{2L}.$$

Combining these two equations we get

$$\frac{(880 \text{ Hz})}{n} = \frac{(1320 \text{ Hz})}{n+1},$$
$$(n+1)(880 \text{ Hz}) = n(1320 \text{ Hz}),$$
$$n = \frac{(880 \text{ Hz})}{(1320 \text{ Hz}) - (880 \text{ Hz})} = 2.$$

There is actually a very important lesson to be learned here. For a string fixed at both ends (or a pipe closed or opened at both ends), the frequency difference between two adjacent harmonics is the same as the fundamental frequency. In mathematical terms, $f_{n+1} - f_n = f_1$.

Enough of that. Now that we know n (and f_1) we can find v,

$$v = 2(0.300 \text{ m})\frac{(880 \text{ Hz})}{2} = 264 \text{ m/s}$$

And, finally, we are in a position to find the tension, since

$$F = \mu v^2 = (0.652 \times 10^{-3} \text{kg/m})(264 \text{ m/s})^2 = 45.4 \text{ N}.$$

A little tense, are we?

P19-17 The sonic boom that you hear is not from the sound given off by the plane when it is overhead, it is from the sound given off *before* the plane was overhead. So this problem isn't as simple as distance = velocity × time. It is *very* useful to sketch a picture. This figure is my version.

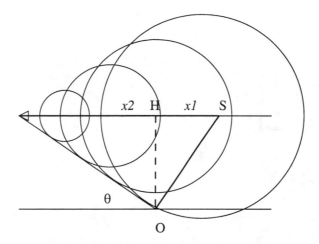

We can find the angle θ from the figure, we'll get Eq. 19-45, so

$$\sin\theta = \frac{v}{v_s} = \frac{(330\,\text{m/s})}{(396\,\text{m/s})} = 0.833 \text{ or } \theta = 56.4°$$

Note that v_s is the speed of the source, not the speed of sound!

Unfortunately $t = 12\,\text{s}$ is *not* the time between when the sonic boom leaves the plane and when it arrives at the observer. It is the time between when the plane is overhead and when the sonic boom arrives at the observer. That's why there are so many marks and variables on the figure. x_1 is the distance from where the sonic boom which is heard by the observer is emitted to the point directly overhead; x_2 is the distance from the point which is directly overhead to the point where the plane is when the sonic boom is heard by the observer. We do have $x_2 = v_s(12.0\,\text{s})$. This length forms one side of a right triangle HSO, the opposite side of this triangle is the side HO, which is the height of the plane above the ground, so

$$h = x_2 \tan\theta = (343\,\text{m/s})(12.0\,\text{s}) \tan(56.4°) = 7150\,\text{m}.$$

P19-19 (a)   We apply Eq. 19-44

$$f' = f\frac{v + v_O}{v - v_S} = (1030\,\text{Hz})\frac{(5470\,\text{km/h}) + (94.6\,\text{km/h})}{(5470\,\text{km/h}) - (20.2\,\text{km/h})} = 1050\,\text{Hz}$$

(b)   The reflected signal has a frequency equal to that of the signal received by the second sub originally. Applying Eq. 19-44 again,

$$f' = f\frac{v + v_O}{v - v_S} = (1050\,\text{Hz})\frac{(5470\,\text{km/h}) + (20.2\,\text{km/h})}{(5470\,\text{km/h}) - (94.6\,\text{km/h})} = 1070\,\text{Hz}$$

Chapter 20

Relativity

E20-3 We are interested in the length contraction, which is given by Eq. 20-8.

$$L = L_0\sqrt{1 - u^2/c^2} = (1.68\,\text{m})\sqrt{1 - (0.632)^2} = 1.30\,\text{m}.$$

One of the biggest challenges in special relativity is trying to figure out which length (or time interval) corresponds to a measurement made on an object at rest or an object which is moving. In other words, which measurement is L, and which is L_0. One rule of thumb: L_0 will always be larger than L. L_0 always corresponds to the measurement of the length of an object which is at rest with respect to the measurer.

E20-5 We can apply $\Delta x = v\Delta t$ to find the time the particle existed before it decayed. Then

$$\Delta t = \frac{x}{v} \frac{(1.05 \times 10^{-3}\,\text{m})}{(0.992)(3.00 \times 10^8\,\text{m/s})} = 3.53 \times 10^{-12}\,\text{s}.$$

We'll then put this into Eq. 20-3. But is this Δt or Δt_0? Since this is the measurement of a time interval which occurs on an object which is moving with respect to the measurer, it is Δt. Then the *proper lifetime* of the particle is

$$\Delta t_0 = \Delta t\sqrt{1 - u^2/c^2} = (3.53 \times 10^{-12}\,\text{s})\sqrt{1 - (0.992)^2} = 4.46 \times 10^{-13}\,\text{s}.$$

E20-11 The distance traveled by the particle is $(6.0\,\text{y})c$; the time required for the particle to travel this distance is 8.0 years. Then the speed of the particle is

$$v = \frac{\Delta x}{\Delta t} = \frac{(6.0\,\text{y})c}{(8.0\,\text{y})} = \frac{3}{4}c.$$

The speed parameter β is given by

$$\beta = \frac{v}{c} = \frac{\frac{3}{4}c}{c} = \frac{3}{4}.$$

E20-17 (a) The first part is easy; we appear to be moving away from A at the same speed as A appears to be moving away from us: $0.347c$.

(b) Using the velocity transformation formula, Eq. 20-18,

$$v'_x = \frac{v_x - u}{1 - uv_x/c^2} = \frac{(0.347c) - (-0.347c)}{1 - (-0.347c)(0.347c)/c^2} = 0.619c.$$

The negative sign reflects the fact that these two velocities are in *opposite* directions.

E20-23 This exercise is a little sneaky. The length of the ship as measured in the "certain" reference frame is

$$L = L_0\sqrt{1 - v^2/c^2} = (358\,\text{m})\sqrt{1 - (0.728)^2} = 245\,\text{m}.$$

In a time Δt the ship will move a distance $x_1 = v_1\Delta t$ while the micrometeorite will move a distance $x_2 = v_2\Delta t$; since they are moving toward each other then the micrometeorite will pass then ship when $x_1 + x_2 = L$. Then

$$\Delta t = L/(v_1 + v_2) = (245\,\text{m})/[(0.728 + 0.817)(3.00 \times 10^8\,\text{m/s})] = 5.29 \times 10^{-7}\,\text{s}.$$

That doesn't look like the answer in the back of the book! This answer is the time measured in the "certain" reference frame. We can't use Eq. 20-3 to boost this time difference to the ship, because the two events: crossing the front of the ship and crossing the back of the ship did not happen in the same place! We can, however, use Eq. 20-21,

$$\Delta t = \frac{\Delta t' + u\Delta x'/c^2}{\sqrt{1 - u^2/c^2}} = \frac{(5.29 \times 10^{-7}\,\text{s}) + (0.728c)(116\,\text{m})/c^2}{\sqrt{1 - (0.728)^2}} = 1.23 \times 10^{-6}\,\text{s}.$$

Note that $\Delta x' = (0.817c)\Delta t' = 116\,\text{m}$, because this is how far the micrometeorite moved in the "certain" reference frame.

E20-29 The magnitude of the momentum of a relativistic particle in terms of the magnitude of the velocity is given by Eq. 20-23,

$$p = \frac{mv}{\sqrt{1 - v^2/c^2}}.$$

The speed parameter, β, is what we are looking for, so we need to rearrange the above expression for the quantity v/c.

$$p/c = \frac{mv/c}{\sqrt{1 - v^2/c^2}},$$

$$\frac{p}{c} = \frac{m\beta}{\sqrt{1-\beta^2}},$$

$$\frac{mc}{p} = \frac{\sqrt{1-\beta^2}}{\beta},$$

$$\frac{mc}{p} = \sqrt{1/\beta^2 - 1}.$$

We aren't done yet. Before we go much farther, we are going to multiply the left hand side by c/c; we do this because both mc^2 and pc are useful ways of viewing these two quantities. Finishing the rearranging,

$$\frac{mc^2}{pc} = \sqrt{1/\beta^2 - 1},$$

$$\left(\frac{mc^2}{pc}\right)^2 = \frac{1}{\beta^2} - 1,$$

$$\sqrt{\left(\frac{mc^2}{pc}\right)^2 + 1} = \frac{1}{\beta},$$

$$\frac{pc}{\sqrt{m^2c^4 + p^2c^2}} = \beta$$

This last expression can also be written as $E\beta = pc$, which is a nice formula to remember.

(a) For the electron,

$$\beta = \frac{(12.5 \text{ MeV}/c)c}{\sqrt{(0.511 \text{ MeV}/c^2)^2 c^4 + (12.5 \text{ MeV}/c)^2 c^2}} = 0.999.$$

(b) For the proton,

$$\beta = \frac{(12.5 \text{ MeV}/c)c}{\sqrt{(938 \text{ MeV}/c^2)^2 c^4 + (12.5 \text{ MeV}/c)^2 c^2}} = 0.0133.$$

E20-31 The kinetic energy (not total relativistic energy) is given by Eq. 20-27,

$$K = \frac{mc^2}{\sqrt{1-v^2/c^2}} - mc^2.$$

We rearrange this to solve for $\beta = v/c$,

$$\frac{K + mc^2}{mc^2} = \frac{1}{\sqrt{1-\beta^2}},$$

$$\left(\frac{mc^2}{K + mc^2}\right)^2 = 1 - \beta^2,$$

160

which can then be written as

$$\beta = \sqrt{1 - \left(\frac{mc^2}{K + mc^2}\right)^2},$$

and yes, that looks ugly.

It is actually *much* easier to find γ, since

$$\gamma = \frac{1}{\sqrt{1 - v^2/c^2}},$$

so

$$K = \gamma mc^2 - mc^2$$

or

$$\gamma = \frac{K + mc^2}{mc^2}.$$

(a) For the electron,

$$\beta = \sqrt{1 - \left(\frac{(0.511 \text{ MeV}/c^2)c^2}{(10 \text{ MeV}) + (0.511 \text{ MeV}/c^2)c^2}\right)^2} = 0.9988,$$

and

$$\gamma = \frac{(10 \text{ MeV}) + (0.511 \text{ MeV}/c^2)c^2}{(0.511 \text{ MeV}/c^2)c^2} = 20.6.$$

(b) For the proton,

$$\beta = \sqrt{1 - \left(\frac{(938 \text{ MeV}/c^2)c^2}{(10 \text{ MeV}) + (938 \text{ MeV}/c^2)c^2}\right)^2} = 0.0145,$$

and

$$\gamma = \frac{(10 \text{ MeV}) + (938 \text{ MeV}/c^2)c^2}{(938 \text{ MeV}/c^2)c^2} = 1.01.$$

(b) For the alpha particle,

$$\beta = \sqrt{1 - \left(\frac{4(938 \text{ MeV}/c^2)c^2}{(10 \text{ MeV}) + 4(938 \text{ MeV}/c^2)c^2}\right)^2} = 0.73,$$

and

$$\gamma = \frac{(10 \text{ MeV}) + 4(938 \text{ MeV}/c^2)c^2}{4(938 \text{ MeV}/c^2)c^2} = 1.0027.$$

E20-35 (a) The kinetic energy is given by Eq. 20-27,

$$K = \frac{mc^2}{\sqrt{1 - v^2/c^2}} - mc^2 = mc^2\left(\left(1 - \beta^2\right)^{-1/2} - 1\right).$$

We want to expand the $1 - \beta^2$ part for small β,

$$\left(1 - \beta^2\right)^{-1/2} = 1 + \frac{1}{2}\beta^2 + \frac{3}{8}\beta^4 + \cdots$$

Inserting this into the kinetic energy expression,

$$K = \frac{1}{2}mc^2\beta^2 + \frac{3}{8}mc^2\beta^4 + \cdots$$

But $\beta = v/c$, so

$$K = \frac{1}{2}mv^2 + \frac{3}{8}m\frac{v^4}{c^2} + \cdots$$

(b) We want to know when the error because of neglecting the second (and higher) terms is 1%; or

$$0.01 = \left(\frac{3}{8}m\frac{v^4}{c^2}\right)/\left(\frac{1}{2}mv^2\right) = \frac{3}{4}\left(\frac{v}{c}\right)^2.$$

This will happen when $v/c = \sqrt{(0.01)4/3} = 0.115$.

E20-37 Start with Eq. 20-34 (an equation that you should commit to memory), in the form

$$E^2 = (pc)^2 + (mc^2)^2$$

The rest energy is mc^2, and if the total energy is three times this then $E = 3mc^2$, so

$$\begin{aligned}
(3mc^2)^2 &= (pc)^2 + (mc^2)^2, \\
8(mc^2)^2 &= (pc)^2, \\
\sqrt{8}\,mc &= p.
\end{aligned}$$

E20-41 Work is change in energy, so

$$W = mc^2/\sqrt{1 - (v_f/c)^2} - mc^2/\sqrt{1 - (v_i/c)^2}.$$

(a) Plug in the numbers,

$$W = (0.511\text{ MeV})(1/\sqrt{1 - (0.19)^2} - 1/\sqrt{1 - (0.18)^2}) = 0.996\text{ keV}.$$

Plug in the numbers,

$$W = (0.511\text{ MeV})(1/\sqrt{1 - (0.99)^2} - 1/\sqrt{1 - (0.98)^2}) = 1.05\text{ MeV}.$$

P20-5 We can choose our coordinate system so that u is directed along the x axis without any loss of generality. Then, according to Table 20-2,

$$
\begin{aligned}
\Delta x' &= \gamma(\Delta x - u\Delta t), \\
\Delta y' &= \Delta y, \\
\Delta z' &= \Delta z, \\
c\Delta t' &= \gamma(c\Delta t - u\Delta x/c).
\end{aligned}
$$

Square these expressions,

$$
\begin{aligned}
(\Delta x')^2 &= \gamma^2(\Delta x - u\Delta t)^2 = \gamma^2\left((\Delta x)^2 - 2u(\Delta x)(\Delta t) + (\Delta t)^2\right), \\
(\Delta y')^2 &= (\Delta y)^2, \\
(\Delta z')^2 &= (\Delta z)^2, \\
c^2(\Delta t')^2 &= \gamma^2(c\Delta t - u\Delta x/c)^2 = \gamma^2\left(c^2(\Delta t)^2 - 2u(\Delta t)(\Delta x) + u^2(\Delta x)^2/c^2\right).
\end{aligned}
$$

We'll add the first three equations and then subtract the fourth. The left hand side is the equal to

$$
(\Delta x')^2 + (\Delta y')^2 + (\Delta z')^2 - c^2(\Delta t')^2,
$$

while the right hand side will equal

$$
\gamma^2\left((\Delta x)^2 + u^2(\Delta t)^2 - c^2(\Delta t)^2 - u^2/c^2(\Delta x)^2\right) + (\Delta y)^2 + (\Delta z)^2,
$$

which can be rearranged as

$$
\gamma^2\left(1 - u^2/c^2\right)(\Delta x)^2 + \gamma^2\left(u^2 - c^2\right)(\Delta t)^2 + (\Delta y)^2 + (\Delta z)^2,
$$
$$
\gamma^2\left(1 - u^2/c^2\right)(\Delta x)^2 + (\Delta y)^2 + (\Delta z)^2 - c^2\gamma^2\left(1 - u^2/c^2\right)(\Delta t)^2.
$$

But

$$
\gamma^2 = \frac{1}{1 - u^2/c^2},
$$

so the previous expression will simplify to

$$
(\Delta x)^2 + (\Delta y)^2 + (\Delta z)^2 - c^2(\Delta t)^2,
$$

and we are done!

P20-7 We want to repeatedly add relativistic velocities starting at $\beta = 0.5$ until we get to $\beta = 0.999$. The first boost is easy,

$$
\beta_1 = 0.5.
$$

The second boost is still easy, apply Eq. 20-12, and

$$
\beta_2 = \frac{\beta_1 + 0.5}{1 + 0.5\beta_1} = \frac{0.5 + 0.5}{1 + .25} = 0.8.
$$

Re-apply the boost equation to find the speed after the third boost,

$$\beta_3 = \frac{\beta_2 + 0.5}{1 + 0.5\beta_2} = \frac{0.8 + 0.5}{1 + 0.4} = 0.929.$$

And then we apply the fourth boost,

$$\beta_4 = \frac{\beta_3 + 0.5}{1 + 0.5\beta_3} = \frac{0.929 + 0.5}{1 + 0.465} = 0.975.$$

We are getting closer, but the closer we get, the slower the approach. There must be an easier way! If we look back at the boost equation we might notice that it looks *very* similar to the rule for the tangent of the sum of two angles. It is exactly the same as the rule for the *hyperbolic* tangent, that "tanh" button on your calculator that you might never have used before. The rule is

$$\tanh(\alpha_1 + \alpha_2) = \frac{\tanh \alpha_1 + \tanh \alpha_2}{1 + \tanh \alpha_1 \tanh \alpha_2}.$$

This means that each boost of $\beta = 0.5$ is the same as a "hyperbolic" rotation of α_r where $\tanh \alpha_r = 0.5$. We need only add these rotations together until we get to α_f, where $\tanh \alpha_f = 0.999$.

$\alpha_f = 3.800$ (check your calculator for the inverse hyperbolic tangent function, *atanh*), and $\alpha_R = 0.5493$. We can fit $(3.800)/(0.5493) = 6.92$ boosts, but we need an integral number, so there are seven boosts required. The final speed after these seven boosts will be $0.9991c$.

P20-13 (a)   Start with Eq. 20-34,

$$E^2 = (pc)^2 + (mc^2)^2,$$

and substitute into this $E = K + mc^2$,

$$K^2 + 2Kmc^2 + (mc^2)^2 = (pc)^2 + (mc^2)^2.$$

We can rearrange this, and then

$$K^2 + 2Kmc^2 = (pc)^2,$$
$$m = \frac{(pc)^2 - K^2}{2Kc^2}$$

(b)   As $v/c \to 0$ we have $K \to \frac{1}{2}mv^2$ and $p \to mv$, the classical limits. Then the above expression becomes

$$m = \frac{m^2 v^2 c^2 - \frac{1}{4}m^2 v^4}{mv^2 c^2},$$
$$= m\frac{v^2 c^2 - \frac{1}{4}v^4}{v^2 c^2},$$
$$= m\left(1 - \frac{1}{4}\frac{v^2}{c^2}\right)$$

164

But $v/c \to 0$, so this expression reduces to $m = m$ in the classical limit, which is a *good thing*.

(c)   We get

$$m = \frac{(121 \text{ MeV})^2 - (55.0 \text{ MeV})^2}{2(55.0 \text{ MeV})c^2} = 1.06 \text{ MeV}/c^2,$$

which is $(1.06 \text{ MeV}/c^2)/(0.511 \text{ MeV}/c^2) = 207m_e$. Sounds like a muon to me.

P20-15 (a)   A completely inelastic collision means the two particles, each of mass m_1, stick together after the collision, in effect becoming a new particle of mass m_2. We'll use the subscript 1 for moving particle of mass m_1, the subscript 0 for the particle which is originally at rest, and the subscript 2 for the new particle after the collision. We need to conserve momentum,

$$
\begin{aligned}
p_1 + p_0 &= p_2, \\
\gamma_1 m_1 u_1 + (0) &= \gamma_2 m_2 u_2,
\end{aligned}
$$

and we need to conserve total energy,

$$
\begin{aligned}
E_1 + E_0 &= E_2, \\
\gamma_1 m_1 c^2 + m_1 c^2 &= \gamma_2 m_2 c^2,
\end{aligned}
$$

Divide the momentum equation by the energy equation and then

$$\frac{\gamma_1 u_1}{\gamma_1 + 1} = u_2.$$

But $u_1 = c\sqrt{1 - 1/\gamma_1^2}$, so

$$
\begin{aligned}
u_2 &= c\frac{\gamma_1 \sqrt{1 - 1/\gamma_1^2}}{\gamma_1 + 1}, \\
&= c\frac{\sqrt{\gamma_1^2 - 1}}{\gamma_1 + 1}, \\
&= c\frac{\sqrt{(\gamma_1 + 1)(\gamma_1 - 1)}}{\gamma_1 + 1}, \\
&= c\sqrt{\frac{\gamma_1 - 1}{\gamma_1 + 1}}.
\end{aligned}
$$

(b) Using the momentum equation,

$$m_2 = m_1 \frac{\gamma_1 u_1}{\gamma_2 u_2},$$

$$= m_1 \frac{c\gamma_1 \sqrt{1 - 1/\gamma_1^2}}{u_2/\sqrt{1 - (u_2/c)^2}},$$

$$= m_1 \frac{\sqrt{\gamma_1^2 - 1}}{1/\sqrt{(c/u_2)^2 - 1}},$$

$$= m_1 \frac{\sqrt{\gamma_1^2 - 1}}{1/\sqrt{(\gamma_1 + 1)/(\gamma_1 - 1) - 1}},$$

$$= m_1 \frac{\sqrt{(\gamma_1 + 1)(\gamma_1 - 1)}}{\sqrt{(\gamma_1 - 1)/2}},$$

$$= m_1 \sqrt{2(\gamma_1 + 1)}.$$

Without a doubt, relativity problems are the canonical example of a problem that starts simple, becomes horrific, and then collapses to something simple.

Chapter 21

Temperature

E21-1 (a) We'll assume that the new temperature scale is related to the Celsius scale by a linear transformation; then

$$T_S = mT_C + b,$$

where m and b are constants to be determined, T_S is the temperature measurement in the "new" scale, and T_C is the temperature measurement in Celsius degrees.

One of our known points is absolute zero;

$$\begin{aligned}
T_S &= mT_C + b, \\
(0) &= m(-273.15°C) + b.
\end{aligned}$$

We have two other points, the melting and boiling points for water,

$$\begin{aligned}
(T_S)_{bp} &= m(100°C) + b, \\
(T_S)_{mp} &= m(0°C) + b;
\end{aligned}$$

we can subtract the top equation from the bottom equation to get

$$(T_S)_{bp} - (T_s)S_{mp} = 100\ C°m.$$

We are told this is 180 S°, so $m = 1.8$ S°/C°. Put this into the first equation and then find b,

$$b = 273.15°Cm = 491.67°S.$$

The conversion is then

$$T_S = (1.8\ S°/C°)T_C + (491.67°S).$$

(b) The melting point for water is 491.67°S; the boiling point for water is 180 S° above this, or 671.67°S.

167

E21-7 If the temperature (in Kelvin) is directly proportional to the resistance then $T = kR$, where k is a constant of proportionality. We are given one point, $T = 273.16$ K when $R = 90.35\ \Omega$, but that is okay; we only have one unknown, k. Then $(273.16$ K$) = k(90.35\ \Omega)$ or $k = 3.023$ K$/\Omega$.

If the resistance is measured to be $R = 96.28\ \Omega$, we have a temperature of

$$T = kR = (3.023\text{ K}/\Omega)(96.28\ \Omega) = 291.1 \text{ K}.$$

E21-9 We must first find the equation which relates gain to temperature, and then find the gain at the specified temperature. If we let G be the gain we can write this linear relationship as

$$G = mT + b,$$

where m and b are constants to be determined. We have two known points:

$$(30.0) = m(20.0°\text{ C}) + b,$$
$$(35.2) = m(55.0°\text{ C}) + b.$$

If we subtract the top equation from the bottom we get

$$5.2 = m(35.0°\text{ C}),$$

or $m = 1.49\,\text{C}^{-1}$. Put this into either of the first two equations and

$$(30.0) = (0.149\,\text{C}^{-1})(20.0°\text{ C}) + b,$$

which has a solution $b = 27.0$

Now to find the gain when $T = 28.0\,°$C:

$$G = mT + b = (0.149\,\text{C}^{-1})(28.0°\text{ C}) + (27.0) = 31.2$$

E21-15 We want to focus on the temperature change, not the absolute temperature. In this case, $\Delta T = T_\text{f} - T_\text{i} = (42°\text{C}) - (-5.0°\text{C}) = 47\,\text{C}°$.

Then we apply Eq. 21-8,
$$\Delta L = \alpha L \Delta T$$
where for steel $\alpha = 11 \times 10^{-6}\,\text{C}^{-1}$. Gluing it all together,

$$\Delta L = (11 \times 10^{-6}\,\text{C}^{-1})(12.0\,\text{m})(47\,\text{C}°) = 6.2 \times 10^{-3}\,\text{m}.$$

This may not seem like much, but it is. Failure to include the expansion spaces will cause the tracks to bulge to the side by 20 centimeters, which is more than enough to cause a derailment.

E21-17 (a) We'll apply Eq. 21-10. The surface area of a cube is six times the area of one face, which is the edge length squared. So

$$A = 6(0.332\,\text{m})^2 = 0.661\,\text{m}^2.$$

The temperature change is $\Delta T = (75.0°\text{C}) - (20.0°\text{C}) = 55.0\,\text{C}°$. Then the increase in surface area is

$$\Delta A = 2\alpha A \Delta T = 2(19 \times 10^{-6}\,\text{C}^{-1})(0.661\,\text{m}^2)(55.0\,\text{C}°) = 1.38 \times 10^{-3}\,\text{m}^2$$

(b) We'll now apply Eq. 21-11. The volume of the cube is the edge length cubed (hence the term *cube*), so

$$V = (0.332\,\text{m})^3 = 0.0366\,\text{m}^3.$$

and then from Eq. 21-11,

$$\Delta V = 2\alpha V \Delta T = 3(19 \times 10^{-6}\,\text{C}^{-1})(0.0366\,\text{m}^3)(55.0\,\text{C}°) = 1.15 \times 10^{-4}\,\text{m}^3,$$

is the change in volume of the cube.

E21-21 I like this question. We'll assume that the steel ruler measures length correctly at room temperature. Then the 20.05 cm measurement of the rod is correct. But both the rod and the ruler will expand in the oven, so the 20.11 cm measurement of the rod is *not* the actual length of the rod in the oven. What is the actual length of the rod in the oven? We can only answer that after figuring out how the 20.11 cm mark on the ruler moves when the ruler expands.

Let $L = 20.11$ cm correspond to the ruler mark at room temperature. Then

$$\Delta L = \alpha_{\text{steel}} L \Delta T = (11 \times 10^{-6}\,\text{C}^{-1})(20.11\,\text{cm})(250\,\text{C}°) = 5.5 \times 10^{-2}\,\text{cm}$$

is the shift in position of the mark as the ruler is raised to the higher temperature. Then the change in length of the rod is *not* $(20.11\,\text{cm}) - (20.05\,\text{cm}) = 0.06\,\text{cm}$, because the 20.11 cm mark is shifted out. We need to add 0.055 cm to this; the rod changed length by 0.115 cm.

The coefficient of thermal expansion for the rod is

$$\alpha = \frac{\Delta L}{L \Delta T} = \frac{(0.115\,\text{cm})}{(20.05\,\text{cm})(250\,\text{C}°)} = 23 \times 10^{-6}\,\text{C}^{-1}.$$

E21-23 Most people solve this problem (okay, I exaggerate, most people don't solve this problem) by assuming the solid is in the form of a cube. Not all solids are cubes, but you can carefully dice up the solid into various cubes of different sizes, so we'll assume the approach is valid.

If the length of one side of a cube is originally L_0, then the volume is originally $V_0 = L_0^3$. After heating, the volume of the cube will be $V = L^3$, where $L = L_0 + \Delta L$. Then

$$
\begin{aligned}
V &= L^3, \\
&= (L_0 + \Delta L)^3, \\
&= (L_0 + \alpha L_0 \Delta T)^3, \\
&= L_0^3(1 + \alpha \Delta T)^3.
\end{aligned}
$$

As long as the quantity $\alpha \Delta T$ is much less than one we can expand the last line in a binomial expansion as

$$V \approx V_0(1 + 3\alpha \Delta T + \cdots),$$

so the change in volume is $\Delta V \approx 3\alpha V_0 \Delta T$.

E21-27 The diameter of the rod as a function of temperature is

$$d_s = d_{s,0}(1 + \alpha_s \Delta T),$$

The diameter of the ring as a function of temperature is

$$d_b = d_{b,0}(1 + \alpha_b \Delta T).$$

We are interested in the temperature when the diameters are equal,

$$
\begin{aligned}
d_{s,0}(1 + \alpha_s \Delta T) &= d_{b,0}(1 + \alpha_b \Delta T), \\
\alpha_s d_{s,0} \Delta T - \alpha_b d_{b,0} \Delta T &= d_{b,0} - d_{s,0}, \\
\Delta T &= \frac{d_{b,0} - d_{s,0}}{\alpha_s d_{s,0} - \alpha_b d_{b,0}}, \\
\Delta T &= \frac{(2.992 \text{ cm}) - (3.000 \text{ cm})}{(11 \times 10^{-6}/\text{C}^\circ)(3.000 \text{ cm}) - (19 \times 10^{-6}/\text{C}^\circ)(2.992 \text{ cm})}, \\
&= 335 \text{ C}^\circ.
\end{aligned}
$$

The final temperature is then $T_f = (25^\circ) + 335 \text{ C}^\circ = 360^\circ$.

E21-31 This problem is related to objects which expand when heated, but we never actually need to calculate any temperature changes. We will, however, be interested in the change in rotational inertia. Rotational inertia is directly proportional to the square of the (appropriate) linear dimension, so

$$I_f/I_i = (r_f/r_i)^2.$$

(a) If the bearings are frictionless then there are no external torques, so the angular momentum is constant.

(b)   If the angular momentum is constant, then

$$L_i = L_f,$$
$$I_i\omega_i = I_f\omega_f.$$

We are interested in the percent change in the angular velocity, which is

$$\frac{\omega_f - \omega_i}{\omega_i} = \frac{\omega_f}{\omega_i} - 1 = \frac{I_i}{I_f} - 1 = \left(\frac{r_i}{r_f}\right)^2 - 1 = \left(\frac{1}{1.0018}\right)^2 - 1 = -0.36\%.$$

(c)   The rotational kinetic energy is proportional to $I\omega^2 = (I\omega)\omega = L\omega$, but L is constant, so

$$\frac{K_f - K_i}{K_i} = \frac{\omega_f - \omega_i}{\omega_i} = -0.36\%.$$

E21-39 (a)   Using Eq. 21-17,

$$n = \frac{pV}{RT} = \frac{(108 \times 10^3\,\text{Pa})(2.47\,\text{m}^3)}{(8.31\,\text{J/mol·K})([12 + 273]\,\text{K})} = 113\ \text{mol}.$$

Notice that we converted to Kelvins!

(b)   Use the same expression again,

$$V = \frac{nRT}{p} = \frac{(113\,\text{mol})(8.31\,\text{J/mol·K})([31 + 273]\,\text{K})}{(316 \times 10^3\,\text{Pa})} = 0.903\,\text{m}^3.$$

P21-1 (a)   The dimensions of A must be $[\text{time}]^{-1}$, as can be seen with a quick inspection of the equation. We would expect that A would depend on the **surface area** at the very least; however, that means that it must also depend on some other factor to fix the dimensionality of A.

(b)   Rearrange and integrate,

$$\int_{\Delta T_0}^{T} \frac{d\Delta T}{\Delta T} = -\int_0^t A\,dt,$$
$$\ln(\Delta T/\Delta T_0) = -At,$$
$$\Delta T = \Delta T_0 e^{-At}.$$

P21-5 Start with a differential form for Eq. 21-8, $dL/dT = \alpha L_0$, rearrange, and integrate:

$$\int_{L_0}^{L} dL = \int_{T_0}^{T} \alpha L_0 \, dT,$$

$$L - L_0 = L_0 \int_{T_0}^{T} \alpha \, dT,$$

$$L = L_0 \left(1 + \int_{T_0}^{T} \alpha \, dT \right).$$

P21-7 (a) Consider the work that was done for Ex. 21-27. The length of rod a is

$$L_a = L_{a,0}(1 + \alpha_a \Delta T),$$

while the length of rod b is

$$L_b = L_{b,0}(1 + \alpha_b \Delta T).$$

The difference is

$$L_a - L_b = L_{a,0}(1 + \alpha_a \Delta T) - L_{b,0}(1 + \alpha_b \Delta T),$$
$$= L_{a,0} - L_{b,0} + (L_{a,0}\alpha_a - L_{b,0}\alpha_b)\Delta T,$$

which will be a constant is $L_{a,0}\alpha_a = L_{b,0}\alpha_b$ or

$$L_{i,0} \propto 1/\alpha_i.$$

(b) We want $L_{a,0} - L_{b,0} = 0.30\,\text{m}$ so

$$k/\alpha_a - k/\alpha_b = 0.30\,\text{m},$$

where k is a constant of proportionality;

$$k = (0.30\,\text{m})/\left(1/(11 \times 10^{-6}/\text{C}^\circ) - 1/(19 \times 10^{-6}/\text{C}^\circ)\right) = 7.84 \times 10^{-6}\text{m}/\text{C}^\circ.$$

The two lengths are

$$L_a = (7.84 \times 10^{-6}\text{m}/\text{C}^\circ)/(11 \times 10^{-6}/\text{C}^\circ) = 0.713\,\text{m}$$

for steel and

$$L_b = (7.84 \times 10^{-6}\text{m}/\text{C}^\circ)/(19 \times 10^{-6}/\text{C}^\circ) = 0.413\,\text{m}$$

for brass.

172

P21-13 The volume of the block which is beneath the surface of the mercury displaces a mass of mercury equal to the mass of the block. The mass of the block is independent of the temperature but the volume of the displaced mercury changes according to

$$V_m = V_{m,0}(1 + \beta_m \Delta T).$$

This volume is equal to the depth which the block sinks times the cross sectional area of the block (which *does* change with temperature). Then

$$h_s h_b{}^2 = h_{s,0} h_{b,0}{}^2 (1 + \beta_m \Delta T),$$

where h_s is the depth to which the block sinks and $h_{b,0} = 20$ cm is the length of the side of the block. But

$$h_b = h_{b,0}(1 + \alpha_b \Delta T),$$

so

$$h_s = h_{s,0} \frac{1 + \beta_m \Delta T}{(1 + \alpha_b \Delta T)^2}.$$

Since the changes are small we can expand the right hand side using the binomial expansion; keeping terms only in ΔT we get

$$h_s \approx h_{s,0}(1 + (\beta_m - 2\alpha_b)\Delta T),$$

which means the block will sink a distance $h_s - h_{s,0}$ given by

$$h_{s,0}(\beta_m - 2\alpha_b)\Delta T = h_{s,0}\left[(1.8 \times 10^{-4}/\text{C}°) - 2(23 \times 10^{-6}/\text{C}°)\right](50\,\text{C}°) = (6.7 \times 10^{-3})h_{s,0}.$$

In order to finish we need to know how much of the block was submerged in the first place. Sine the fraction submerged is equal to the ratio of the densities, we have

$$h_{s,0}/h_{b,0} = \rho_b/\rho_m = (2.7 \times 10^3 \text{kg/m}^3)/(1.36 \times 10^4 \text{kg/m}^3),$$

so $h_{s,0} = 3.97$ cm, and the change in depth is 0.27 mm.

P21-17 Call the containers one and two so that $V_1 = 1.22$ L and $V_2 = 3.18$ L. Then the initial number of moles in the two containers are

$$n_{1,i} = \frac{p_i V_1}{RT_i} \quad \text{and} \quad n_{2,i} = \frac{p_i V_2}{RT_i}.$$

The total is

$$n = p_i(V_1 + V_2)/(RT_i).$$

Later the temperatures are changed and then the number of moles of gas in each container is

$$n_{1,f} = \frac{p_f V_1}{RT_{1,f}} \quad \text{and} \quad n_{2,f} = \frac{p_f V_2}{RT_{2,f}}.$$

173

The total is still n, so

$$\frac{p_{\mathrm{f}}}{R}\left(\frac{V_1}{T_{1,\mathrm{f}}} + \frac{V_2}{T_{2,\mathrm{f}}}\right) = \frac{p_{\mathrm{i}}(V_1 + V_2)}{RT_{\mathrm{i}}}.$$

We can solve this for the final pressure, so long as we remember to convert all temperatures to Kelvins,

$$p_{\mathrm{f}} = \frac{p_{\mathrm{i}}(V_1 + V_2)}{T_{\mathrm{i}}}\left(\frac{V_1}{T_{1,\mathrm{f}}} + \frac{V_2}{T_{2,\mathrm{f}}}\right)^{-1},$$

or

$$p_{\mathrm{f}} = \frac{(1.44\ \mathrm{atm})(1.22\mathrm{L} + 3.18\ \mathrm{L})}{(289\,\mathrm{K})}\left(\frac{(1.22\ \mathrm{L})}{(289\,\mathrm{K})} + \frac{(3.18\ \mathrm{L})}{(381\,\mathrm{K})}\right)^{-1} = 1.74\ \mathrm{atm}.$$

Chapter 22

Molecular Properties of Gases

E22-3 (a) We want to use the equation from Section 21-5: $pV = nRT$. We are given the pressure in the tank and the temperature, we want to find the volume. We could if we knew the number of moles of ammonia present, so we first need to calculate the molar mass of ammonia. This is

$$M = M(N) + 3M(H) = (14.0 \text{ g/mol}) + 3(1.01 \text{ g/mol}) = 17.0 \text{ g/mol}$$

The number of moles of nitrogen present is

$$n = m/M_r = (315 \text{ g})/(17.0 \text{ g/mol}) = 18.5 \text{ mol}.$$

The volume of the tank is

$$V = nRT/p = (18.5 \text{ mol})(8.31 \text{ J/mol} \cdot \text{K})(350 \text{ K})/(1.35 \times 10^6 \text{Pa}) = 3.99 \times 10^{-2} \text{m}^3.$$

(b) After the tank is checked the number of moles of gas in the tank is

$$n = pV/(RT) = (8.68 \times 10^5 \text{Pa})(3.99 \times 10^{-2} \text{m}^3)/[(8.31 \text{ J/mol} \cdot \text{K})(295 \text{ K})] = 14.1 \text{ mol}.$$

In that case, 4.4 mol must have escaped; that corresponds to a mass of

$$m = nM_r = (4.4 \text{ mol})(17.0 \text{ g/mol}) = 74.8 \text{ g}.$$

E22-7 (a) From Eq. 22-9,

$$v_{\text{rms}} = \sqrt{\frac{3p}{\rho}}.$$

Be careful with units, however; since atmospheres, grams, and centimeters don't mix well. We'll convert everything to MKSA. Then

$$p = 1.23 \times 10^{-3} \text{ atm} \left(\frac{1.01 \times 10^5 \text{ Pa}}{1 \text{ atm}} \right) = 124 \text{ Pa}$$

and

$$\rho = 1.32 \times 10^{-5} \text{ g/cm}^3 \left(\frac{1 \text{kg}}{1000 \text{ g}}\right) \left(\frac{100 \text{ cm}}{1 \text{ m}}\right)^3 = 1.32 \times 10^{-2} \text{ kg/m}^3.$$

Finally,

$$v_{\text{rms}} = \sqrt{\frac{3(1240 \text{ Pa})}{(1.32 \times 10^{-2} \text{ kg/m}^3)}} = 531 \text{ m/s}.$$

But did the units really work out correctly?

$$\sqrt{\frac{\text{Pa}}{\text{kg/m}^3}} = \sqrt{\frac{\text{N/m}^2}{\text{kg/m}^3}} = \sqrt{\frac{\text{kg m/s}^2}{\text{kg/m}}} = \sqrt{\frac{\text{m}^2}{\text{s}^2}} = \frac{\text{m}}{\text{s}},$$

so yes, they did.

(b) The molar density of the gas is just n/V; but this can be found quickly from the ideal gas law as

$$\frac{n}{V} = \frac{p}{RT} = \frac{(1240 \text{ Pa})}{(8.31 \text{ J/mol} \cdot \text{K})(317 \text{ K})} = 4.71 \times 10^{-1} \text{ mol/m}^3.$$

Note that we used Kelvin in the equation, not Celsius; the conversion was done by adding 273. This is not something that you want to forget to do!

(c) The molar mass is the mass per mole. In part (b) we found the moles per volume. On part (a) we found the distance per time. If you don't see the connection, don't worry, neither do I. But we were given the density, which is mass per volume, so we could find the molar mass from

$$\frac{\rho}{n/V} = \frac{(1.32 \times 10^{-2} \text{ kg/m}^3)}{(4.71 \times 10^{-1} \text{ mol/m}^3)} = 28.0 \text{ g/mol}.$$

But what gas is it? It could contain any atom lighter than silicon; trial and error is the way to go. Some of my guesses include C_2H_4 (ethene), CO (carbon monoxide), and N_2. There's no way to tell which is correct at this point, in fact, the gas could be a mixture of all three.

E22-11 This exercise sounds more confusing than it really is. We have $v = f\lambda$, where λ is the wavelength (which we will set equal to the mean free path), and v is the speed of sound. The mean free path is, from Eq. 22-13,

$$\lambda = \frac{kT}{\sqrt{2}\pi d^2 p}$$

so

$$f = \frac{\sqrt{2}\pi d^2 p v}{kT} = \frac{\sqrt{2}\pi (315 \times 10^{-12} \text{m})^2 (1.02 \times 1.01 \times 10^5 \text{Pa})(343 \text{ m/s})}{(1.38 \times 10^{-23} \text{J/K})(291 \text{ K})} = 3.88 \times 10^9 \text{Hz}.$$

E22-15 (a) Spreadsheets are a wonderful way to calculate averages and root-mean-square values. The average is

$$\frac{4(200\,\text{s}) + 2(500\,\text{m/s}) + 4(600\,\text{m/s})}{4 + 2 + 4} = 420\,\text{m/s}.$$

The mean-square value is

$$\frac{4(200\,\text{s})^2 + 2(500\,\text{m/s})^2 + 4(600\,\text{m/s})^2}{4 + 2 + 4} = 2.1 \times 10^5\,\text{m}^2/\text{s}^2.$$

The root-mean-square value is the square root of this, or 458 m/s.

(b) I'll be lazy. Nine particles are not moving, and the tenth has a speed of 10 m/s. Then the average speed is 1 m/s, and the root-mean-square speed is 3.16 m/s. Look, v_{rms} is larger than v_{av}!

(c) Can $v_{\text{rms}} = v_{\text{av}}$? Assume that the speeds are *not* all the same. Transform to a frame of reference where $v_{\text{av}} = 0$, then some of the individual speeds must be greater than zero, and some will be less than zero. Squaring these speeds will result in positive, non-zero, numbers; the mean square will necessarily be greater than zero, so $v_{\text{rms}} > 0$.

Only if *all* of the particles have the same speed will $v_{\text{rms}} = v_{\text{av}}$.

E22-19 We want to integrate

$$
\begin{aligned}
v_{\text{av}} &= \frac{1}{N} \int_0^\infty N(v) v\, dv, \\
&= \frac{1}{N} \int_0^\infty 4\pi N \left(\frac{m}{2\pi kT}\right)^{3/2} v^2 e^{-mv^2/2kT} v\, dv, \\
&= 4\pi \left(\frac{m}{2\pi kT}\right)^{3/2} \int_0^\infty v^2 e^{-mv^2/2kT} v\, dv.
\end{aligned}
$$

The easiest way to attack this is first with a change of variables— let $x = mv^2/2kT$, then $kT\, dx = mv\, dv$. The limits of integration don't change, since $\sqrt{\infty} = \infty$. Then

$$
\begin{aligned}
v_{\text{av}} &= 4\pi \left(\frac{m}{2\pi kT}\right)^{3/2} \int_0^\infty \frac{2kT}{m} x e^{-x} \frac{kT}{m} dx, \\
&= 2 \left(\frac{2kT}{\pi m}\right)^{1/2} \int_0^\infty x e^{-\alpha x} dx
\end{aligned}
$$

The factor of α that was introduced in the last line is a Feynman trick; we'll set it equal to one when we are finished, so it won't change the result.

Feynman's trick looks like

$$\frac{d}{d\alpha} \int e^{-\alpha x} dx = \int \frac{\partial}{\partial \alpha} e^{-\alpha x} dx = \int (-x) e^{-\alpha x} dx.$$

Applying this to our original problem,

$$
\begin{aligned}
v_{\text{av}} &= 2 \left(\frac{2kT}{\pi m}\right)^{1/2} \int_0^\infty x e^{-\alpha x} dx, \\
&= -\frac{d}{d\alpha} 2 \left(\frac{2kT}{\pi m}\right)^{1/2} \int_0^\infty e^{-\alpha x} dx, \\
&= -2 \left(\frac{2kT}{\pi m}\right)^{1/2} \frac{d}{d\alpha} \left(\frac{-1}{\alpha} e^{-\alpha x} \Big|_0^\infty \right), \\
&= -2 \left(\frac{2kT}{\pi m}\right)^{1/2} \frac{d}{d\alpha} \left(\frac{1}{\alpha}\right), \\
&= -2 \left(\frac{2kT}{\pi m}\right)^{1/2} \frac{-1}{\alpha^2}.
\end{aligned}
$$

We promised, however, that we would set $\alpha = 1$ in the end, so this last line is

$$
\begin{aligned}
v_{\text{av}} &= 2 \left(\frac{2kT}{\pi m}\right)^{1/2}, \\
&= \sqrt{\frac{8kT}{\pi m}}.
\end{aligned}
$$

Now wasn't that fun?

E22-25 According to the equation directly beneath Fig. 22-8,

$$\omega = v\phi/L = (212 \, \text{m/s})(0.0841 \, \text{rad})/(0.204 \, \text{m}) = 87.3 \, \text{rad/s}.$$

E22-29 The fraction of particles that interests us is

$$\frac{2}{\sqrt{\pi}} \frac{1}{(kT)^{3/2}} \int_{0.01kT}^{0.03kT} E^{1/2} e^{-E/kT} \, dE.$$

Because we want this to be easier, we'll change variables according to $E/kT = x$, so that $dE = kT \, dx$. The integral is then

$$\frac{2}{\sqrt{\pi}} \int_{0.01}^{0.03} x^{1/2} e^{-x} \, dx.$$

178

Since the value of x is so small compared to 1 throughout the range of integration, we can expand according to

$$e^{-x} \approx 1 - x \text{ for } x \ll 1.$$

The integral then simplifies to

$$\frac{2}{\sqrt{\pi}} \int_{0.01}^{0.03} x^{1/2}(1-x)\,dx = \frac{2}{\sqrt{\pi}} \left[\frac{2}{3}x^{3/2} - \frac{2}{5}x^{5/2} \right]_{0.01}^{0.03} = 3.09 \times 10^{-3}.$$

E22-31 The volume correction is on page 508; we need first to find d. If we assume that the particles in water are arranged in a cubic lattice (a bad guess, but we'll use it anyway), then 18 grams of water has a volume of $18 \times 10^{-6} \text{m}^3$, and

$$d^3 = \frac{(18 \times 10^{-6}\,\text{m}^3)}{(6.02 \times 10^{23})} = 3.0 \times 10^{-29} \text{m}^3$$

is the volume allocated to each water molecule. In this case $d = 3.1 \times 10^{-10}$m. Then

$$b = \frac{1}{2}(6.02 \times 10^{23})(\frac{4}{3}\pi(3.1 \times 10^{-10}\text{m})^3) = 3.8\,\text{m}^3/\text{mol}.$$

P22-3 The only thing that matters is the total number of moles of gas (2.5) and the number of moles of the second gas (0.5). Since 1/5 of the total number of moles of gas is associated with the second gas, then 1/5 of the total pressure is associated with the second gas.

P22-7 What is important here is the temperature; since the temperatures are the same then the average kinetic energies per particle are the same. Then

$$\frac{1}{2}m_1(v_{\text{rms},1})^2 = \frac{1}{2}m_2(v_{\text{rms},2})^2.$$

We are given in the problem that $v_{\text{av},2} = 2v_{\text{rms},1}$. According to Eqs. 22-18 and 22-20 we have

$$v_{\text{rms}} = \sqrt{\frac{3RT}{M}} = \sqrt{\frac{3\pi}{8}}\sqrt{\frac{8RT}{\pi M}} = \sqrt{\frac{3\pi}{8}}v_{\text{av}}.$$

Combining this with the kinetic energy expression above,

$$\frac{m_1}{m_2} = \left(\frac{v_{\text{rms},2}}{v_{\text{rms},1}} \right)^2 = \left(2\sqrt{\frac{3\pi}{8}} \right)^2 = 4.71.$$

P22-9 (a)   We need to first find the number of particles by integrating

$$N = \int_0^\infty N(v)\,dv,$$

$$= \int_0^{v_0} Cv^2\,dv + \int_{v_0}^\infty (0)\,dv = C\int_0^{v_0} v^2\,dv = \frac{C}{3}v_0^3.$$

Invert, then $C = 3N/v_0^3$.

(b)   The average velocity is found from

$$v_{\text{av}} = \frac{1}{N}\int_0^\infty N(v)v\,dv.$$

Using our result from above,

$$v_{\text{av}} = \frac{1}{N}\int_0^{v_0}\left(\frac{3N}{v_0^3}v^2\right)v\,dv,$$

$$= \frac{3}{v_0^3}\int_0^{v_0} v^3\,dv = \frac{3}{v_0^3}\frac{v_0^4}{4} = \frac{3}{4}v_0.$$

As expected, the average speed is less than the maximum speed. We can make a prediction about the root mean square speed; it will be larger than the average speed (see Exercise 22-15 above) but smaller than the maximum speed.

(c)   The root-mean-square velocity is found from

$$v_{\text{rms}}^2 = \frac{1}{N}\int_0^\infty N(v)v^2\,dv.$$

Using our results from above,

$$v_{\text{rms}}^2 = \frac{1}{N}\int_0^{v_0}\left(\frac{3N}{v_0^3}v^2\right)v^2\,dv,$$

$$= \frac{3}{v_0^3}\int_0^{v_0} v^4\,dv = \frac{3}{v_0^3}\frac{v_0^5}{5} = \frac{3}{5}v_0^2.$$

Then, taking the square root,

$$v_{\text{rms}}^2 = \sqrt{\frac{3}{5}}\,v_0$$

Is $\sqrt{3/5} > 3/4$? It had better be.

P22-15 The mass of air displaced by 2180 m^3 is $m = (1.22 \, \text{kg/m}^3)(2180 \, \text{m}^3) = 2660$ kg. The mass of the balloon and basket is 249 kg and we want to lift 272 kg; this leaves a remainder of 2140 kg for the mass of the air inside the balloon. This corresponds to $(2140 \, \text{kg})/(0.0289 \, \text{kg/mol}) = 7.4 \times 10^4$ mol.

The temperature of the gas inside the balloon is then

$$T = (pV)/(nR) = [(1.01 \times 10^5 \text{Pa})(2180 \, \text{m}^3)]/[(7.4 \times 10^4 \text{mol})(8.31 \, \text{J/mol} \cdot \text{K}) = 358 \, \text{K}.$$

That's 85°C.

Chapter 23

The First Law of Thermodynamics

E23-1 Thankfully, we can ignore the curvature of the Earth. This is justifiable because the thickness of the the crust is much less than the radius of the Earth. We apply Eq. 23-1,

$$H = kA\frac{\Delta T}{\Delta x}$$

The rate at which heat flows out is given as a power per area (mW/m^2), so the quantity given is really H/A. Then the temperature difference is

$$\Delta T = \frac{H}{A}\frac{\Delta x}{k} = (0.054\,W/m^2)\frac{(33,000\,m)}{(2.5\,W/m\cdot K)} = 710\,K$$

The heat flow is out, so that the temperature is higher at the base of the crust. The temperature there is then

$$710 + 10 = 720\,°C.$$

Yes, it is perfectly acceptable to add a temperature difference in Kelvin to a temperature in Celsius.

E23-5 There are three possible arrangements: a sheet of type 1 with a sheet of type 1; a sheet of type 2 with a sheet of type 2; and a sheet of type 1 with a sheet of type 2. We can look back on Sample Problem 23-1 to see how to start the problem; the heat flow will be

$$H_{12} = \frac{A\Delta T}{(L/k_1) + (L/k_2)}$$

for substances of different types; and

$$H_{11} = \frac{A\Delta T/L}{(L/k_1) + (L/k_1)} = \frac{1}{2}\frac{A\Delta Tk_1}{L}$$

for a double layer if substance 1. There is a similar expression for a double layer of substance 2.

For configuration (a) we then have

$$H_{11} + H_{22} = \frac{1}{2}\frac{A\Delta T k_1}{L} + \frac{1}{2}\frac{A\Delta T k_2}{L} = \frac{A\Delta T}{2L}(k_1 + k_2),$$

while for configuration (b) we have

$$H_{12} + H_{21} = 2\frac{A\Delta T}{(L/k_1) + (L/k_2)} = \frac{2A\Delta T}{L}\left((1/k_1) + (1/k_2)\right)^{-1}.$$

We want to compare these, so expanding the relevant part of the second configuration

$$\left((1/k_1) + (1/k_2)\right)^{-1} = \left((k_1 + k_2)/(k_2 k_2)\right)^{-1} = \frac{k_1 k_2}{k_1 + k_2}.$$

Then which is larger

$$(k_1 + k_2)/2 \text{ or } \frac{2k_1 k_2}{k_1 + k_2} \text{ ?}$$

If $k_1 \gg k_2$ then the expression become

$$k_1/2 \text{ and } 2k_2,$$

so the first expression is larger, and therefore configuration (b) has the lower heat flow. Notice that we get the same result if $k_1 \ll k_2$!

E23-9 (a)   This exercise has a distraction: it asks about the heat flow through the window, but what you need to find first is the heat flow through the air near the window. We are given the temperature gradient both inside and outside the window. Inside,

$$\frac{\Delta T}{\Delta x} = \frac{(20°C) - (5°C)}{(0.08\,\text{m})} = 190\,\text{C}°/\text{m};$$

a similar expression exists for outside.

From Eq. 23-1 we find the heat flow *through the air*;

$$H = kA\frac{\Delta T}{\Delta x} = (0.026\,\text{W/m}\cdot\text{K})(0.6\,\text{m})^2(190\,\text{C}°/\text{m}) = 1.8\,\text{W}.$$

It is acceptable to "cancel" the Kelvins with the Celsius in the above expression, because both refer to a temperature difference. Although the area is that of the window, we'll assume that it applies to the air as well.

The value that we arrived at is the rate that heat flows through the air across an area the size of the window on either side of the window. This heat flow had to occur through the window as well, so

$$H = 1.8\,\text{W}$$

answers the window question.

(b) Now that we know the rate that heat flows through the window, we are in a position to find the temperature difference across the window. Rearranging Eq. 32-1,

$$\Delta T = \frac{H\Delta x}{kA} = \frac{(1.8\,\text{W})(0.005\,\text{m})}{(1.0\,\text{W/m}\cdot\text{K})(0.6\,\text{m})^2} = 0.025\,\text{C}^\circ,$$

so we were well justified in our approximation that the temperature drop across the glass is very small.

E23-13 We don't need to know the outside temperature because the amount of heat energy required is explicitly stated: 5.22 GJ. We just need to know how much water is required to transfer this amount of heat energy. Use Eq. 23-11, and then

$$m = \frac{Q}{c\Delta T} = \frac{(5.22 \times 10^9\,\text{J})}{(4190\,\text{J/kg}\cdot\text{K})(50.0^\circ\text{C} - 22.0^\circ\text{C})} = 4.45 \times 10^4\,\text{kg}.$$

This is the mass of the water, we want to know the volume, so we'll use the density, and then

$$V = \frac{m}{\rho} = \frac{(4.45 \times 10^4\,\text{kg})}{(998\,\text{kg/m}^3)} = 44.5\,\text{m}^3.$$

E23-17 There are three "things" in this problem: the copper bowl (b), the water (w), and the copper cylinder (c). The total internal energy changes must add up to zero, so

$$\Delta E_{\text{int},b} + \Delta E_{\text{int},w} + \Delta E_{\text{int},c} = 0.$$

As in Sample Problem 23-3, no work is done on any object, so

$$Q_b + Q_w + Q_c = 0.$$

The heat transfers for these three objects are

$$
\begin{aligned}
Q_b &= m_b c_b (T_{f,b} - T_{i,b}),\\
Q_w &= m_w c_w (T_{f,w} - T_{i,w}) + L_v m_2,\\
Q_c &= m_c c_c (T_{f,c} - T_{i,c}).
\end{aligned}
$$

For the most part, this looks exactly like the presentation in Sample Problem 23-3; but there is an extra term in the second line. This term reflects the extra heat required to vaporize $m_2 = 4.70$ g of water at 100°C into steam 100°C.

Some of the initial temperatures are specified in the exercise: $T_{i,b} = T_{i,w} = 21.0^\circ\text{C}$ and $T_{f,b} = T_{f,w} = T_{f,c} = 100^\circ\text{C}$.

(a) The heat transferred to the water, then, is

$$Q_{\mathrm{w}} = (0.223\,\mathrm{kg})(4190\,\mathrm{J/kg\cdot K})\,((100^\circ\mathrm{C}) - (21.0^\circ\mathrm{C})),$$
$$+(2.26\times10^6\,\mathrm{J/kg})(4.70\times10^{-3}\mathrm{kg}),$$
$$= 8.44\times10^4\,\mathrm{J}.$$

This answer differs from the back of the book. I think that they (or was it me) used the latent heat of fusion when they should have used the latent heat of vaporization!

(b) The heat transfered to the bowl, then, is

$$Q_{\mathrm{w}} = (0.146\,\mathrm{kg})(387\,\mathrm{J/kg\cdot K})\,((100^\circ\mathrm{C}) - (21.0^\circ\mathrm{C})) = 4.46\times10^3\,\mathrm{J}.$$

(c) The heat transfered from the cylinder was transfered into the water and bowl, so

$$Q_{\mathrm{c}} = -Q_{\mathrm{b}} - Q_{\mathrm{w}} = -(4.46\times10^3\,\mathrm{J}) - (8.44\times10^4\,\mathrm{J}) = -8.89\times10^4\,\mathrm{J}.$$

The initial temperature of the cylinder is then given by

$$T_{\mathrm{i,c}} = T_{\mathrm{f,c}} - \frac{Q_{\mathrm{c}}}{m_{\mathrm{c}}c_{\mathrm{c}}} = (100^\circ\mathrm{C}) - \frac{(-8.89\times10^4\,\mathrm{J})}{(0.314\,\mathrm{kg})(387\,\mathrm{J\cdot K})} = 832^\circ\mathrm{C}.$$

E23-21 The linear dimensions of the ring and sphere change with the temperature change according to

$$\Delta d_{\mathrm{r}} = \alpha_{\mathrm{r}} d_{\mathrm{r}}(T_{\mathrm{f,r}} - T_{\mathrm{i,r}}),$$
$$\Delta d_{\mathrm{s}} = \alpha_{\mathrm{s}} d_{\mathrm{s}}(T_{\mathrm{f,s}} - T_{\mathrm{i,s}}).$$

When the ring and sphere are at the same (final) temperature the ring and the sphere have the same diameter. This means that

$$d_{\mathrm{r}} + \Delta d_{\mathrm{r}} = d_{\mathrm{s}} + \Delta d_{\mathrm{s}}$$

when $T_{\mathrm{f,s}} = T_{\mathrm{f,r}}$. We'll solve these expansion equations first, and then go back to the heat equations.

$$d_{\mathrm{r}} + \Delta d_{\mathrm{r}} = d_{\mathrm{s}} + \Delta d_{\mathrm{s}},$$
$$d_{\mathrm{r}}\left(1 + \alpha_{\mathrm{r}}(T_{\mathrm{f,r}} - T_{\mathrm{i,r}})\right) = d_{\mathrm{s}}\left(1 + \alpha_{\mathrm{s}}(T_{\mathrm{f,s}} - T_{\mathrm{i,s}})\right),$$

which can be rearranged to give

$$\alpha_{\mathrm{r}} d_{\mathrm{r}} T_{\mathrm{f,r}} - \alpha_{\mathrm{s}} d_{\mathrm{s}} T_{\mathrm{f,s}} = d_{\mathrm{s}}\left(1 - \alpha_{\mathrm{s}} T_{\mathrm{i,s}}\right) - d_{\mathrm{r}}\left(1 - \alpha_{\mathrm{r}} T_{\mathrm{i,r}}\right),$$

but since the final temperatures are the same,

$$T_{\mathrm{f}} = \frac{d_{\mathrm{s}}\left(1 - \alpha_{\mathrm{s}} T_{\mathrm{i,s}}\right) - d_{\mathrm{r}}\left(1 - \alpha_{\mathrm{r}} T_{\mathrm{i,r}}\right)}{\alpha_{\mathrm{r}} d_{\mathrm{r}} - \alpha_{\mathrm{s}} d_{\mathrm{s}}}$$

Putting in the numbers,

$$T_{\mathrm{f}} = $$
$$\frac{(2.54533\mathrm{cm})[1 - (23 \times 10^{-6}/\mathrm{C}°)(100°\mathrm{C})] - (2.54000\mathrm{cm})[1 - (17 \times 10^{-6}/\mathrm{C}°)(0°\mathrm{C})]}{(2.54000\mathrm{cm})(17 \times 10^{-6}/\mathrm{C}°) - (2.54533\mathrm{cm})(23 \times 10^{-6}/\mathrm{C}°)},$$
$$= 34.1°\mathrm{C}.$$

Now for the heat part of the problem.

No work is done, so we only have the issue of heat flow, then

$$Q_{\mathrm{r}} + Q_{\mathrm{s}} = 0.$$

Where "r" refers to the copper ring and "s" refers to the aluminum sphere. The heat equations are

$$
\begin{aligned}
Q_{\mathrm{r}} &= m_{\mathrm{r}} c_{\mathrm{r}} (T_{\mathrm{f}} - T_{\mathrm{i,r}}), \\
Q_{\mathrm{s}} &= m_{\mathrm{s}} c_{\mathrm{s}} (T_{\mathrm{f}} - T_{\mathrm{i,s}}).
\end{aligned}
$$

Equating and rearranging,

$$m_{\mathrm{s}} = \frac{m_{\mathrm{r}} c_{\mathrm{r}} (T_{\mathrm{i,r}} - T_{\mathrm{f}})}{c_{\mathrm{s}} (T_{\mathrm{f}} - T_{\mathrm{i,s}})}$$

or

$$m_{\mathrm{s}} = \frac{(21.6\ \mathrm{g})(387\ \mathrm{J/kg{\cdot}K})(0°\mathrm{C} - 34.1°\mathrm{C})}{(900\ \mathrm{J/kg{\cdot}K})(34.1°\mathrm{C} - 100°\mathrm{C})} = 4.81\ \mathrm{g}.$$

I wish there was an easier way.

E23-25 Net work done on the gas is given by Eq. 23-15,

$$W = - \int p\, dV.$$

But integrals are just the area under the curve; and that's the easy way to solve this problem. In the case of closed paths, it becomes the area inside the curve, with a clockwise sense giving a positive value for the integral.

The magnitude of the area is the same for either path, since it is a rectangle divided in half by a square. The area of the rectangle is

$$(15 \times 10^{3}\mathrm{Pa})(6\,\mathrm{m}^{3}) = 90 \times 10^{3}\mathrm{J},$$

so the area of path 1 (counterclockwise) is -45 kJ; this means the work done on the gas is -(-45 kJ) or 45 kJ. The work done on the gas for path 2 is the negative of this because the path is clockwise.

E23-29 (a) According to Eq. 23-20,

$$p_f = \frac{p_i V_i^\gamma}{V_f^\gamma} = \frac{(1.17 \text{ atm})(4.33 \text{ L})^{(1.40)}}{(1.06 \text{ L})^{(1.40)}} = 8.39 \text{ atm.}$$

(b) The final temperature can be found from the ideal gas law,

$$T_f = T_i \frac{p_f V_f}{p_i V_i} = (310 \text{ K}) \frac{(8.39 \text{ atm})(1.06 \text{ L})}{(1.17 \text{ atm})(4.33 \text{ L})} = 544 \text{ K.}$$

(c) The work done (for an adiabatic process) is given by Eq. 23-22,

$$\begin{aligned} W &= \frac{1}{(1.40) - 1} \Big[(8.39 \times 1.01 \times 10^5 \text{Pa})(1.06 \times 10^{-3} \text{m}^3) \\ &\quad -(1.17 \times 1.01 \times 10^5 \text{Pa})(4.33 \times 10^{-3} \text{m}^3) \Big], \\ &= 966 \text{ J.} \end{aligned}$$

E23-33 (a) Invert Eq. 32-20,

$$\gamma = \frac{\ln(p_1/p_2)}{\ln(V_2/V_1)} = \frac{\ln(122 \text{ kPa}/1450 \text{ kPa})}{\ln(1.36 \text{ m}^3/10.7 \text{ m}^3)} = 1.20.$$

(b) The final temperature is found from the ideal gas law,

$$T_f = T_i \frac{p_f V_f}{p_i V_i} = (250 \text{ K}) \frac{(1450 \times 10^3 \text{Pa})(1.36 \text{ m}^3)}{(122 \times 10^3 \text{Pa})(10.7 \text{ m}^3)} = 378 \text{ K,}$$

which is the same as 105°C.

(c) Ideal gas law, again:

$$n = [pV]/[RT] = [(1450 \times 10^3 \text{Pa})(1.36 \text{ m}^3)]/[(8.31 \text{ J/mol} \cdot \text{K})(378 \text{ K})] = 628 \text{ mol.}$$

(d) From Eq. 23-24,

$$E_{\text{int}} = \frac{3}{2} nRT = \frac{3}{2}(628 \text{ mol})(8.31 \text{ J/mol} \cdot \text{K})(250 \text{ K}) = 1.96 \times 10^6 \text{J}$$

before the compression and

$$E_{\text{int}} = \frac{3}{2} nRT = \frac{3}{2}(628 \text{ mol})(8.31 \text{ J/mol} \cdot \text{K})(378 \text{ K}) = 2.96 \times 10^6 \text{J}$$

after the compression.

(e) The ratio of the rms speeds will be proportional to the square root of the ratio of the internal energies,

$$\sqrt{(1.96 \times 10^6 \,\mathrm{J})/(2.96 \times 10^6 \,\mathrm{J})} = 0.813;$$

we can do this because the number of particles is the same before and after, hence the ratio of the energies per particle is the same as the ratio of the total energies.

E23-37 (a) From Eq. 23-37,

$$Q = nc_{\mathrm{p}}\Delta T = (4.34 \,\mathrm{mol})(29.1 \,\mathrm{J/mol \cdot K})(62.4 \,\mathrm{K}) = 7880 \,\mathrm{J}.$$

(b) From Eq. 23-28,

$$E_{\mathrm{int}} = \frac{5}{2}nR\Delta T = \frac{5}{2}(4.34 \,\mathrm{mol})(8.31 \,\mathrm{J/mol \cdot K})(62.4 \,\mathrm{K}) = 5630 \,\mathrm{J}.$$

(c) From Eq. 23-23,

$$K_{\mathrm{trans}} = \frac{3}{2}nR\Delta T = \frac{5}{2}(4.34 \,\mathrm{mol})(8.31 \,\mathrm{J/mol \cdot K})(62.4 \,\mathrm{K}) = 3380 \,\mathrm{J}.$$

E23-41 The first law of thermodynamics is

$$Q + W = \Delta E_{\mathrm{int}}.$$

According to Eq. 23-25 (which is specific to ideal gases),

$$\Delta E_{\mathrm{int}} = \frac{3}{2}nR\Delta T,$$

and for an isothermal process $\Delta T = 0$, so for an ideal gas $\Delta E_{\mathrm{int}} = 0$. Consequently,

$$Q + W = 0$$

for an ideal gas which undergoes an isothermal process.
   But we know W for an isotherm, Eq. 23-18 shows

$$W = -nRT \ln \frac{V_{\mathrm{f}}}{V_{\mathrm{i}}}$$

Then finally

$$Q = -W = nRT \ln \frac{V_{\mathrm{f}}}{V_{\mathrm{i}}}$$

E23-45 If the pressure and volume are both doubled along a straight line then the process can be described by

$$p = \frac{p_1}{V_1} V$$

The final point involves the doubling of both the pressure and the volume, so according to the ideal gas law, $pV = nRT$, the final temperature T_2 will be *four* times the initial temperature T_1.

Now for the exercises.

(a) The work done on the gas is

$$W = -\int_1^2 p\, dV = -\int_1^2 \frac{p_1}{V_1} V\, dV = -\frac{p_1}{V_1} \left(\frac{V_2^2}{2} - \frac{V_1^2}{2} \right)$$

We want to express our answer in terms of T_1. First we take advantage of the fact that $V_2 = 2V_1$, then

$$W = -\frac{p_1}{V_1} \left(\frac{4V_1^2}{2} - \frac{V_1^2}{2} \right) = -\frac{3}{2} p_1 V_1 = -\frac{3}{2} nRT_1$$

(b) The nice thing about ΔE_{int} is that it is path independent, we care only of the initial and final points. From Eq. 23-25,

$$\Delta E_{\text{int}} = \frac{3}{2} nR\Delta T = \frac{3}{2} nR (T_2 - T_1) = \frac{9}{2} nRT_1$$

(c) Finally we are in a position to find Q by applying the first law,

$$Q = \Delta E_{\text{int}} - W = \frac{9}{2} nRT_1 + \frac{3}{2} nRT_1 = 6nRT_1.$$

(d) If we define specific heat as heat divided by temperature change, then

$$c = \frac{Q}{n\Delta T} = \frac{6RT_1}{4T_1 - T_1} = 2R.$$

P23-3 Follow the example in Sample Problem 23-2. We start with Eq. 23-1:

$$H = kA\frac{dT}{dr},$$

$$H = k(4\pi r^2)\frac{dT}{dr},$$

$$\int_{r_1}^{r_2} H \frac{dr}{4\pi r^2} = \int_{T_1}^{T_2} k\, dT,$$

$$\frac{H}{4pi}\left(\frac{1}{r_1} - \frac{1}{r_2}\right) = k(T_1 - T_2),$$

$$H\left(\frac{r_2 - r_1}{r_1 r_2}\right) = 4\pi k(T_1 - T_2),$$

$$H = \frac{4\pi k(T_1 - T_2)r_1 r_2}{r_2 - r_1}.$$

P23-7 The key point here is that how you solve the problem depends on what the answer is! If there is any ice left then the final temperature will be zero; if the final temperature is *not* zero then there will be no ice left. We will assume that all of the ice melts, and find out what happens.

(a) Start with the heat equation:

$$Q_t + Q_i + Q_w = 0,$$

where Q_t is the heat from the tea, Q_i is the heat from the ice when it melts, and Q_w is the heat from the water (which used to be ice). Then

$$m_t c_t(T_f - T_{t,i}) + m_i L_f + m_w c_w(T_f - T_{w,i}) = 0,$$

which, since we have assumed all of the ice melts and the masses are all equal, can be solved for T_f as

$$T_f = \frac{c_t T_{t,i} + c_w T_{w,i} - L_f}{c_t + c_w},$$
$$= \frac{(4190 \text{J/kg} \cdot \text{K})(90°\text{C}) + (4190 \text{J/kg} \cdot \text{K})(0°\text{C}) - (333 \times 10^3 \text{J/kg})}{(4190 \text{J/kg} \cdot \text{K}) + (4190 \text{J/kg} \cdot \text{K})},$$
$$= 5.3°\text{C}.$$

Look, we guessed right, all of the ice did melt! But how do we know? Stayed tuned for part (b), and I'll show you.

(b) Once again, assume all of the ice melted. Then we can do the same steps, and we get

$$T_f = \frac{c_t T_{t,i} + c_w T_{w,i} - L_f}{c_t + c_w},$$
$$= \frac{(4190 \text{J/kg} \cdot \text{K})(70°\text{C}) + (4190 \text{J/kg} \cdot \text{K})(0°\text{C}) - (333 \times 10^3 \text{J/kg})}{(4190 \text{J/kg} \cdot \text{K}) + (4190 \text{J/kg} \cdot \text{K})},$$
$$= -4.7°\text{C}.$$

What? A negative temperature? Such a temperature is possible, but it is not likely that putting ice at 0°C into hot water will result in a final temperature which is below the initial temperature of the ice. So we must have guessed wrong when we assumed that all of the ice melted. The heat equation then simplifies to

$$m_t c_t (T_f - T_{t,i}) + m_i L_f = 0,$$

and then

$$
\begin{aligned}
m_i &= \frac{m_t c_t (T_{t,i} - T_f)}{L_f}, \\
&= \frac{(0.520\,\text{kg})(4190\,\text{J/kg}\cdot\text{K})(90°\text{C} - 0°)}{(333\times 10^3\,\text{J/kg})}, \\
&= 0.458\,\text{kg}.
\end{aligned}
$$

If we had in part (a) assumed that all of the ice had *not* melted, then we would have found that the mass would have been larger than the mass of ice available.

P23-11 We can use Eq. 23-10, but we will need to approximate c first. If we assume that the line is straight then we use $c = mT + b$. I approximate m from

$$m = \frac{(14\,\text{J/mol}\cdot\text{K}) - (3\,\text{J/mol}\cdot\text{K})}{(500\,\text{K}) - (200\,\text{K})} = 3.67\times 10^{-2}\,\text{J/mol}.$$

Then I find b from those same data points,

$$b = (3\,\text{J/mol}\cdot\text{K}) - (3.67\times 10^{-2}\,\text{J/mol})(200\,\text{K}) = -4.34\,\text{J/mol}\cdot\text{K}.$$

Then from Eq. 23-10,

$$
\begin{aligned}
Q &= n\int_{T_i}^{T_f} c\,dT, \\
&= n\int_{T_i}^{T_f} (mT + b)\,dT, \\
&= n\left[\frac{m}{2}T^2 + bT\right]_{T_i}^{T_f}, \\
&= n\left(\frac{m}{2}(T_f{}^2 - T_i{}^2) + b(T_f - T_i)\right), \\
&= (0.45\text{mol})\left(\frac{(3.67\times 10^{-2}\,\text{J/mol})}{2}((500\,\text{K})^2 - (200\,\text{K})^2)\right. \\
&\qquad \left. + (-4.34\,\text{J/mol}\cdot\text{K})(500\,\text{K} - 200\,\text{K})\right), \\
&= 1.15\times 10^3\,\text{J}.
\end{aligned}
$$

P23-15 When the tube is horizontal there are two regions filled with gas, one at $p_{1,i}$, $V_{1,i}$; the other at $p_{2,i}$, $V_{2,i}$. Originally $p_{1,i} = p_{2,i} = 1.01 \times 10^5 \text{Pa}$ and $V_{1,i} = V_{2,i} = (0.45\,\text{m})A$, where A is the cross sectional area of the tube.

When the tube is held so that region 1 is on top then the mercury has three force on it: the force of gravity, mg; the force from the gas above pushing down $p_{2,f}A$; and the force from the gas below pushing up $p_{1,f}A$. The balanced force expression is

$$p_{1,f}A = p_{2,f}A + mg.$$

If we write $m = \rho l_{\text{m}}A$ where $l_{\text{m}} = 0.10\,\text{m}$, then

$$p_{1,f} = p_{2,f} + \rho g l_{\text{m}}.$$

Finally, since the tube has uniform cross section, we can write $V = Al$ everywhere.

(a) For an isothermal process $p_i l_i = p_f l_f$, where we have used $V = Al$, and then

$$p_{1,i}\frac{l_{1,i}}{l_{1,f}} - p_{2,i}\frac{l_{2,i}}{l_{2,f}} = \rho g l_{\text{m}}.$$

But we can factor out $p_{1,i} = p_{2,i}$ and $l_{1,i} = l_{2,i}$, and we can apply $l_{1,f} + l_{2,f} = 0.90\,\text{m}$. Then

$$\frac{1}{l_{1,f}} - \frac{1}{0.90\,\text{m} - l_{1,f}} = \frac{\rho g l_{\text{m}}}{p_i l_i}.$$

Put in some numbers and rearrange,

$$0.90\,\text{m} - 2l_{1,f} = (0.294\,\text{m}^{-1})l_{1,f}(0.90\,\text{m} - l_{1,f}),$$

which can be written as an ordinary quadratic,

$$(0.294\,\text{m}^{-1})l_{1,f}{}^2 - (2.265)l_{1,f} + (0.90\,\text{m}) = 0$$

The solutions are $l_{1,f} = 7.284\,\text{m}$ and $0.421\,\text{m}$. Only one of these solutions is reasonable, so the mercury shifted down $0.450 - 0.421 = 0.029\,\text{m}$.

(b) The math is a wee bit uglier here, but we can start with $p_i l_i{}^\gamma = p_f l_f{}^\gamma$, and this means that everywhere we had a $l_{1,f}$ in the previous derivation we need to replace it with $l_{1,f}{}^\gamma$. Then we have

$$\frac{1}{l_{1,f}{}^\gamma} - \frac{1}{(0.90\,\text{m} - l_{1,f})^\gamma} = \frac{\rho g l_{\text{m}}}{p_i l_i{}^\gamma}.$$

This can be written as

$$(0.90\,\text{m} - l_{1,f})^\gamma - l_{1,f}{}^\gamma - = (0.404\,\text{m}^{-\gamma})l_{1,f}{}^\gamma(0.90\,\text{m} - l(0._{1,f})^\gamma,$$

which looks nasty to me! I'll use Maple to get the answer, and find $l_{1,f} = 0.429$, so the mercury shifted down $0.450 - 0.429 = 0.021\,\text{m}$.

Which is more likely? Turn the tube fast, and the adiabatic approximation works. Eventually the system will return to room temperature, and then the isothermal approximation is valid.

Chapter 24

Entropy and The Second Law of Thermodynamics

E24-1   For isothermal processes the entropy expression is almost trivial,

$$\Delta S = \frac{Q}{T},$$

where if Q is positive (heat flow into system) the entropy increases.

Then

$$Q = T\Delta S = (405 \, \text{K})(46.2 \, \text{J/K}) = 1.87 \times 10^4 \text{J}$$

We *must* use Kelvins (or some other absolute temperature scale) in this expression!

E24-5 (a)   We want to find the heat absorbed, so

$$Q = mc\Delta T = (1.22 \, \text{kg})(387 \, \text{J/mol} \cdot \text{K}) \left((105^\circ\text{C}) - (25.0^\circ\text{C}) \right) = 3.77 \times 10^4 \text{J}.$$

(b)   We want to find the entropy change, so, according to Eq. 24-1,

$$\Delta S = \int_{T_\text{i}}^{T_\text{f}} \frac{dQ}{T}.$$

We won't worry about the mathematical imprecision of writing dQ. When you take additional physics courses you will learn that writing dQ is a "bad thing", because dQ is not a true differential. If we substitute $dQ = mc\,dT$, then suddenly (mathemagically?) everything is okay, so we write

$$
\begin{aligned}
\Delta S &= \int_{T_\text{i}}^{T_\text{f}} \frac{mc\,dT}{T}, \\
&= mc\ln\frac{T_\text{f}}{T_\text{i}}.
\end{aligned}
$$

This expression is quite important; it occurs as part of Sample Problem 24-1 on Page 547, but isn't referenced with an equation number. We'll call it "that really important equation that isn't referenced with an equation number" when we need to use it.

The entropy change of the copper block is then

$$\Delta S = mc\ln\frac{T_f}{T_i} = (1.22\,\text{kg})(387\,\text{J/mol}\cdot\text{K})\ln\frac{(378\,\text{K})}{(298\,\text{K})} = 112\,\text{J/K}.$$

It was crucial that we remember to use Kelvins in that expression!

Incidentally, we could combine the heat transfer expression, $mc = Q/\Delta T$, with the entropy change expression and write

$$\Delta S = \frac{Q}{\Delta T}\ln\frac{T_f}{T_i}.$$

E24-9 (a) If the rod is in a steady state we wouldn't expect the entropy of the rod to change. Heat energy is flowing out of the hot reservoir into the rod, but this process happens at a fixed temperature, so the entropy change in the hot reservoir is

$$\Delta S_H = \frac{Q_H}{T_H} = \frac{(-1200\,\text{J})}{(403\,\text{K})} = -2.98\,\text{J/K}.$$

The negative sign is because heat is flowing out of the hot reservoir.

The heat energy flows into the cold reservoir, so

$$\Delta S_C = \frac{Q_H}{T_H} = \frac{(1200\,\text{J})}{(297\,\text{K})} = 4.04\,\text{J/K}.$$

The total change in entropy of the system is the sum of these two terms

$$\Delta S = \Delta S_H + \Delta S_C = 1.06\,\text{J/K}.$$

(b) Since the rod is in a steady state, nothing is changing, not even the entropy.

E24-11 The total mass of ice and water is 2.04 kg. If eventually the ice and water have the same mass, then the final state will have 1.02 kg of each. This means that $1.78\,\text{kg} - 1.02\,\text{kg} = 0.76\,\text{kg}$ of water changed into ice.

(a) The change of water at 0°C to ice at 0°C is isothermal, so the entropy change is

$$\Delta S = \frac{Q}{T} = \frac{-mL}{T} = \frac{(0.76\,\text{kg})(333\times10^3\,\text{J/kg})}{(273\,\text{K})} = -927\,\text{J/K}.$$

(b) The entropy change is now $+927\,\text{J/K}$.

E24-15 One hour's worth of coal, when burned, will provide energy equal to

$$(382 \times 10^3 \text{kg})(28.0 \times 10^6 \text{J/kg}) = 1.07 \times 10^{13} \text{J}.$$

In this hour, however, the plant only generates

$$(755 \times 10^6 \text{ W})(3600\,s) = 2.72 \times 10^{12} \text{J}.$$

The efficiency is then

$$e = (2.72 \times 10^{12} \text{J})/(1.07 \times 10^{13} \text{J}) = 25.4\%.$$

E24-19 The BC and DA processes are both adiabatic; so if we could find an expression for work done during an adiabatic process we might be almost done. But what is an adiabatic process? It is a process for which $Q = 0$, so according to the first law

$$\Delta E_{\text{int}} = W.$$

But for an ideal gas

$$\Delta E_{\text{int}} = nC_V \Delta T,$$

as was pointed out in Table 23-5. So we have

$$|W| = nC_V |\Delta T|$$

and since the adiabatic paths BC and DA operate between the same two isotherms, we can conclude that the magnitude of the work is the same for both paths.

E24-23 Note that since the coefficient of performance is given, we don't care about the temperatures, except to ask that age-old question: "is this refrigerator physically possible?"

Since, according to Eq. 24-15,

$$K = \frac{T_L}{T_H - T_L} = \frac{(261\,\text{K})}{(299\,\text{K}) - (261\,\text{K})} = 6.87$$

Note that we used Kelvin temperature, not Celsius. I can't stress this enough! But since the coefficient of performance of this fridge is less than the theoretical maximum, this fridge is physically possible.

Now we solve the question out of order.

(b) Since $K = |Q_L|/|W|$, it makes more sense to do part (b) first. The work required to run the freezer is

$$|W| = |Q_L|/K = (185 \text{ kJ})/(5.70) = 32.5 \text{ kJ}.$$

(a)   The freezer will discharge heat into the room equal to

$$|Q_L| + |W| = (185 \text{ kJ}) + (32.5 \text{ kJ}) = 218 \text{ kJ}.$$

E24-33   The Carnot engine has an efficiency

$$\epsilon = 1 - \frac{T_2}{T_1} = \frac{|W|}{|Q_1|}.$$

The Carnot refrigerator has a coefficient of performance

$$K = \frac{T_4}{T_3 - T_4} = \frac{|Q_4|}{|W|}.$$

Lastly, $|Q_4| = |Q_3| - |W|$. We just need to combine these three expressions into one. Starting with the first, and solving for $|W|$,

$$|W| = |Q_1| \frac{T_1 - T_2}{T_1}.$$

Then we combine the last two expressions, and

$$\frac{T_4}{T_3 - T_4} = \frac{|Q_3| - |W|}{|W|} = \frac{|Q_3|}{|W|} - 1.$$

Finally, combine them all,

$$\frac{T_4}{T_3 - T_4} = \frac{|Q_3|}{|Q_1|} \frac{T_1}{T_1 - T_2} - 1.$$

Now, we rearrange,

$$
\begin{aligned}
\frac{|Q_3|}{|Q_1|} &= \left(\frac{T_4}{T_3 - T_4} + 1 \right) \frac{T_1 - T_2}{T_1}, \\
&= \left(\frac{T_3}{T_3 - T_4} \right) \frac{T_1 - T_2}{T_1}, \\
&= (1 - T_2/T_1)/(1 - T_4/T_3).
\end{aligned}
$$

E24-35 (a)   For this problem we don't care how the particles are arranged inside a section, we only care how they are divided up between the two sides.

Consequently, there is only one way to arrange the particles: you put them all on one side, and you have no other choices. So the multiplicity in this case is one, or $w_1 = 1$.

(b) Once the particles are allowed to mix we have more work in computing the multiplicity. Using Eq. 24-19, we have

$$w_2 = \frac{N!}{(N/2)!(N/2)!} = \frac{N!}{((N/2)!)^2}$$

Don't try to factor out the 1/2, you can't do this with factorials!

(c) The entropy of a state of multiplicity w is given by Eq. 24-20,

$$S = k \ln w$$

For part (a), with a multiplicity of 1, $S_1 = 0$. That was easy. Now for part (b),

$$S_2 = k \ln \left(\frac{N!}{((N/2)!)^2} \right) = k \ln N! - 2k \ln(N/2)!$$

and we need to expand each of those terms with Stirling's approximation,

$$\ln N! \approx N \ln N - N.$$

Combining,

$$
\begin{aligned}
S_2 &= = k \left(N \ln N - N \right) - 2k \left((N/2) \ln(N/2) - (N/2) \right), \\
&= kN lnN - kN - kN \ln N + kN \ln 2 + kN, \\
&= kN \ln 2
\end{aligned}
$$

Finally, $\Delta S = S_2 - S_1 = kN \ln 2$.

(d) The answer should be the same; it is a free expansion problem in both cases!

P24-1 We want to evaluate

$$
\begin{aligned}
\Delta S &= \int_{T_i}^{T_f} \frac{n C_v \, dT}{T}, \\
&= \int_{T_i}^{T_f} \frac{n A T^3 \, dT}{T}, \\
&= \int_{T_i}^{T_f} n A T^2 \, dT, \\
&= \frac{n A}{3} \left(T_f{}^3 - T_i{}^3 \right).
\end{aligned}
$$

Into this last expression, which is true for many substances at sufficiently low temperatures, we substitute the given numbers.

$$\Delta S = \frac{(4.8 \text{ mol})(3.15 \times 10^{-5} \text{J/mol} \cdot \text{K}^4)}{3} \left((10 \text{ K})^3 - (5.0 \text{ K})^3 \right) = 4.41 \times 10^{-2} \text{J/K}.$$

P24-7 (a) This is a problem where the total internal energy of the two objects doesn't change, but since no work is done during the process, we can start with the simpler expression

$$Q_1 + Q_2 = 0$$

The heat transfers by the two objects are

$$Q_1 = m_1 c_1 (T_1 - T_{1,i}),$$
$$Q_2 = m_2 c_2 (T_2 - T_{2,i}).$$

Note that we don't call the final temperature T_f here, because we *are not* assuming that the two objects are at equilibrium.

We combine these three equations,

$$m_2 c_2 (T_2 - T_{2,i}) = -m_1 c_1 (T_1 - T_{1,i}),$$
$$m_2 c_2 T_2 = m_2 c_2 T_{2,i} + m_1 c_1 (T_{1,i} - T_1),$$
$$T_2 = T_{2,i} + \frac{m_1 c_1}{m_2 c_2} (T_{1,i} - T_1)$$

As object 1 "cools down", object 2 "heats up", as expected.

(b) The entropy change of *one* object is given by

$$\Delta S = \int_{T_i}^{T_f} \frac{mc\, dT}{T} = mc \ln \frac{T_f}{T_i},$$

and the total entropy change for the system will be the sum of the changes for each object, so

$$\Delta S = m_1 c_1 \ln \frac{T_1}{T_{i,1}} + m_2 c_2 \ln \frac{T_2}{T_{i,2}}.$$

Into the this last equation we need to substitute the expression for T_2 in as a function of T_1. There's no new physics in doing this, just a mess of algebra.

(c) We want to evaluate $d(\Delta S)/dT_1$. To save on algebra we will work with the last expression, remembering that T_2 is a function, not a variable. Then

$$\frac{d(\Delta S)}{dT_1} = \frac{m_1 c_1}{T_1} + \frac{m_2 c_2}{T_2} \frac{dT_2}{dT_1}.$$

We've saved on algebra, but now we need to evaluate dT_2/dT_1. Starting with the results from part (a),

$$\frac{dT_2}{dT_1} = \frac{d}{dT_1} \left(T_{2,i} + \frac{m_1 c_1}{m_2 c_2} (T_{1,i} - T_1) \right),$$
$$= -\frac{m_1 c_1}{m_2 c_2}.$$

198

Now we collect the two results and write

$$\frac{d(\Delta S)}{dT_1} = \frac{m_1 c_1}{T_1} + \frac{m_2 c_2}{T_2}\left(-\frac{m_1 c_1}{m_2 c_2}\right),$$

$$= m_1 c_1 \left(\frac{1}{T_1} - \frac{1}{T_2}\right).$$

We could consider writing T_2 out in all of its glory, but what would it gain us? Nothing. There is actually considerably more physics in the expression as written, because...

(d) ...we get a maximum for ΔS when $d(\Delta S)/dT_1 = 0$, and this can only occur when $T_1 = T_2$ according to the expression.

What is the significance? Equilibrium occurs when two objects are at the same temperature is a consequence of entropy principles: when two objects in thermal contact reach the same temperature they have maximized the change in entropy!

Chapter 25

Electric Charge and Coulomb's Law

E25-1 Later we'll learn the "official" formula, but for now we consider the units of current: Coulombs per second. Then the charge transferred is

$$Q = (2.5 \times 10^4 \, \text{C/s})(20 \times 10^{-6} \, \text{s}) = 5.0 \times 10^{-1} \, \text{C}.$$

E25-5 (a) The magnitude of the electric force is found from a quick application of Coulomb's Law, Eq. 25-4,

$$F = \frac{1}{4\pi\epsilon_0} \frac{q_1 q_2}{r_{12}^2} = \frac{1}{4\pi(8.85 \times 10^{-12} \, \text{C}^2/\text{N} \cdot \text{m}^2)} \frac{(21.3 \, \mu\text{C})(21.3 \, \mu\text{C})}{(1.52 \, \text{m})^2} = 1.77 \, \text{N}$$

We weren't asked, but it is worth pointing out that this force on q_1 is directed up.

(b) In part (a) we found F_{12}; to solve part (b) we need to first find F_{13}. Let's see if we can do this without any work. Since $q_3 = q_2$ and $r_{13} = r_{12}$, we can immediately conclude that $F_{13} = F_{12}$.

We must assess the direction of the force of q_3 on q_1; it will be directed along the line which connects the two charges, and will be directed away from q_3. The diagram below shows the directions.

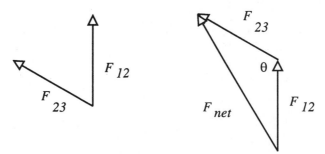

From this diagram we want to find the magnitude of the *net* force on q_1. The cosine law is appropriate here:

$$
\begin{aligned}
F_{\text{net}}{}^2 &= F_{12}^2 + F_{13}^2 - 2F_{12}F_{13}\cos\theta, \\
&= (1.77\,\text{N})^2 + (1.77\,\text{N})^2 - 2(1.77\,\text{N})(1.77\,\text{N})\cos(120°), \\
&= 9.40\,\text{N}^2, \\
F_{\text{net}} &= 3.07\,\text{N}.
\end{aligned}
$$

E25-7 At first glance it might appear odd that we weren't asked what q_3 is. It turns out that not only the value but also the sign of q_3 is unimportant. The forces on q_3 are $\vec{F}_{31}$ and $\vec{F}_{32}$. These forces are given by the vector form of Coulomb's Law, Eq. 25-5,

$$
\begin{aligned}
\vec{F}_{31} &= \frac{1}{4\pi\epsilon_0}\frac{q_3 q_1}{r_{31}^2}\hat{r}_{31} = \frac{1}{4\pi\epsilon_0}\frac{q_3 q_1}{(2d)^2}\hat{r}_{31}, \\
\vec{F}_{32} &= \frac{1}{4\pi\epsilon_0}\frac{q_3 q_2}{r_{32}^2}\hat{r}_{32} = \frac{1}{4\pi\epsilon_0}\frac{q_3 q_2}{(d)^2}\hat{r}_{32}.
\end{aligned}
$$

These two forces are the only forces which act on q_3, so in order to have q_3 in equilibrium the forces must be equal in magnitude, but opposite in direction. In short,

$$
\begin{aligned}
\vec{F}_{31} &= -\vec{F}_{32}, \\
\frac{1}{4\pi\epsilon_0}\frac{q_3 q_1}{(2d)^2}\hat{r}_{31} &= -\frac{1}{4\pi\epsilon_0}\frac{q_3 q_2}{(d)^2}\hat{r}_{32}, \\
\frac{q_1}{4}\hat{r}_{31} &= -\frac{q_2}{1}\hat{r}_{32}.
\end{aligned}
$$

As was foretold, q_3 cancels out of this expression. Note that $\hat{r}_{31}$ and $\hat{r}_{32}$ both point in the same direction and are both of unit length. This means they are the same thing. We then get

$$
q_1 = -4q_2.
$$

The negative sign is very important. Regardless of the sign of q_3, we must have q_1 and q_2 be opposite in sign.

E25-11 This problem is similar to Ex. 25-7. There are some additional issues, however. It is easy enough to write expressions for the forces on the third charge

$$
\begin{aligned}
\vec{F}_{31} &= \frac{1}{4\pi\epsilon_0}\frac{q_3 q_1}{r_{31}^2}\hat{r}_{31}, \\
\vec{F}_{32} &= \frac{1}{4\pi\epsilon_0}\frac{q_3 q_2}{r_{32}^2}\hat{r}_{32}.
\end{aligned}
$$

As with the previous worked exercise we write

$$\vec{\mathbf{F}}_{31} = -\vec{\mathbf{F}}_{32},$$

$$\frac{1}{4\pi\epsilon_0}\frac{q_3 q_1}{r_{31}^2}\hat{\mathbf{r}}_{31} = -\frac{1}{4\pi\epsilon_0}\frac{q_3 q_2}{r_{32}^2}\hat{\mathbf{r}}_{32},$$

$$\frac{q_1}{r_{31}^2}\hat{\mathbf{r}}_{31} = -\frac{q_2}{r_{32}^2}\hat{\mathbf{r}}_{32}.$$

The only way to satisfy the *vector* nature of the above expression is to have $\hat{\mathbf{r}}_{31} = \pm\hat{\mathbf{r}}_{32}$; this means that q_3 must be collinear with q_1 and q_2. q_3 could be between q_1 and q_2, or it could be on either side. Let's resolve this issue now by putting the values for q_1 and q_2 into the expression:

$$\frac{(1.07\,\mu\text{C})}{r_{31}^2}\hat{\mathbf{r}}_{31} = -\frac{(-3.28\,\mu\text{C})}{r_{32}^2}\hat{\mathbf{r}}_{32},$$

$$r_{32}^2\hat{\mathbf{r}}_{31} = (3.07)r_{31}^2\hat{\mathbf{r}}_{32}.$$

Since squared quantities are positive, we can only get this to work if $\hat{\mathbf{r}}_{31} = \hat{\mathbf{r}}_{32}$, so q_3 is *not* between q_1 and q_2. We are then left with

$$r_{32}^2 = (3.07)r_{31}^2,$$

so that q_3 is closer to q_1 than it is to q_2. Then $r_{32} = r_{31} + r_{12} = r_{31} + 0.618\,\text{m}$, and if we take the square root of both sides of the above expression,

$$r_{31} + (0.618\,\text{m}) = \sqrt{(3.07)}r_{31},$$

$$(0.618\,\text{m}) = \sqrt{(3.07)}r_{31} - r_{31},$$

$$(0.618\,\text{m}) = 0.752r_{31},$$

$$0.822\,\text{m} = r_{31}$$

E25-13   On any corner charge there are seven forces; one from each of the other seven charges. The net force will be the sum. Since all eight charges are the same all of the forces will be repulsive. We need to sketch a diagram to show how the charges are labeled.

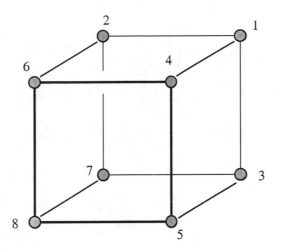

The magnitude of the force of charge 2 on charge 1 is

$$F_{12} = \frac{1}{4\pi\epsilon_0} \frac{q^2}{r_{12}^2},$$

where $r_{12} = a$, the length of a side. Since both charges are the same we wrote q^2. By symmetry we expect that the magnitudes of F_{12}, F_{13}, and F_{14} will all be the same and they will all be at right angles to each other directed along the edges of the cube. Written in terms of vectors the forces would be

$$\vec{\mathbf{F}}_{12} = \frac{1}{4\pi\epsilon_0} \frac{q^2}{a^2}\hat{\mathbf{i}},$$

$$\vec{\mathbf{F}}_{13} = \frac{1}{4\pi\epsilon_0} \frac{q^2}{a^2}\hat{\mathbf{j}},$$

$$\vec{\mathbf{F}}_{14} = \frac{1}{4\pi\epsilon_0} \frac{q^2}{a^2}\hat{\mathbf{k}}.$$

The force from charge 5 is

$$F_{15} = \frac{1}{4\pi\epsilon_0} \frac{q^2}{r_{15}^2},$$

and is directed along the side diagonal away from charge 5. The distance r_{15} is also the side diagonal distance, and can be found from

$$r_{15}^2 = a^2 + a^2 = 2a^2,$$

then

$$F_{15} = \frac{1}{4\pi\epsilon_0} \frac{q^2}{2a^2}.$$

By symmetry we expect that the magnitudes of F_{15}, F_{16}, and F_{17} will all be the same and they will all be directed along the diagonals of the faces of the cube. In terms of components we would have

$$\vec{F}_{15} = \frac{1}{4\pi\epsilon_0} \frac{q^2}{2a^2} \left(\hat{j}/\sqrt{2} + \hat{k}/\sqrt{2}\right),$$

$$\vec{F}_{16} = \frac{1}{4\pi\epsilon_0} \frac{q^2}{2a^2} \left(\hat{i}/\sqrt{2} + \hat{k}/\sqrt{2}\right),$$

$$\vec{F}_{17} = \frac{1}{4\pi\epsilon_0} \frac{q^2}{2a^2} \left(\hat{i}/\sqrt{2} + \hat{j}/\sqrt{2}\right).$$

The last force is the force from charge 8 on charge 1, and is given by

$$F_{18} = \frac{1}{4\pi\epsilon_0} \frac{q^2}{r_{18}^2},$$

and is directed along the cube diagonal away from charge 8. The distance r_{18} is also the cube diagonal distance, and can be found from

$$r_{18}^2 = a^2 + a^2 + a^2 = 3a^2,$$

then in term of components

$$\vec{F}_{18} = \frac{1}{4\pi\epsilon_0} \frac{q^2}{3a^2} \left(\hat{i}/\sqrt{3} + \hat{j}/\sqrt{3} + \hat{k}/\sqrt{3}\right).$$

We can add the components together. By symmetry we expect the same answer for each components, so we'll just do one. How about $\hat{i}$. This component has contributions from charge 2, 6, 7, and 8:

$$\frac{1}{4\pi\epsilon_0} \frac{q^2}{a^2} \left(\frac{1}{1} + \frac{2}{2\sqrt{2}} + \frac{1}{3\sqrt{3}}\right),$$

or

$$\frac{1}{4\pi\epsilon_0} \frac{q^2}{a^2} (1.90)$$

The three components add according to Pythagoras to pick up a final factor of $\sqrt{3}$, so

$$F_{\text{net}} = (0.262) \frac{q^2}{\epsilon_0 a^2}.$$

E25-19 We can make considerable headway by considering some simple symmetry arguments. The net force could have components in both the x and y directions. But any y component from the top half of the rod will be canceled out by the y component from the bottom half, because the two components would point in different direction since the charges densities are opposite in sign.

We can start with the work that was done for us on Page 577, except since we are concerned with $\sin \theta = z/r$ we would have

$$dF_x = dF \sin \theta = \frac{1}{4\pi\epsilon_0} \frac{q_0 \lambda\, dz}{(y^2 + z^2)} \frac{z}{\sqrt{y^2 + z^2}}.$$

We will need to take into consideration that λ changes sign for the two halves of the rod. Then

$$
\begin{aligned}
F_x &= \frac{q_0 \lambda}{4\pi\epsilon_0} \left(\int_{-L/2}^{0} \frac{-z\,dz}{(y^2 + z^2)^{3/2}} + \int_{0}^{L/2} \frac{+z\,dz}{(y^2 + z^2)^{3/2}} \right), \\
&= \frac{q_0 \lambda}{2\pi\epsilon_0} \int_{0}^{L/2} \frac{z\,dz}{(y^2 + z^2)^{3/2}}, \\
&= \frac{q_0 \lambda}{2\pi\epsilon_0} \left. \frac{-1}{\sqrt{y^2 + z^2}} \right|_{0}^{L/2}, \\
&= \frac{q_0 \lambda}{2\pi\epsilon_0} \left(\frac{1}{y} - \frac{1}{\sqrt{y^2 + (L/2)^2}} \right).
\end{aligned}
$$

That's simple enough for me.

E25-21 In each case we conserve charge by making sure that the total number of protons is the same on both sides of the expression. In the last few chapters we will learn how to conserve charge while allowing the number of protons to change. But not here. We also need to conserve the number of neutrons.

(a) Hydrogen has one proton, Beryllium has four, so X must have five protons. Then X must be Boron, B.

(b) Carbon has six protons, Hydrogen has one, so X must have seven. Then X is Nitrogen, N.

(c) Nitrogen has seven protons, Hydrogen has one, but Helium has two, so X has $7 + 1 - 2 = 6$ protons. This means X is Carbon, C.

E25-25 The electrostatic repulsion is given by Coulomb's law, Eq. 25-4,

$$F = \frac{1}{4\pi\epsilon_0} \frac{q_1 q_2}{r_{12}^2} = \frac{(\frac{1}{3} 1.6 \times 10^{-19}\,\text{C})(\frac{1}{3} 1.6 \times 10^{-19}\,\text{C})}{4\pi (8.85 \times 10^{-12}\,\text{C}^2/\text{N} \cdot \text{m}^2)(2.6 \times 10^{-15}\,\text{m})^2} = 3.8\,\text{N}.$$

This might not seem like a large force, but at the level of subatomic particles, it is huge.

E25-31 (a) The gravitational force of attraction between the Moon and the Earth is

$$F_{\rm G} = \frac{GM_{\rm E}M_{\rm M}}{R^2},$$

where R is the distance between them. If both the Earth and the moon are provided a charge q, then the electrostatic repulsion would be

$$F_{\rm E} = \frac{1}{4\pi\epsilon_0}\frac{q^2}{R^2}.$$

Setting these two expression equal to each other,

$$\frac{q^2}{4\pi\epsilon_0} = GM_{\rm E}M_{\rm M},$$

which has solution

$$
\begin{aligned}
q &= \sqrt{4\pi\epsilon_0 GM_{\rm E}M_{\rm M}}, \\
&= \sqrt{4\pi(8.85\times10^{-12}{\rm C}^2/{\rm Nm}^2)(6.67\times10^{-11}{\rm Nm}^2/{\rm kg}^2)(5.98\times10^{24}{\rm kg})(7.36\times10^{22}{\rm kg})}, \\
&= 5.71\times10^{13}\,{\rm C}.
\end{aligned}
$$

Notice that the distance canceled out of this expression. The same charge would be required to offset gravity no matter where the moon is.

(b) We need

$$(5.71\times10^{13}\,{\rm C})/(1.60\times10^{-19}\,{\rm C}) = 3.57\times10^{32}$$

protons on each body. The mass of protons needed is then

$$(3.57\times10^{32})(1.67\times10^{-27}\,{\rm kg}) = 5.97\times10^{65}\,{\rm kg}.$$

Ignoring the mass of the electron (why not?) we can assume that hydrogen is all protons, so we need that much hydrogen.

P25-1 Assume that the spheres initially have charges q_1 and q_2. Note that one of these two must be negative, we just don't know which! The force of attraction between them is

$$F_1 = \frac{1}{4\pi\epsilon_0}\frac{q_1q_2}{r_{12}^2} = -0.108\,{\rm N},$$

where $r_{12} = 0.500\,{\rm m}$. The negative sign for attraction *is* important.

When the spheres are connected by a conducting wire the net charge on the spheres is conserved, but it is also rearranged. The net charge is $q_1 + q_2$, and after the conducting wire is connected each sphere will get *half* of the total. The spheres will have the same charge, and repel with a force of

$$F_2 = \frac{1}{4\pi\epsilon_0} \frac{\frac{1}{2}(q_1 + q_2)\frac{1}{2}(q_1 + q_2)}{r_{12}^2} = 0.0360\,\text{N}.$$

Since we know the separation of the spheres we can find $q_1 + q_2$ quickly,

$$q_1 + q_2 = 2\sqrt{4\pi\epsilon_0 r_{12}^2 (0.0360\,\text{N})} = 2.00\,\mu\text{C}$$

We'll put this back into the first expression and solve for q_2.

$$-0.108\,\text{N} = \frac{1}{4\pi\epsilon_0} \frac{(2.00\,\mu\text{C} - q_2)q_2}{r_{12}^2},$$
$$-3.00 \times 10^{-12}\,\text{C}^2 = (2.00\,\mu\text{C} - q_2)q_2,$$
$$0 = -q_2^2 + (2.00\,\mu\text{C})q_2 + (1.73\,\mu\text{C})^2.$$

The solution is $q_2 = 3.0\,\mu\text{C}$ or $q_2 = -1.0\,\mu\text{C}$. Then $q_1 = -1.0\,\mu\text{C}$ or $q_1 = 3.0\,\mu\text{C}$.

P25-3 We've already done this problem twice, with the appropriate variations. Here is yet another version, and we must additionally show that the solution is unstable. Like Ex. 25-11 we were given the location of two charges and we looked for the location of a third, but now the first two particles have charges of the same sign, so the equilibrium position for the third charge will be *between* the first two charges.

(a) The third charge, q_3, will be between the first two. The net force on the third charge will be zero if

$$\frac{1}{4\pi\epsilon_0} \frac{q\,q_3}{r_{31}^2} = \frac{1}{4\pi\epsilon_0} \frac{4q\,q_3}{r_{32}^2},$$

which will occur if

$$\frac{1}{r_{31}} = \frac{2}{r_{32}}.$$

The total distance is L, so $r_{31} + r32 = L$, or $r_{31} = L/3$ and $r_{32} = 2L/3$.

Now that we have found the position of the third charge we need to find the magnitude. The second and third charges both exert a force on the first charge; we want this net force on the first charge to be zero, so

$$\frac{1}{4\pi\epsilon_0} \frac{q\,q_3}{r_{13}^2} = \frac{1}{4\pi\epsilon_0} \frac{q\,4q}{r_{12}^2},$$

or

$$\frac{q_3}{(L/3)^2} = \frac{4q}{L^2},$$

which has solution $q_3 = -4q/9$. The negative sign is because the force between the first and second charge must be in the opposite direction to the force between the first and third charge.

(b) Consider what happens to the net force on the middle charge if is is displaced a small distance z. If the charge 3 is moved toward charge 1 then the force of attraction with charge 1 will increase. But moving charge 3 closer to charge 1 means moving charge 3 away from charge 2, so the force of attraction between charge 3 and charge 2 will decrease. So charge 3 experiences more attraction to ward the charge that it moves toward, and less attraction to the charge it moves away from. Sounds unstable to me.

P25-7 The force between the two charges is

$$F = \frac{1}{4\pi\epsilon_0} \frac{(Q-q)q}{r_{12}^2}.$$

We want to maximize this force with respect to variation in q, this means finding dF/dq and setting it equal to 0. Then

$$\frac{dF}{dq} = \frac{d}{dq}\left(\frac{1}{4\pi\epsilon_0}\frac{(Q-q)q}{r_{12}^2}\right) = \frac{1}{4\pi\epsilon_0}\frac{Q-2q}{r_{12}^2}.$$

This will vanish if $Q - 2q = 0$, or $q = \frac{1}{2}Q$.

P25-11 We can pretend that this problem is in a single plane containing all three charges. The magnitude of the force on the test charge q_0 from the charge q on the left is

$$F_1 = \frac{1}{4\pi\epsilon_0}\frac{q\,q_0}{(a^2 + R^2)}.$$

A force of identical magnitude exists from the charge on the right. we need to add these two forces as vectors. Only the components along R will survive, and each force will contribute an amount

$$F_1 \sin\theta = F_1\frac{R}{\sqrt{R^2 + a^2}},$$

so the net force on the test particle will be

$$\frac{2}{4\pi\epsilon_0}\frac{q\,q_0}{(a^2 + R^2)}\frac{R}{\sqrt{R^2 + a^2}}.$$

We want to find the maximum value as a function of R. This means take the derivative, and set it equal to zero. The derivative is

$$\frac{2q\,q_0}{4\pi\epsilon_0}\left(\frac{1}{(a^2 + R^2)^{3/2}} - \frac{3R^2}{(a^2 + R^2)^{5/2}}\right),$$

which will vanish when

$$a^2 + R^2 = 3R^2,$$

a *simple* quadratic equation with solutions $R = \pm a/\sqrt{2}$.

Chapter 26

The Electric Field

E26-3 Balancing the weight means that the electric force is equal in magnitude to the gravitational force, or $F_E = W = mg$. But the magnitude of the electric force is given by the *scalar* form of Eq. 26-4, $F_E = Eq$. Combining these two expressions,

$$E = \frac{mg}{q} = \frac{(6.64 \times 10^{-27}\,\text{kg})(9.81\,\text{m/s}^2)}{2(1.60 \times 10^{-19}\,\text{C})} = 2.03 \times 10^{-7}\,\text{N/C}.$$

The alpha particle has a positive charge, this means that it will experience an electric force which is in the same direction as the electric field. Since the gravitational force is down, the electric force, and consequently the electric field, must be directed up.

Although working out the direction of the electric field might appear to be trivial, learn it well now before things get nasty with the magnetic field.

E26-7 The electric field far from an electric dipole is given by Eq. 26-12 for points along the perpendicular bisector. Then

$$E = \frac{1}{4\pi\epsilon_0}\frac{p}{x^3} = (8.99 \times 10^9\,\text{N}\cdot\text{m}^2/\text{C}^2)\frac{(3.56 \times 10^{-29}\,\text{C}\cdot\text{m})}{(25.4 \times 10^{-9}\,\text{m})^3} = 1.95 \times 10^4\,\text{N/C}.$$

Even if your instructor doesn't assign Problem 1, you should do it.

E26-11 Treat the two charges on the left as one dipole and treat the two charges on the right as a second dipole. Point P is on the perpendicular bisector of both dipoles, so we can use Eq. 26-12 to find the two fields.

For the dipole on the left $p = 2aq$ and the electric field due to this dipole at P has magnitude

$$E_1 = \frac{1}{4\pi\epsilon_0}\frac{2aq}{(x+a)^3}$$

and is directed *up*.

For the dipole on the right $p = 2aq$ and the electric field due to this dipole at P has magnitude

$$E_r = \frac{1}{4\pi\epsilon_0} \frac{2aq}{(x-a)^3}$$

and is directed *down*.

The net electric field at P is the sum of these two fields, but since the two component fields point in opposite directions we must actually subtract these values,

$$\begin{aligned} E &= E_r - E_l, \\ &= \frac{2aq}{4\pi\epsilon_0} \left(\frac{1}{(x-a)^3} - \frac{1}{(x+a)^3} \right), \\ &= \frac{aq}{2\pi\epsilon_0} \frac{1}{x^3} \left(\frac{1}{(1-a/x)^3} - \frac{1}{(1+a/x)^3} \right). \end{aligned}$$

We can use the binomial expansion on the terms containing $1 \pm a/x$,

$$\begin{aligned} E &\approx \frac{aq}{2\pi\epsilon_0} \frac{1}{x^3} \left((1+3a/x) - (1-3a/x) \right), \\ &= \frac{aq}{2\pi\epsilon_0} \frac{1}{x^3} (6a/x), \\ &= \frac{3(2qa^2)}{2\pi\epsilon_0 x^4}. \end{aligned}$$

E26-15 (a) The electric field strength just above the center surface of a charged disk is given by Eq. 26-19, but with $z = 0$,

$$E = \frac{\sigma}{2\epsilon_0}$$

The surface charge density is $\sigma = q/A = q/(\pi R^2)$. Combining,

$$q = 2\epsilon_0 \pi R^2 E = 2(8.85 \times 10^{-12}\,\text{C}^2/\text{N·m}^2)\pi(2.5\times10^{-2}\text{m})^2(3\times10^6\,\text{N/C}) = 1.04\times10^{-7}\text{C}.$$

Notice we used an electric field strength of $E = 3 \times 10^6\,\text{N/C}$, which is the field at air breaks down and sparks happen.

(b) This question is *not* an electric field question. We want to find out how many atoms are on the surface; if a is the cross sectional area of one atom, and N the number of atoms, then $A = Na$ is the surface area of the disk. The number of atoms is

$$N = \frac{A}{a} = \frac{\pi(0.0250\,\text{m})^2}{(0.015 \times 10^{-18}\,\text{m}^2)} = 1.31 \times 10^{17}$$

(c)   The total charge on the disk is 1.04×10^{-7}C, this corresponds to

$$(1.04 \times 10^{-7}\text{C})/(1.6 \times 10^{-19}\text{C}) = 6.5 \times 10^{11}$$

electrons. (We are ignoring the sign of the charge here.) If each surface atoms can have at most one excess electron, then the fraction of atoms which are charged is

$$(6.5 \times 10^{11})/(1.31 \times 10^{17}) = 4.96 \times 10^{-6},$$

which isn't very many.

E26-17   We want to fit the data to Eq. 26-19,

$$E_z = \frac{\sigma}{2\epsilon_0}\left(1 - \frac{z}{\sqrt{z^2 + R^2}}\right).$$

There are only two variables, R and q, with $q = \sigma \pi R^2$.

We can find σ *very* easily if we assume that the measurements have no error because then at the surface (where $z = 0$), the expression for the electric field simplifies to

$$E = \frac{\sigma}{2\epsilon_0}.$$

Then $\sigma = 2\epsilon_0 E = 2(8.854 \times 10^{-12}\,\text{C}^2/\text{N} \cdot \text{m}^2)(2.043 \times 10^7\,\text{N/C}) = 3.618 \times 10^{-4}\,\text{C/m}^2$.

Finding the radius will take a little more work. We can choose one point, and make that the reference point, and then solve for R. Starting with

$$E_z = \frac{\sigma}{2\epsilon_0}\left(1 - \frac{z}{\sqrt{z^2 + R^2}}\right),$$

and then rearranging,

$$\frac{2\epsilon_0 E_z}{\sigma} = 1 - \frac{z}{\sqrt{z^2 + R^2}},$$

$$\frac{2\epsilon_0 E_z}{\sigma} = 1 - \frac{1}{\sqrt{1 + (R/z)^2}},$$

$$\frac{1}{\sqrt{1 + (R/z)^2}} = 1 - \frac{2\epsilon_0 E_z}{\sigma},$$

$$1 + (R/z)^2 = \frac{1}{(1 - 2\epsilon_0 E_z/\sigma)^2},$$

$$\frac{R}{z} = \sqrt{\frac{1}{(1 - 2\epsilon_0 E_z/\sigma)^2} - 1}.$$

That was ugly. Using $z = 0.03\,\text{m}$ and $E_z = 1.187 \times 10^7\,\text{N/C}$, along with our value of $\sigma = 3.618 \times 10^{-4}\,\text{C/m}^2$, we find

$$\frac{R}{z} = \sqrt{\frac{1}{\left(1 - 2(8.854 \times 10^{-12}\text{C}^2/\text{Nm}^2)(1.187 \times 10^7\text{N/C})/(3.618 \times 10^{-4}\text{C/m}^2)\right)^2} - 1},$$

$$R = 2.167(0.03\,\text{m}) = 0.065\,\text{m}.$$

(b) And now we can find the charge from the charge density and the radius,

$$q = \pi R^2 \sigma = \pi (0.065\,\text{m})^2 (3.618 \times 10^{-4}\,\text{C/m}^2) = 4.80\,\mu\text{C}.$$

Really, we should check to see if this result is consistent with the other measurements.

E26-27 (a) The electric field does (negative) work on the electron. The magnitude of this work is $W = Fd$, where $F = Eq$ is the magnitude of the electric force on the electron and d is the distance through which the electron moves. Combining, but with a vector notation for extra fun,

$$W = \vec{\mathbf{F}} \cdot \vec{\mathbf{d}} = q\vec{\mathbf{E}} \cdot \vec{\mathbf{d}},$$

which gives the work done by the electric field on the electron. The electron originally possessed a kinetic energy of $K = \frac{1}{2}mv^2$, since we want to bring the electron to a rest, the work done must be negative. The charge q of the electron is negative, so $\vec{\mathbf{E}}$ and $\vec{\mathbf{d}}$ are pointing in the same direction, and $\vec{\mathbf{E}} \cdot \vec{\mathbf{d}} = Ed$.

By the work energy theorem,

$$W = \Delta K = 0 - \frac{1}{2}mv^2.$$

We put all of this together and find d,

$$d = \frac{W}{qE} = \frac{-mv^2}{2qE} = \frac{-(9.11 \times 10^{-31}\text{kg})(4.86 \times 10^6\,\text{m/s})^2}{2(-1.60 \times 10^{-19}\text{C})(1030\,\text{N/C})} = 0.0653\,\text{m}.$$

(b) We can resort to fundamental kinematics: there is a net force on an electron, that force causes an acceleration, the acceleration (in this case) is opposite to the direction of the initial velocity. Then

$$Eq = ma$$

gives the magnitude of the acceleration, and

$$v_\text{f} = v_\text{i} + at$$

gives the time. But $v_\text{f} = 0$. Combining these expressions,

$$t = -\frac{mv_\text{i}}{Eq} = -\frac{(9.11 \times 10^{-31}\text{kg})(4.86 \times 10^6\,\text{m/s})}{(1030\,\text{N/C})(-1.60 \times 10^{-19}\text{C})} = 2.69 \times 10^{-8}\,\text{s}.$$

(c) We will apply the work energy theorem again, except now we don't assume the final kinetic energy is zero. Instead,

$$W = \Delta K = K_f - K_i,$$

and dividing through by the initial kinetic energy to get the fraction lost,

$$\frac{W}{K_i} = \frac{K_f - K_i}{K_i} = \text{ fractional change of kinetic energy.}$$

But $K_i = \frac{1}{2}mv^2$, and $W = qEd$, so the fractional change is

$$\frac{W}{K_i} = \frac{qEd}{\frac{1}{2}mv^2} = \frac{(-1.60\times10^{-19}\text{C})(1030\,\text{N/C})(7.88\times10^{-3}\text{m})}{\frac{1}{2}(9.11\times10^{-31}\text{kg})(4.86\times10^6\,\text{m/s})^2} = -12.1\%.$$

E26-31 The drop is balanced if the electric force is equal to the force of gravity, or

$$Eq = mg$$

The mass of the drop is given in terms of the density by

$$m = \rho V = \rho\frac{4}{3}\pi r^3.$$

We could have found the radius of this drop by watching it fall through the air and measuring the terminal velocity. Combining,

$$q = \frac{mg}{E} = \frac{4\pi\rho r^3 g}{3E} = \frac{4\pi(851\,\text{kg/m}^3)(1.64\times10^{-6}\text{m})^3(9.81\,\text{m/s}^2)}{3(1.92\times10^5\text{N/C})} = 8.11\times10^{-19}\text{C}.$$

We want the charge in terms of e, so we divide, and get

$$\frac{q}{e} = \frac{(8.11\times10^{-19}\text{C})}{(1.60\times10^{-19}\text{C})} = 5.07 \approx 5.$$

E26-33 If each value of q measured by Millikan was a multiple of e, then the difference between any two values of q must also be a multiple of q. The smallest difference would be the smallest multiple, and this multiple might be unity. The differences are 1.641, 1.63, 1.60, 1.63, 3.30, 3.35, 3.18, 3.24, all times 10^{-19} C. This is a pretty clear indication that the fundamental charge is on the order of 1.6×10^{-19} C. If so, the likely number of fundamental charges on each of the drops is shown below in a table arranged like the one in the book:

$$
\begin{array}{ccc}
4 & 8 & 12 \\
5 & 10 & 14 \\
7 & 11 & 16 \\
\end{array}
$$

The total number of charges is 87, while the total charge is 142.69×10^{-19} C, so the average charge per quanta is 1.64×10^{-19} C.

It is of interest to find the uncertainty in this result.

E26-37 The torque on a dipole is given by Eq. 26-27, $\vec{\tau} = \vec{p} \times \vec{E}$. If we are interested in only the magnitude we could instead write $\tau = pE \sin\theta$, where θ is the angle between $\vec{p}$ and $\vec{E}$. For this dipole $p = qd = 2ed$ or $p = 2(1.6 \times 10^{-19}\,C)(0.78 \times 10^{-9}\,m) = 2.5 \times 10^{-28}\,C \cdot m$. For all three cases

$$pE = (2.5 \times 10^{-28}\,C \cdot m)(3.4 \times 10^{6}N/C) = 8.5 \times 10^{-22}\,N \cdot m.$$

The only thing we care about is the angle.

    (a) For the parallel case $\theta = 0$, so $\sin\theta = 0$, and $\tau = 0$.

    (b) For the perpendicular case $\theta = 90°$, so $\sin\theta = 1$, and $\tau = 8.5 \times 10^{-22}\,N \cdot m..$

    (c) For the anti-parallel case $\theta = 180°$, so $\sin\theta = 0$, and $\tau = 0$.

E26-39 This exercise isn't near as bad as it looks. There are two sources to the electric field– the positively charged point-like nucleus, and the negatively charged uniform electron "cloud".

The point-like nucleus contributes an electric field

$$E_+ = \frac{1}{4\pi\epsilon_0} \frac{Ze}{r^2},$$

while the uniform sphere of negatively charged electron cloud of radius R contributes an electric field given by Eq. 26-24,

$$E_- = \frac{1}{4\pi\epsilon_0} \frac{-Zer}{R^3}.$$

The net electric field is just the sum,

$$E = \frac{Ze}{4\pi\epsilon_0} \left(\frac{1}{r^2} - \frac{r}{R^3} \right)$$

P26-1 I must confess that I'm not sure why this is a problem while Exercise 11 is an exercise. We solve them much the same way. Let the positive charge be located *closer* to the point in question, then the electric field from the positive charge is

$$E_+ = \frac{1}{4\pi\epsilon_0} \frac{q}{(x - d/2)^2}$$

and is directed *away from* the dipole.

The negative charge is located farther from the point in question, so

$$E_- = \frac{1}{4\pi\epsilon_0} \frac{q}{(x + d/2)^2}$$

and is directed *toward* the dipole.

The net electric field is the sum of these two fields, but since the two component fields point in opposite direction we must actually subtract these values,

$$E = E_+ - E_-,$$
$$= \frac{1}{4\pi\epsilon_0}\frac{q}{(z-d/2)^2} - \frac{1}{4\pi\epsilon_0}\frac{q}{(z+d/2)^2},$$
$$= \frac{1}{4\pi\epsilon_0}\frac{q}{z^2}\left(\frac{1}{(1-d/2z)^2} - \frac{1}{(1+d/2z)^2}\right)$$

We can use the binomial expansion on the terms containing $1 \pm d/2z$,

$$E \approx \frac{1}{4\pi\epsilon_0}\frac{q}{z^2}\left((1+d/z)-(1-d/z)\right),$$
$$= \frac{1}{2\pi\epsilon_0}\frac{qd}{z^3}$$

(b) The electric field is directed away from the positive charge when you are closer to the positive charge; the electric field is directed toward the negative charge when you are closer to the negative charge. In short, along the axis the electric field is directed in the same direction as the dipole moment.

P26-5 A monopole field falls off as $1/r^2$. A dipole field falls off as $1/r^3$, and consists of two oppositely charge monopoles close together. A quadrupole field (see Exercise 11 above or read Problem 4) falls off as $1/r^4$ and (can) consist of two otherwise identical dipoles arranged with anti-parallel dipole moments. Just taking a leap of faith it seems as if we can construct a $1/r^6$ field behavior by extending the reasoning.

First we need an *octopole* which is constructed from a quadrupole. We want to keep things as simple as possible, so the construction steps are

1. The monopole is a charge $+q$ at $x = 0$.

2. The dipole is a charge $+q$ at $x = 0$ and a charge $-q$ at $x = a$. We'll call this a dipole at $x = a/2$

3. The quadrupole is the dipole at $x = a/2$, and a second dipole pointing the other way at $x = -a/2$. The charges are then $-q$ at $x = -a$, $+2q$ at $x = 0$, and $-q$ at $x = a$.

4. The octopole will be two stacked, offset quadrupoles. There will be $-q$ at $x = -a$, $+3q$ at $x = 0$, $-3q$ at $x = a$, and $+q$ at $x = 2a$.

5. Finally, our distribution with a far field behavior of $1/r^6$. There will be $+q$ at $x = 2a$, $-4q$ at $x = -a$, $+6q$ at $x = 0$, $-4q$ at $x = a$, and $+q$ at $x = 2a$.

Yes, there is a pattern, and it is intimately associated with Pascal's triangle.

P26-9 The key statement is the second italicized paragraph on page 595; the number of field lines through a unit cross-sectional area is proportional to the electric field strength. If the exponent is n, then the electric field strength a distance r from a point charge is

$$E = \frac{kq}{r^n},$$

and the *total* cross sectional area at a distance r is the area of a spherical shell, $4\pi r^2$. Then the number of field lines through the shell is proportional to

$$EA = \frac{kq}{r^n} 4\pi r^2 = 4\pi kq r^{2-n}.$$

Note that the number of field lines varies with r if $n \neq 2$. This means that as we go farther from the point charge we need more and more field lines (or fewer and fewer). But the field lines can only start on charges, and we don't have any except for the point charge. We have a problem; we really do need $n = 2$ if we want workable field lines.

P26-15 Use the a variation of the *exact* result from Problem 26-1. The two charge are positive, but since we will eventually focus on the area between the charges we must *subtract* the two field contributions, since they point in opposite directions. Then

$$E_z = \frac{q}{4\pi\epsilon_0} \left(\frac{1}{(z - a/2)^2} - \frac{1}{(z + a/2)^2} \right)$$

and then take the derivative,

$$\frac{dE_z}{dz} = -\frac{q}{2\pi\epsilon_0} \left(\frac{1}{(z - a/2)^3} - \frac{1}{(z + a/2)^3} \right).$$

Applying the binomial expansion for points $z \ll a$,

$$\begin{aligned}
\frac{dE_z}{dz} &= -\frac{8q}{2\pi\epsilon_0} \frac{1}{a^3} \left(\frac{1}{(2z/a - 1)^3} - \frac{1}{(2z/a + 1)^3} \right), \\
&\approx -\frac{8q}{2\pi\epsilon_0} \frac{1}{a^3} \left(-(1 + 6z/a) - (1 - 6z/a) \right), \\
&= \frac{8q}{\pi\epsilon_0} \frac{1}{a^3}.
\end{aligned}$$

There were some fancy sign flips in the second line, so review those steps carefully!

(b) The electrostatic force on a dipole is the difference in the magnitudes of the electrostatic forces on the two charges that make up the dipole. Near the center of the above charge arrangement the electric field behaves as

$$E_z \approx E_z(0) + \left. \frac{dE_z}{dz} \right|_{z=0} z + \text{ higher ordered terms.}$$

The net force on a dipole is

$$F_+ - F_- = q(E_+ - E_-) = q \left(E_z(0) + \left. \frac{dE_z}{dz} \right|_{z=0} z_+ - E_z(0) - \left. \frac{dE_z}{dz} \right|_{z=0} z_- \right)$$

where the "+" and "-" subscripts refer to the locations of the positive and negative charges. This last line can be simplified to yield

$$q \left. \frac{dE_z}{dz} \right|_{z=0} (z_+ - z_-) = qd \left. \frac{dE_z}{dz} \right|_{z=0} .$$

Chapter 27

Gauss' Law

E27-3 (a) The flat base is easy enough, since according to Eq. 27-7,

$$\Phi_E = \int \vec{\mathbf{E}} \cdot d\vec{\mathbf{A}}.$$

There are two important facts to consider in order to integrate this expression. $\vec{\mathbf{E}}$ is parallel to the axis of the hemisphere, $\vec{\mathbf{E}}$ points inward while $d\vec{\mathbf{A}}$ points outward on the flat base. $\vec{\mathbf{E}}$ is uniform, so it is everywhere the same on the flat base. Since $\vec{\mathbf{E}}$ is anti-parallel to $d\vec{\mathbf{A}}$, $\vec{\mathbf{E}} \cdot d\vec{\mathbf{A}} = -E\,dA$, then

$$\Phi_E = \int \vec{\mathbf{E}} \cdot d\vec{\mathbf{A}} = -\int E\,dA.$$

Since $\vec{\mathbf{E}}$ is uniform, it is constant in the region of integration (it is constant *everywhere*, eh?), so we can bring it out of the integral,

$$\Phi_E = -\int E\,dA = -E\int dA = -EA = -\pi R^2 E.$$

The last steps are just substituting the area of a circle for the flat side of the hemisphere.

 These steps that we went through are fairly standard for evaluating the vector integrals of electricity and magnetism. First we try to change the vector nature of the problem into a scalar problem, and then we try to pull any constants out of the problem. We'll *try* to choose a region of integration where we expect $\vec{\mathbf{E}}$ to be constant or zero, and then life will be very easy.

(b) If we knew Gauss' Law (we'll know it soon), then we would already be done. We don't, so we need to integrate the hard way. As before, we must first sort out the dot product

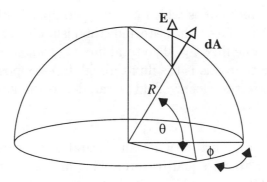

We can simplify the vector part of the problem with $\vec{\mathbf{E}} \cdot d\vec{\mathbf{A}} = \cos\theta\, E\, dA$, so

$$\Phi_E = \int \vec{\mathbf{E}} \cdot d\vec{\mathbf{A}} = \int \cos\theta\, E\, dA$$

Once again, $\vec{\mathbf{E}}$ is uniform, so we can take it out of the integral,

$$\Phi_E = \int \cos\theta\, E\, dA = E \int \cos\theta\, dA$$

Finally, $dA = (R\, d\theta)(R\sin\theta\, d\phi)$ on the surface of a sphere centered on $R = 0$. You would be well advised to remember this differential expression. I'll point out several others in the next few chapters!

We'll integrate ϕ around the axis, from 0 to 2π. We'll integrate θ from the axis to the equator, from 0 to $\pi/2$. Then

$$\Phi_E = E \int \cos\theta\, dA = E \int_0^{2\pi} \int_0^{\pi/2} R^2 \cos\theta \sin\theta\, d\theta\, d\phi.$$

Pulling out the constants, doing the *trivial* ϕ integration, and then writing $2\cos\theta\sin\theta$ as $\sin(2\theta)$,

$$\Phi_E = 2\pi R^2 E \int_0^{\pi/2} \cos\theta\sin\theta\, d\theta = \pi R^2 E \int_0^{\pi/2} \sin(2\theta)\, d\theta,$$

Change variables and let $\beta = 2\theta$, then we have

$$\Phi_E = \pi R^2 E \int_0^{\pi} \sin\beta \frac{1}{2} d\beta = \pi R^2 E.$$

E27-5 The size or shape of the surface doesn't matter, only the fact that the charge is inside does. Then by Eq. 27-8,

$$\Phi_E = \frac{q}{\epsilon_0} = \frac{(1.84\,\mu\text{C})}{(8.85\times10^{-12}\,\text{C}^2/\text{N}\cdot\text{m}^2)} = 2.08\times10^5\,\text{N}\cdot\text{m}^2/\text{C}.$$

We didn't need to do any more work than this. Don't be tempted to make problems harder than they really are.

E27-11 There are *eight* cubes which can be "wrapped" around the charge. Each cube has three external faces that are indistinguishable for a total of twenty-four faces, each with the same flux Φ_E. The total flux is q/ϵ_0, so the flux through one face is $\Phi_E = q/24\epsilon_0$. Note that this is the flux through faces opposite the charge; for faces which touch the charge the electric field is parallel to the surface, so the flux would be zero.

E27-17 We don't really need to write an integral, we just need the charge per unit length in the cylinder to be equal to zero. This means that the positive charge in cylinder must be $+3.60nC/m$. This positive charge is uniformly distributed in a circle of radius $R = 1.50$ cm, so

$$\rho = \frac{3.60nC/m}{\pi R^2} = \frac{3.60nC/m}{\pi (0.0150\,\text{m})^2} = 5.09\mu C/m^3.$$

E27-19 The proton orbits with a speed v, so the centripetal force on the proton is $F_C = mv^2/r$. This centripetal force is from the electrostatic attraction with the sphere; so long as the proton is outside the sphere the electric field is equivalent to that of a point charge Q (Eq. 27-15),

$$E = \frac{1}{4\pi\epsilon_0} \frac{Q}{r^2}.$$

If q is the charge on the proton we can write $F = Eq$, or

$$\frac{mv^2}{r} = q\frac{1}{4\pi\epsilon_0} \frac{Q}{r^2}$$

Solving for Q,

$$
\begin{aligned}
Q &= \frac{4\pi\epsilon_0 m v^2 r}{q}, \\
&= \frac{4\pi(8.85\times 10^{-12}\,\text{C}^2/\text{N}\cdot\text{m}^2)(1.67\times 10^{-27}\text{kg})(294\times 10^3\text{m/s})^2(0.0113\,\text{m})}{(1.60\times 10^{-19}\text{C})}, \\
&= -1.13\times 10^{-9}\text{C}.
\end{aligned}
$$

Where'd the negative sign come from? The proton is positive so to stay in orbit it does need to be attracted to the sphere...

E27-25 (a) We want to follow steps similar to the infinite line of charge on page 617. We use the same cylindrical Gaussian surface, and we will have no contributions from the end caps because $\vec{\mathbf{E}}$ will be parallel to those surfaces. Starting with Gauss' Law,

$$\frac{q_{\text{enc}}}{\epsilon_0} = \oint \vec{\mathbf{E}} \cdot d\vec{\mathbf{A}} = \int_{\text{tube}} \vec{\mathbf{E}} \cdot d\vec{\mathbf{A}} + \int_{\text{ends}} \vec{\mathbf{E}} \cdot d\vec{\mathbf{A}},$$

but only the tube part is non-zero. $\vec{E}$ is directly out from the surface, so the dot product simplifies, then

$$\frac{q_{enc}}{\epsilon_0} = \int_{tube} E\, dA.$$

By symmetry, E is everywhere the same on the surface of the tube (but E is *not* constant; it varies with r). We can bring it out of the integral, because we are only integrating on the surface. Then

$$\int_{tube} E\, dA = E \int_{tube} dA = E2\pi rl,$$

where l is the length of the Gaussian surface.

Now for the q_{enc} part. If the (uniform) volume charge density is ρ, then the charge enclosed in the Gaussian cylinder is

$$q_{enc} = \int \rho dV = \rho \int dV = \rho V = \pi r^2 l\rho.$$

Combining,

$$\pi r^2 l\rho/\epsilon_0 = E2\pi rl$$

or

$$E = \rho r/2\epsilon_0.$$

(b) Outside the charged cylinder the charge enclosed in the Gaussian surface is just the charge in the cylinder. Then

$$q_{enc} = \int \rho dV = \rho \int dV = \rho V = \pi R^2 l\rho.$$

and

$$\pi R^2 l\rho/\epsilon_0 = E2\pi rl,$$

and then finally

$$E = \frac{R^2\rho}{2\epsilon_0 r}.$$

E27-29 (a) The near field is given by Eq. 27-12, $E = \sigma/2\epsilon_0$, so

$$E \approx \frac{(6.0\times10^{-6}\text{C})/(8.0\times10^{-2}\,\text{m})^2}{2(8.85\times10^{-12}\,\text{C}^2/\text{N}\cdot\text{m}^2)} = 5.3\times10^7\text{N/C}.$$

(b) Very far from *any* object a point charge approximation is valid. Then

$$E = \frac{1}{4\pi\epsilon_0}\frac{q}{r^2} = \frac{1}{4\pi(8.85\times10^{-12}\,\text{C}^2/\text{N}\cdot\text{m}^2)}\frac{(6.0\times10^{-6}\text{C})}{(30\,\text{m})^2} = 60\text{N/C}.$$

Actually, the above approximation is only valid if the distance from the object is much larger than the linear dimensions of the object, even if the distance from the object is small.

P27-3 The net force on the small sphere is zero; this force is the vector sum of the force of gravity W, the electric force F_E, and the tension T.

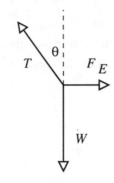

These forces are related by $F_E/W = \tan\theta$, or

$$Eq = mg\tan\theta.$$

We also have $E = \sigma/2\epsilon_0$, so

$$
\begin{aligned}
\sigma &= \frac{2\epsilon_0 mg\tan\theta}{q}, \\
&= \frac{2(8.85\times 10^{-12}\,\mathrm{C}^2/\mathrm{N}\cdot\mathrm{m}^2)(1.12\times 10^{-6}\mathrm{kg})(9.81\,\mathrm{m/s}^2)\tan(27.4^\circ)}{(19.7\times 10^{-9}\mathrm{C})}, \\
&= 5.11\times 10^{-9}\mathrm{C/m}^2.
\end{aligned}
$$

P27-7 This problem is closely related to Ex. 27-25, except for the part concerning q_{enc}. We'll set up the problem the same way: the Gaussian surface will be a (imaginary) cylinder centered on the axis of the physical cylinder. For Gaussian surfaces of radius $r < R$, there is *no* charge enclosed while for Gaussian surfaces of radius $r > R$, $q_{\mathrm{enc}} = \lambda l$.

We've already worked out the integral

$$\int_{\mathrm{tube}} \vec{\mathbf{E}}\cdot d\vec{\mathbf{A}} = 2\pi r l E,$$

for the cylinder, and then from Gauss' law,

$$q_{\mathrm{enc}} = \epsilon_0 \int_{\mathrm{tube}} \vec{\mathbf{E}}\cdot d\vec{\mathbf{A}} = 2\pi\epsilon_0 r l E.$$

(a) When $r < R$ there is no enclosed charge, so the left hand vanishes and consequently $E = 0$ inside the physical cylinder.

(b) When $r > R$ there is a charge λl enclosed, so

$$E = \frac{\lambda}{2\pi\epsilon_0 r}.$$

Yes, it should look familiar!

P27-11 (a) Put a spherical Gaussian surface inside the shell centered on the point charge. We can use Gauss' law here because there is spherical symmetry in the entire problem, both inside and outside the Gaussian surface. If the point charge inside the shell has been off-center we would *not* be able to apply Gauss' law to find the electric field. But back to the problem at hand. Gauss' law states

$$\oint \vec{E} \cdot d\vec{A} = \frac{q_{enc}}{\epsilon_0}.$$

Since there is spherical symmetry the electric field is normal to the spherical Gaussian surface, and it is everywhere the same on this surface. The dot product simplifies to $\vec{E} \cdot d\vec{A} = E\, dA$, while since E is a constant on the surface we can pull it out of the integral, and we end up with

$$E \oint dA = \frac{q}{\epsilon_0},$$

where q is the point charge in the center. Now $\oint dA = 4\pi r^2$, where r is the radius of the Gaussian surface, so

$$E = \frac{q}{4\pi\epsilon_0 r^2},$$

which should look vaguely familiar. We just reinvented the wheel! But that's okay, it was good for us. Note that this is the electric field inside the metallic shell.

(b) Repeat the above steps, except put the Gaussian surface outside the conducting shell. Keep it centered on the charge. Two things are different from the above derivation: (1) r is bigger, and (2) there is an uncharged spherical conducting shell inside the Gaussian surface. Neither change will affect the surface integral or q_{enc}, so the electric field outside the shell is still

$$E = \frac{q}{4\pi\epsilon_0 r^2},$$

(c) This is a subtle question. With all the symmetry here it appears as if the shell has no effect; the field just looks like a point charge field. If, however, the charge were moved off center the field inside the shell would become distorted, and we wouldn't be able to use Gauss' law to find it. So the shell does make a difference.

Outside the shell, however, we can't tell what is going on inside the shell. So the electric field outside the shell looks like a point charge field originating from the center of the shell *regardless of where inside the shell the point charge is placed!*

(d) Yes, q induces surface charges on the shell. There will be a charge $-q$ on the inside surface and a charge q on the outside surface.

(e) Yes, as there is an electric field from the shell, isn't there?

(f) No, as the electric field from the outside charge won't make it through a conducting shell. The conductor acts as a shield.

(g) No, this is not a contradiction, because the outside charge never experienced any electrostatic attraction or repulsion from the inside charge. The force is between the shell and the outside charge.

P27-15 If a point is an equilibrium point then the electric field at that point should be zero. If it is a stable point then moving the test charge (assumed positive) a small distance from the equilibrium point should result in a restoring force directed back toward the equilibrium point. In other words, there will be a point where the electric field is zero, and around this point there will be an electric field pointing inward. Applying Gauss' law to a small surface surrounding our point P, we have a net inward flux, so there must be a negative charge *inside* the surface. But there should be nothing inside the surface except an empty point P, so we have a contradiction.

P27-17 (a) The total charge is the volume integral over the whole sphere,

$$Q = \int \rho \, dV.$$

This is actually a three dimensional integral, and

$$dV = (dr)(r d\theta)(r \sin \theta d\phi).$$

This is another *good* differential to memorize. We do have spherical symmetry here, so it is easier to write this as $dV = A \, dr$, where A is the surface area of a spherical shell, or $A = 4\pi r^2$. Then

$$
\begin{aligned}
Q &= \int \rho \, dV, \\
&= \int_0^R \left(\frac{\rho_s r}{R} \right) 4\pi r^2 \, dr, \\
&= \frac{4\pi \rho_s}{R} \frac{1}{4} R^4, \\
&= \pi \rho_s R^3.
\end{aligned}
$$

(b) Put a spherical Gaussian surface inside the sphere centered on the center. We can use Gauss' law here because there is spherical symmetry in the entire problem, both inside and outside the Gaussian surface. Gauss' law states

$$\oint \vec{\mathbf{E}} \cdot d\vec{\mathbf{A}} = \frac{q_{enc}}{\epsilon_0}.$$

Since there is spherical symmetry the electric field is normal to the spherical Gaussian surface, and it is everywhere the same on this surface. The dot product simplifies to $\vec{\mathbf{E}} \cdot d\vec{\mathbf{A}} = E\, dA$, while since E is a constant on the surface we can pull it out of the integral, and we end up with

$$E \oint dA = \frac{q_{enc}}{\epsilon_0},$$

Now $\oint dA = 4\pi r^2$, where r is the radius of the Gaussian surface, so

$$E = \frac{q_{enc}}{4\pi\epsilon_0 r^2},$$

which should look vaguely familiar. We aren't done yet, because the charge enclosed depends on the radius of the Gaussian surface. We need to do part (a) again, except this time we don't want to do the whole volume of the sphere, we only want to go out as far as the Gaussian surface. Then

$$
\begin{aligned}
q_{enc} &= \int \rho\, dV, \\
&= \int_0^r \left(\frac{\rho_S r}{R}\right) 4\pi r^2\, dr, \\
&= \frac{4\pi\rho_S}{R}\frac{1}{4} r^4, \\
&= \pi\rho_S \frac{r^4}{R}.
\end{aligned}
$$

Combine these last two results and

$$
\begin{aligned}
E &= \frac{\pi\rho_S}{4\pi\epsilon_0 r^2}\frac{r^4}{R}, \\
&= \frac{\pi\rho_S}{4\pi\epsilon_0}\frac{r^2}{R}, \\
&= \frac{Q}{4\pi\epsilon_0}\frac{r^2}{R^4}.
\end{aligned}
$$

In the last line we used the results of part (a) to eliminate ρ_S from the expression.

Note that the electric field outside the sphere should be that of a point charge, and that the electric field should be continuous across the surface. You should be able to verify that

$$\frac{Q}{4\pi\epsilon_0}\frac{r^2}{R^4} \quad \text{for } r = R$$

is the same as

$$\frac{Q}{4\pi\epsilon_0 r^2} \text{ for } r = R.$$

Chapter 28

Electric Potential Energy and Potential

Throughout this chapter we will use the convention that $V(\infty) = 0$ unless explicitly stated otherwise. Then the potential in the vicinity of a point charge will be given by Eq. 28-18, $V = q/4\pi\epsilon_0 r$.

E28-3 (a) We build the electron one part at a time; each part has a charge $q = e/3$. Moving the first part from infinity to the location where we want to construct the electron is easy and takes no work at all. Moving the second part in requires work to change the potential energy to

$$U_{12} = \frac{1}{4\pi\epsilon_0}\frac{q_1 q_2}{r},$$

which is basically Eq. 28-7. The separation $r = 2.82 \times 10^{-15}$ m.

Bringing in the third part requires work against the force of repulsion between the third charge and both of the other two charges. Potential energy then exists in the form U_{13} and U_{23}, where all three charges are the same, and all three separations are the same. Then $U_{12} = U_{13} = U_{12}$, so the total potential energy of the system is

$$U = 3\frac{1}{4\pi\epsilon_0}\frac{(e/3)^2}{r} = \frac{3}{4\pi(8.85\times 10^{-12}\,\mathrm{C^2/N\cdot m^2})}\frac{(1.60\times 10^{-19}\,\mathrm{C}/3)^2}{(2.82\times 10^{-15}\,\mathrm{m})} = 2.72\times 10^{-14}\,\mathrm{J}$$

(b) Dividing our answer by the speed of light squared to find the mass,

$$m = \frac{2.72 \times 10^{-14}\,\mathrm{J}}{(3.00 \times 10^8\,\mathrm{m/s})^2} = 3.02 \times 10^{-31}\,\mathrm{kg}.$$

This answer is remarkably close to the accepted electron mass. It should be; the classical radius of the electron is defined as the radius of a sphere of fundamental charge which has a mass energy equal to that of the electron. In effect we have picked ourselves up by our belt and carried ourselves to the solution.

E28-7 (a) The energy released is equal to the charges times the potential through which the charge was moved (Eq. 28-14, more or less). Then

$$\Delta U = q\Delta V = (30\,\text{C})(1.0 \times 10^9\,\text{V}) = 3.0 \times 10^{10}\,\text{J}.$$

(b) Although the problem mentions acceleration, we want to focus on energy. The energy will change the kinetic energy of the car from 0 to $K_f = 3.0 \times 10^{10}\,\text{J}$. The speed of the car is then

$$v = \sqrt{\frac{2K}{m}} = \sqrt{\frac{2(3.0 \times 10^{10}\,\text{J})}{(1200\,\text{kg})}} = 7100\,\text{m/s}.$$

Rather fast.

(c) The energy required to melt ice is given by $Q = mL$, where L is the latent heat of fusion. Then

$$m = \frac{Q}{L} = \frac{(3.0 \times 10^{10}\,\text{J})}{(3.33 \times 10^5\,\text{J/kg})} = 90,100\text{kg},$$

which would make a cube of ice about 45 meters on a side!

E28-13 (a) The magnitude of the electric field would be found from

$$E = \frac{F}{q} = \frac{(3.90 \times 10^{-15}\,\text{N})}{(1.60 \times 10^{-19}\,\text{C})} = 2.44 \times 10^4\,\text{N/C}.$$

(b) The potential difference between the plates is found by evaluating Eq. 28-15,

$$\Delta V = -\int_a^b \vec{\mathbf{E}} \cdot d\vec{\mathbf{s}}.$$

The electric field between two parallel plates is uniform and perpendicular to the plates (see previous chapters). We can take any path we want between the two plates, since conductors are equipotential surfaces. For convenience we follow the shortest path between the plates; this path is also perpendicular to the plates. Then $\vec{\mathbf{E}} \cdot d\vec{\mathbf{s}} = E\,ds$ along this path, and since E is uniform,

$$\Delta V = -\int_a^b \vec{\mathbf{E}} \cdot d\vec{\mathbf{s}} = -\int_a^b E\,ds = -E\int_a^b ds = E\Delta x,$$

where Δx is the separation between the plates. *This is a relationship worth remembering, and it is true for parallel conducting plates.*

Finally, $\Delta V = (2.44 \times 10^4\,\text{N/C})(0.120\,\text{m}) = 2930\,\text{V}.$

E28-19 (a) We evaluate V_A and V_B individually, and then find the difference.

$$V_A = \frac{1}{4\pi\epsilon_0}\frac{q}{r} = \frac{1}{4\pi(8.85 \times 10^{-12}\,\mathrm{C^2/N\cdot m^2})}\frac{(1.16\mu\mathrm{C})}{(2.06\,\mathrm{m})} = 5060\,\mathrm{V},$$

and

$$V_B = \frac{1}{4\pi\epsilon_0}\frac{q}{r} = \frac{1}{4\pi(8.85 \times 10^{-12}\,\mathrm{C^2/N\cdot m^2})}\frac{(1.16\mu\mathrm{C})}{(1.17\,\mathrm{m})} = 8910\,\mathrm{V},$$

The difference is then $V_A - V_B = -3850\,\mathrm{V}$.

(b) The answer is the same, since when concerning ourselves with electric potential we only care about distances, and not directions.

E28-23 Since the surface of the Earth is outside the Earth we can treat the Earth as a point charge and consider the surface as a point r away from the charge.

The electric field is given by

$$E = \frac{1}{4\pi\epsilon_0}\frac{q}{r^2},$$

the electric potential is given by

$$V = \frac{1}{4\pi\epsilon_0}\frac{q}{r}.$$

The ratio is

$$\frac{V}{E} = \left(\frac{1}{4\pi\epsilon_0}\frac{q}{r}\right)\Big/\left(\frac{1}{4\pi\epsilon_0}\frac{q}{r^2}\right) = r.$$

In this problem r is the radius of the Earth, so at the surface of the Earth the potential is

$$V = Er = (100\,\mathrm{V/m})(6.38\times 10^6\,\mathrm{m}) = 6.38\times 10^8\,\mathrm{V}.$$

That's a potential which is revoltingly large!

E28-25 (a) When finding V_A we need to consider the contribution from both the positive and the negative charge, so

$$V_A = \frac{1}{4\pi\epsilon_0}\left(qa + \frac{-q}{a+d}\right)$$

There will be a similar expression for V_B,

$$V_B = \frac{1}{4\pi\epsilon_0}\left(-qa + \frac{q}{a+d}\right).$$

Now to evaluate the difference.

$$
\begin{aligned}
V_A - V_B &= \frac{1}{4\pi\epsilon_0}\left(qa + \frac{-q}{a+d}\right) - \frac{1}{4\pi\epsilon_0}\left(-qa + \frac{q}{a+d}\right), \\
&= \frac{q}{2\pi\epsilon_0}\left(\frac{1}{a} - \frac{1}{a+d}\right), \\
&= \frac{q}{2\pi\epsilon_0}\left(\frac{a+d}{a(a+d)} - \frac{a}{a(a+d)}\right), \\
&= \frac{q}{2\pi\epsilon_0}\frac{d}{a(a+d)}.
\end{aligned}
$$

(b) Does it do what we expect when $d = 0$? I expect it the difference to go to zero as the two points A and B get closer together. The numerator will go to zero as d gets smaller. The denominator, however, stays finite, which is a good thing. So yes, $V_a - V_B \to 0$ as $d \to 0$.

E28-29 The potential on the axis of a uniform ring of charge is given by Eq. 28-30,

$$
V = \frac{1}{4\pi\epsilon_0}\frac{2\pi\lambda R}{\sqrt{R^2 + x^2}}.
$$

We can find the linear charge density by dividing the charge by the circumference,

$$
\lambda = \frac{Q}{2\pi R},
$$

where Q refers to the charge on the ring. The work done to move a charge q from a point x to the origin will be given by

$$
\begin{aligned}
W &= q\Delta V, \\
W &= q\left(V(0) - V(x)\right), \\
&= q\left(\frac{1}{4\pi\epsilon_0}\frac{Q}{\sqrt{R^2}} - \frac{1}{4\pi\epsilon_0}\frac{Q}{\sqrt{R^2 + x^2}}\right), \\
&= \frac{qQ}{4\pi\epsilon_0}\left(\frac{1}{R} - \frac{1}{\sqrt{R^2 + x^2}}\right).
\end{aligned}
$$

Putting in the numbers,

$$
\frac{(-5.93\times 10^{-12}\mathrm{C})(-9.12\times 10^{-9}\mathrm{C})}{4\pi(8.85\times 10^{-12}\mathrm{C}^2/\mathrm{N}\cdot\mathrm{m}^2)}\left(\frac{1}{1.48\mathrm{m}} - \frac{1}{\sqrt{(1.48\mathrm{m})^2 + (3.07\mathrm{m})^2}}\right) = 1.86\times 10^{-10}\mathrm{J}.
$$

E28-33 The radial potential gradient is just the magnitude of the radial component of the electric field,

$$E_r = -\frac{\partial V}{\partial r}$$

Since we are outside the gold nucleus we can treat the nucleus as a point charge and evaluate the electric field strength at a distance r from the center, where r is the radius of a gold nucleus. Then

$$\frac{\partial V}{\partial r} = -\frac{1}{4\pi\epsilon_0}\frac{q}{r^2},$$

$$= \frac{1}{4\pi(8.85\times 10^{-12}\,\text{C}^2/\text{N}\cdot\text{m}^2)}\frac{79(1.60\times 10^{-19}\text{C})}{(7.0\times 10^{-15}\text{m})^2},$$

$$= -2.32\times 10^{21}\,\text{V/m}.$$

E28-37 Since the magnitude of the work is equal to the magnitude of the charge times the potential difference through which the charge is moved, $|W| = |q\Delta V|$, this problem is fairly easy.

(a) $|V_B - V_A| = |W/q| = |(3.94\times 10^{-19}\,\text{J})/(1.60\times 10^{-19}\,\text{C})| = 2.46\,\text{V}$. The electric field did work on the electron, so the electron was moving from a region of low potential to a region of high potential; or $V_B > V_A$. Consequently, $V_B - V_A = 2.46\,\text{V}$.

(b) V_C is at the same potential as V_B (both points are on the same equipotential line), so $V_C - V_A = V_B - V_A = 2.46\,\text{V}$.

(b) V_C is at the same potential as V_B (both points are on the same equipotential line), so $V_C - V_B = 0\,\text{V}$.

E28-47 (a) The total charge ($Q = 57.2$nC) will be divided up between the two spheres so that they are at the same potential. If q_1 is the charge on one sphere, $q_2 = Q - q_1$ is the charge on the other. Consequently

$$V_1 = V_2,$$

$$\frac{1}{4\pi\epsilon_0}\frac{q_1}{r_1} = \frac{1}{4\pi\epsilon_0}\frac{Q-q_1}{r_2},$$

$$q_1 r_2 = (Q - q_1)r_1,$$

$$q_1 = \frac{Qr_2}{r_2 + r_1}.$$

Putting in the numbers, we find

$$q_1 = \frac{Qr_1}{r_2 + r_1} = \frac{(57.2\text{ nC})(12.2\text{ cm})}{(5.88\text{ cm}) + (12.2\text{ cm})} = 38.6\text{ nC},$$

and

$$q_2 = Q - q_1 = (57.2 \text{ nC}) - (38.6 \text{ nC}) = 18.6 \text{ nC}$$

(b) The potential on each sphere should be the same, so we only need to solve one. Then

$$\frac{1}{4\pi\epsilon_0}\frac{q_1}{r_1} = \frac{1}{4\pi(8.85 \times 10^{-12}\,\text{C}^2/\text{N}\cdot\text{m}^2)}\frac{(38.6 \text{ nC})}{(12.2 \text{ cm})} = 2850 \text{ V}.$$

Note that we were given units in centimeters!

E28-49 (a) Apply the point charge formula again, but solve for the charge. Then

$$\frac{1}{4\pi\epsilon_0}\frac{q}{r} = V,$$
$$q = 4\pi\epsilon_0 rV,$$
$$q = 4\pi(8.85 \times 10^{-12}\,\text{C}^2/\text{N}\cdot\text{m}^2)(1\,\text{m})(10^6\,\text{V}) = 0.11\,\text{mC}.$$

Now that's a fairly small charge. But if the radius were decreased by a factor of 100, so would the charge $(1.10\,\mu\text{C})$. Consequently, smaller metal balls can be raised to higher potentials with less charge.

(b) The electric field near the surface of the ball is a function of the surface charge density, $E = \sigma/\epsilon_0$. But surface charge density depends on the area, and varies as r^{-2}. For a given potential, the electric field near the surface would then be given by

$$E = \frac{\sigma}{\epsilon_0} = \frac{q}{4\pi\epsilon_0 r^2} = \frac{V}{r}.$$

Note that the electric field grows as the ball gets smaller. This means that the break down field is more likely to be exceeded with a low voltage small ball; you'll get sparking. Hence, lighting rods are pointed, and not big-balled.

P28-1 (a) According to Newtonian mechanics we want

$$K = \frac{1}{2}mv^2$$

to be equal to

$$W = q\Delta V$$

which means

$$\Delta V = \frac{mv^2}{2q} = \frac{(0.511 \text{ MeV})}{2e} = 256 \text{ kV}.$$

We were somewhat sneaky here. mc^2 is the rest mass energy of an electron, so we just wrote it in. We *can* cancel the e in eV with the e in the denominator, but only because they both refer to the charge on one electron.

(b) Let's do some rearranging first.

$$K = mc^2\left[\frac{1}{\sqrt{1-\beta^2}} - 1\right],$$

$$\frac{K}{mc^2} = \frac{1}{\sqrt{1-\beta^2}} - 1,$$

$$\frac{K}{mc^2} + 1 = \frac{1}{\sqrt{1-\beta^2}},$$

$$\frac{1}{\frac{K}{mc^2} + 1} = \sqrt{1-\beta^2},$$

$$\frac{1}{\left(\frac{K}{mc^2} + 1\right)^2} = 1-\beta^2,$$

and finally,

$$\beta = \sqrt{1 - \frac{1}{\left(\frac{K}{mc^2} + 1\right)^2}}$$

Putting in the numbers,

$$\sqrt{1 - \frac{1}{\left(\frac{(256\ \text{keV})}{(511\ \text{keV})} + 1\right)^2}} = 0.746,$$

so $v = 0.746c$.

P28-7 (a) First apply Eq. 28-18, but solve for r. Then

$$r = \frac{q}{4\pi\epsilon_0 V} = \frac{(32.0 \times 10^{-12}\,\text{C})}{4\pi(8.85 \times 10^{-12}\,\text{C}^2/\text{N}\cdot\text{m}^2)(512\,\text{V})} = 562\,\mu\text{m},$$

which is not a very big drop.

(b) If two such drops join together the charge doubles, and the volume of water doubles, but the radius of the new drop only increases by a factor of $\sqrt[3]{2} = 1.26$ because volume is proportional to the radius cubed.

The potential on the surface of the new drop will be

$$
\begin{aligned}
V_{\text{new}} &= \frac{1}{4\pi\epsilon_0}\frac{q_{\text{new}}}{r_{\text{new}}}, \\
&= \frac{1}{4\pi\epsilon_0}\frac{2q_{\text{old}}}{\sqrt[3]{2}\,r_{\text{old}}}, \\
&= (2)^{2/3}\frac{1}{4\pi\epsilon_0}\frac{q_{\text{old}}}{r_{\text{old}}} = (2)^{2/3}V_{\text{old}}.
\end{aligned}
$$

The new potential is 813 V.

P28-9 (a) The potential at any point will be the sum of the contribution from each charge,

$$V = \frac{1}{4\pi\epsilon_0}\frac{q_1}{r_1} + \frac{1}{4\pi\epsilon_0}\frac{q_2}{r_2},$$

where r_1 is the distance the point in question from q_1 and r_2 is the distance the point in question from q_2. Pick a point, any point. Call it (x, y). Since q_1 is at the origin,

$$r_1 = \sqrt{x^2 + y^2}.$$

Since q_2 is at $(d, 0)$, where $d = 9.60$ nm,

$$r_2 = \sqrt{(x - d)^2 + y^2}.$$

Define the "Stanley Number" as $S = 4\pi\epsilon_0 V$. Equipotential surfaces are also equistanley surfaces. In particular, when $V = 0$, so does S. We can then write the potential expression in a sightly simplified form

$$S = \frac{q_1}{r_1} + \frac{q_2}{r_2}.$$

If $S = 0$ we can rearrange and square this expression.

$$\frac{q_1}{r_1} = -\frac{q_2}{r_2},$$

$$\frac{r_1^2}{q_1^2} = \frac{r_2^2}{q_2^2},$$

$$\frac{x^2 + y^2}{q_1^2} = \frac{(x - d)^2 + y^2}{q_2^2},$$

Let $\alpha = q_2/q_1$, then we can write

$$\alpha^2\left(x^2 + y^2\right) = (x - d)^2 + y^2,$$
$$\alpha^2 x^2 + \alpha^2 y^2 = x^2 - 2xd + d^2 + y^2,$$
$$(\alpha^2 - 1)x^2 + 2xd + (\alpha^2 - 1)y^2 = d^2.$$

We complete the square for the $(\alpha^2 - 1)x^2 + 2xd$ term by adding $d^2/(\alpha^2 - 1)$ to both sides of the equation. Then

$$(\alpha^2 - 1)\left[\left(x + \frac{d}{\alpha^2 - 1}\right)^2 + y^2\right] = d^2\left(1 + \frac{1}{\alpha^2 - 1}\right).$$

The center of the circle is at

$$-\frac{d}{\alpha^2 - 1} = \frac{(9.60\text{ nm})}{(-10/6)^2 - 1} = -5.4\text{ nm}.$$

(b) The radius of the circle is

$$\sqrt{d^2 \frac{\left(1 + \frac{1}{\alpha^2 - 1}\right)}{\alpha^2 - 1}},$$

which can be simplified to

$$d \frac{\alpha}{\alpha^2 - 1} = (9.6 \text{ nm}) \frac{|(-10/6)|}{(-10/6)^2 - 1} = 9.00 \text{ nm}.$$

P28-13 (a) We follow the work done in Section 28-6 for a uniform line of charge, starting with Eq. 28-26,

$$
\begin{aligned}
dV &= \frac{1}{4\pi\epsilon_0} \frac{\lambda\, dx}{\sqrt{x^2 + y^2}}, \\
dV &= \frac{1}{4\pi\epsilon_0} \int_0^L \frac{kx\, dx}{\sqrt{x^2 + y^2}}, \\
&= \frac{k}{4\pi\epsilon_0} \left. \sqrt{x^2 + y^2} \right|_0^L, \\
&= \frac{k}{4\pi\epsilon_0} \left(\sqrt{L^2 + y^2} - y \right).
\end{aligned}
$$

I can look at this expression and verify that the potential does go to zero as $y \to \infty$. You should do the same; focus on the limiting values as y gets very large compared to L.

(b) The y component of the electric field can be found from

$$E_y = -\frac{\partial V}{\partial y},$$

which (using a computer-aided math program) is

$$E_y = \frac{k}{4\pi\epsilon_0} \left(1 - \frac{y}{\sqrt{L^2 + y^2}} \right).$$

(c) We could find E_x if we knew the x variation of V. But we don't; we only found the values of V along a fixed value of x.

(d) We want to find y such that the ratio

$$\left[\frac{k}{4\pi\epsilon_0}\left(\sqrt{L^2+y^2}-y\right)\right]\Big/\left[\frac{k}{4\pi\epsilon_0}(L)\right]$$

is one-half. Simplifying,

$$\sqrt{L^2+y^2}-y = L/2,$$

which can be written as

$$L^2+y^2 = L^2/4 + Ly + y^2,$$

or

$$3L^2/4 = Ly,$$

with solution $y = 3L/4$.

P28-15 Calculating the fraction of excess electrons is the same as calculating the fraction of excess charge, so we'll skip counting the electrons. This problem is effectively the same as Exercise 28-47; we have a total charge that is divided between two unequal size spheres which are at the same potential on the surface. Using the result from that exercise we have

$$q_1 = \frac{Qr_1}{r_2+r_1},$$

where $Q = -6.2$ nC is the total charge available, and q_1 is the charge left on the sphere. r_1 is the radius of the small ball, r_2 is the radius of Earth. Since the fraction of charge remaining is q_1/Q, we can write

$$\frac{q_1}{Q} = \frac{r_1}{r_2+r_1} \approx \frac{r_1}{r_2} = 2.0 \times 10^{-8}$$

As small as it is, it looks as if we have finished this chapter with no extra charge.

Chapter 29

The Electrical Properties of Materials

E29-1 (a)   The charge which flows through a cross sectional surface area in a time t is given by Eq. 29-2,

$$q = it,$$

where i is the current. For this exercise we have

$$q = (4.82\,\text{A})(4.60 \times 60\,\text{s}) = 1330\,\text{C}$$

as the charge which passes through a cross section of this resistor.

   (b)   The number of electrons is given by $(1330\,\text{C})/(1.60 \times 10^{-19}\,\text{C}) = 8.31 \times 10^{21}$ electrons.

E29-5   The current rating of a fuse of cross sectional area A would be

$$i_{\max} = (440\,\text{A/cm}^2)A,$$

and if the fuse wire is cylindrical $A = \pi d^2/4$. Then

$$d = \sqrt{\frac{4}{\pi}\frac{(0.552\,\text{A})}{(440\,\text{A/m}^2)}} = 4.00 \times 10^{-2}\ \text{cm}.$$

E29-11   The drift velocity is given by Eq. 29-6,

$$v_{\text{d}} = \frac{j}{ne} = \frac{i}{Ane} = \frac{(115\,\text{A})}{(31.2 \times 10^{-6}\text{m}^2)(8.49 \times 10^{28}/m^3)(1.60 \times 10^{-19}\text{C})} = 2.71 \times 10^{-4}\text{m/s}.$$

We used the electron density for copper as was previously found in Sample Problem 29-3.

The time it takes for the electrons to get to the starter motor is

$$t = \frac{x}{v} = \frac{(0.855\,\text{m})}{(2.71 \times 10^{-4}\text{m/s})} = 3.26 \times 10^3\text{s}.$$

That's about 54 minutes. The battery would be dead long before any of the electrons actually completed the journey!

E29-13 The resistance of an object with constant cross section is given by Eq. 29-13,

$$R = \rho\frac{L}{A} = (3.0 \times 10^{-7}\,\Omega \cdot \text{m})\frac{(11,000\,\text{m})}{(0.0056\,\text{m}^2)} = 0.59\,\Omega.$$

This is a small resistance; when considering power transmission lines between states the resistances can become significant. There is certainly a great deal of resistance between California and the rest of the United States while I write this.

E29-17 Start with Eq. 29-16,

$$\rho - \rho_0 = \rho_0\alpha_{\text{av}}(T - T_0),$$

and multiply through by L/A,

$$\frac{L}{A}(\rho - \rho_0) = \frac{L}{A}\rho_0\alpha_{\text{av}}(T - T_0),$$

to get

$$R - R_0 = R_0\alpha_{\text{av}}(T - T_0).$$

This only works if we can ignore variations in L or A with temperature.

E29-19 We'll use Eq. 29-13 again. If the length of each conductor is L and has resistivity ρ, then

$$R_A = \rho\frac{L}{\pi D^2/4} = \rho\frac{4L}{\pi D^2}$$

and

$$R_B = \rho\frac{L}{(\pi 4D^2/4 - \pi D^2/4)} = \rho\frac{4L}{3\pi D^2}.$$

The ratio of the resistances is then

$$\frac{R_A}{R_B} = 3.$$

E29-23 Conductivity is given by Eq. 29-8, $\vec{j} = \sigma\vec{E}$. If the wire is long and thin (but not necessarily straight), then the magnitude of the electric field in the wire will be given by

$$E \approx \Delta V/L = (115\,\text{V})/(9.66\,\text{m}) = 11.9\,\text{V/m}.$$

This is an important approximation. Be prepared to use it again some day!
We can now find the conductivity,

$$\sigma = \frac{j}{E} = \frac{(1.42\times10^4\text{A/m}^2)}{(11.9\,\text{V/m})} = 1.19\times10^3(\Omega\cdot\text{m})^{-1}.$$

E29-27 (a) The resistance is defined as

$$R = \frac{\Delta V}{i} = \frac{(3.55\times10^6\,\text{V/A}^2)i^2}{i} = (3.55\times10^6\,\text{V/A}^2)i.$$

When $i = 2.40$ mA the resistance would be

$$R = (3.55\times10^6\,\text{V/A}^2)(2.40\times10^{-3}\text{A}) = 8.52\,\text{k}\Omega.$$

(b) Invert the above expression, and

$$i = R/(3.55\times10^6\,\text{V/A}^2) = (16.0\,\Omega)/(3.55\times10^6\,\text{V/A}^2) = 4.51\,\mu\text{A}.$$

E29-31 (a) At the surface of a conductor of radius R with charge Q the magnitude of the electric field is given by

$$E = \frac{1}{4\pi\epsilon_0}QR^2,$$

while the potential (assuming $V = 0$ at infinity) is given by

$$V = \frac{1}{4\pi\epsilon_0}QR.$$

The ratio is $V/E = R$.
The dielectric strength of air is found in Chapter 25 and is 3×10^6N/C. The potential on the sphere that would result in "sparking" is

$$V = ER = (3\times10^6\text{N/C})R.$$

(b) It is "easier" to get a spark off of a sphere with a smaller radius, because any potential on the sphere will result in a larger electric field.

(c) The points of a lighting rod are like small hemispheres; the electric field will be large near these points so that this will be the likely place for sparks to form and lightning bolts to strike.

P29-1 If there is more current flowing into the sphere than is flowing out then there must be a change in the net charge on the sphere. The net current is the difference, or $2\,\mu\text{A}$. The potential on the surface of the sphere will be given by the point-charge expression,

$$V = \frac{1}{4\pi\epsilon_0}\frac{q}{r},$$

and the charge will be related to the current by $q = it$. Combining,

$$V = \frac{1}{4\pi\epsilon_0}\frac{it}{r},$$

or

$$t = \frac{4\pi\epsilon_0 V r}{i} = \frac{4\pi(8.85\times 10^{-12}\,\text{C}^2/\text{N}\cdot\text{m}^2)(980\,\text{V})(0.13\,\text{m})}{(2\,\mu\text{A})} = 7.1\,\text{ms}.$$

That did not take very long. We would build up enough charge on the sphere to generate a spark in a second or so with this small current difference. It is therefore reasonable to assume that the current into an object is balanced by the current out of an object under most circumstances; this is definitely true under steady-state conditions.

P29-9 Originally we have a resistance R_1 made out of a wire of length l_1 and cross sectional area A_1. The volume of this wire is $V_1 = A_1 l_1$. When the wire is drawn out to the new length we have $l_2 = 3l_1$, but the volume of the wire should be constant so

$$
\begin{aligned}
A_2 l_2 &= A_1 l_1, \\
A_2 (3l_1) &= A_1 l_1, \\
A_2 &= A_1/3.
\end{aligned}
$$

The original resistance is

$$R_1 = \rho\frac{l_1}{A_1}.$$

The new resistance is

$$R_2 = \rho\frac{l_2}{A_2} = \rho\frac{3l_1}{A_1/3} = 9R_1,$$

or $R_2 = 54\,\Omega$.

P29-13 We will use the results of Exercise 29-17,

$$R - R_0 = R_0\alpha_{\text{av}}(T - T_0).$$

To save on subscripts we will drop the "av" notation, and just specify whether it is carbon "c" or iron "i".

The disks will be effectively in series, so we will add the resistances to get the total. Looking only at *one* disk pair, we have

$$
\begin{aligned}
R_c + R_i &= R_{0,c}\left(\alpha_c(T - T_0) + 1\right) + R_{0,i}\left(\alpha_i(T - T_0) + 1\right), \\
&= R_{0,c} + R_{0,i} + \left(R_{0,c}\alpha_c + R_{0,i}\alpha_i\right)(T - T_0).
\end{aligned}
$$

This last equation will only be constant if the coefficient for the term $(T - T_0)$ vanishes. Then

$$
R_{0,c}\alpha_c + R_{0,i}\alpha_i = 0,
$$

but $R = \rho L/A$, and the disks have the same cross sectional area, so

$$
L_c \rho_c \alpha_c + L_i \rho_i \alpha_i = 0,
$$

or

$$
\frac{L_c}{L_i} = -\frac{\rho_i \alpha_i}{\rho_c \alpha_c} = -\frac{(9.68 \times 10^{-8}\,\Omega \cdot \mathrm{m})(6.5 \times 10^{-3}/\mathrm{C}^\circ)}{(3500 \times 10^{-8}\,\Omega \cdot \mathrm{m})(-0.50 \times 10^{-3}/\mathrm{C}^\circ)} = 0.036.
$$

P29-15 Resistance is defined as $\Delta V/i$. The current is found from Eq. 29-5,

$$
i = \int \vec{\mathbf{j}} \cdot d\vec{\mathbf{A}},
$$

where the region of integration is over a spherical shell concentric with the two conducting shells but between them. The current density is given by Eq. 29-10,

$$
\vec{\mathbf{j}} = \vec{\mathbf{E}}/\rho,
$$

and we will have an electric field which is perpendicular to the spherical shell. Consequently,

$$
i = \frac{1}{\rho} \int \vec{\mathbf{E}} \cdot d\vec{\mathbf{A}} = \frac{1}{\rho} \int E\, dA
$$

By symmetry we expect the electric field to have the same magnitude anywhere on a spherical shell which is concentric with the two conducting shells, so we can bring it out of the integral sign, and then

$$
i = \frac{1}{\rho} E \int dA = \frac{4\pi r^2 E}{\rho},
$$

where E is the magnitude of the electric field on the shell, which has radius r such that $b > r > a$.

Now to sort out the voltage difference between the conducting shells. The above expression can be inverted to give the electric field as a function of radial distance, since the current is a constant in the above expression. Then

$$
E = \frac{i\rho}{4\pi r^2}
$$

The potential is given by

$$\Delta V = -\int_b^a \vec{\mathbf{E}} \cdot d\vec{\mathbf{s}},$$

we will integrate along a radial line, which is parallel to the electric field, so

$$\begin{aligned}
\Delta V &= -\int_b^a E \, dr, \\
&= -\int_b^a \frac{i\rho}{4\pi r^2} \, dr, \\
&= -\frac{i\rho}{4\pi} \int_b^a \frac{dr}{r}, \\
&= \frac{i\rho}{4\pi} \left(\frac{1}{a} - \frac{1}{b} \right).
\end{aligned}$$

We divide this expression by the current to get the resistance. Then

$$R = \frac{\rho}{4\pi} \left(\frac{1}{a} - \frac{1}{b} \right)$$

Chapter 30

Capacitance

E30-1 We apply Eq. 30-1,

$$q = C\Delta V = (50 \times 10^{-12}\,\text{F})(0.15\,\text{V}) = 7.5 \times 10^{-12}\,\text{C};$$

This might not seem like much, but it is still some fifty million electrons.

E30-5 Eq. 30-11 gives the capacitance of a cylinder,

$$C = 2\pi\epsilon_0 \frac{L}{\ln(b/a)} = 2\pi(8.85 \times 10^{-12}\,\text{F/m}) \frac{(0.0238\,\text{m})}{\ln((9.15\text{mm})/(0.81\text{mm}))} = 5.46 \times 10^{-13}\text{F}.$$

E30-9 The potential difference across each capacitor in parallel is the same; it is equal to 110 V. The charge on each of the capacitors is then

$$q = C\Delta V = (1.00 \times 10^{-6}\,\text{F})(110\,\text{V}) = 1.10 \times 10^{-4}\,\text{C}.$$

If there are N capacitors, then the total charge will be Nq, and we want this total charge to be 1.00 C. Then

$$N = \frac{(1.00\,\text{C})}{q} = \frac{(1.00\,\text{C})}{(1.10 \times 10^{-4}\,\text{C})} = 9090.$$

E30-13 (a) The equivalent capacitance is given by Eq. 30-21,

$$\frac{1}{C_{\text{eq}}} = \frac{1}{C_1} + \frac{1}{C_2} = \frac{1}{(4.0\mu\text{F})} + \frac{1}{(6.0\mu\text{F})} = \frac{5}{(12.0\mu\text{F})}$$

or $C_{\text{eq}} = 2.40\mu\text{F}$.

(b) The charge on the equivalent capacitor is $q = C\Delta V = (2.40\mu\text{F})(200\,\text{V}) = 0.480\,\text{mC}$. For parallel capacitors, the charge on the equivalent capacitor is the same as the charge on each of the capacitors.

(c) The potential difference across the equivalent capacitor is *not* the same as the potential difference across each of the individual capacitors. We need to apply $q = C\Delta V$ to each capacitor using the charge from part (b). Then for the $4.0\mu\text{F}$ capacitor,

$$\Delta V = \frac{q}{C} = \frac{(0.480\,\text{mC})}{(4.0\mu\text{F})} = 120\,\text{V};$$

and for the $6.0\mu\text{F}$ capacitor,

$$\Delta V = \frac{q}{C} = \frac{(0.480\,\text{mC})}{(6.0\mu\text{F})} = 80\,\text{V}.$$

Note that the sum of the potential differences across each of the capacitors is equal to the potential difference across the equivalent capacitor.

E30-19 Consider any junction other than A or B. Call this junction point 0; label the four nearest junctions to this as points 1, 2, 3, and 4. The charge on the capacitor that links point 0 to point 1 is

$$q_1 = C\Delta V_{01},$$

where ΔV_{01} is the potential difference across the capacitor, so

$$\Delta V_{01} = V_0 - V_1,$$

where V_0 is the potential at the junction 0, and V_1 is the potential at the junction 1. Similar expressions exist for the other three capacitors.

For the junction 0 the net charge must be zero; there is no way for charge to cross the plates of the capacitors. Then

$$q_1 + q_2 + q_3 + q_4 = 0,$$

and this means

$$C\Delta V_{01} + C\Delta V_{02} + C\Delta V_{03} + C\Delta V_{04} = 0$$

or

$$\Delta V_{01} + \Delta V_{02} + \Delta V_{03} + \Delta V_{04} = 0.$$

Let $\Delta V_{0i} = V_0 - V_i$, and then rearrange,

$$4V_0 = V_1 + V_2 + V_3 + V_4,$$

or

$$V_0 = \frac{1}{4}\left(V_1 + V_2 + V_3 + V_4\right).$$

E30-23 (a) The capacitance of an air filled parallel-plate capacitor is given by Eq. 30-5,

$$C = \frac{\epsilon_0 A}{d} = \frac{(8.85 \times 10^{-12}\text{F/m})(42.0 \times 10^{-4}\text{m}^2)}{(1.30 \times 10^{-3}\text{m})} = 2.86 \times 10^{-11}\,\text{F}.$$

(b) The magnitude of the charge on each plate is given by

$$q = C\Delta V = (2.86 \times 10^{-11}\,\text{F})(625\,\text{V}) = 1.79 \times 10^{-8}\,\text{C}.$$

(c) The stored energy in a capacitor is given by Eq. 30-25, regardless of the type or shape of the capacitor, so

$$U = \frac{1}{2}C(\Delta V)^2 = \frac{1}{2}(2.86 \times 10^{-11}\,\text{F})(625\,\text{V})^2 = 5.59\,\mu\text{J}.$$

(d) Assuming a parallel plate arrangement with *no* fringing effects, the magnitude of the electric field between the plates is given by $Ed = \Delta V$, where d is the separation between the plates. Then

$$E = \Delta V/d = (625\,\text{V})/(0.00130\,\text{m}) = 4.81 \times 10^5\,\text{V/m}.$$

We *did not* need to know the capacitance to answer this question!

(e) The energy density is Eq. 30-28,

$$u = \frac{1}{2}\epsilon_0 E^2 = \frac{1}{2}((8.85 \times 10^{-12}\text{F/m}))(4.81 \times 10^5\,\text{V/m})^2 = 1.02\,\text{J/m}^3.$$

E30-27 There is enough work on this problem without deriving once again the electric field between charged cylinders. I will instead refer you back to Section 26-4, and state

$$E = \frac{1}{2\pi\epsilon_0}\frac{q}{Lr},$$

where q is the magnitude of the charge on a cylinder and L is the length of the cylinders.

The energy density as a function of radial distance is found from Eq. 30-28,

$$u = \frac{1}{2}\epsilon_0 E^2 = \frac{1}{8\pi^2\epsilon_0}\frac{q^2}{L^2r^2}$$

The total energy stored in the electric field is given by Eq. 30-24,

$$U = \frac{1}{2}\frac{q^2}{C} = \frac{q^2}{2}\frac{\ln(b/a)}{2\pi\epsilon_0 L},$$

where we substituted into the last part Eq. 30-11, the capacitance of a cylindrical capacitor.

We want to show that integrating a volume integral from $r = a$ to $r = \sqrt{ab}$ over the energy density function will yield $U/2$. Since we want to do this problem the hard way, we will pretend we don't know the answer, and integrate from $r = a$ to $r = c$, and then find out what c is.

The volume differential is dV, which in cylindrical coordinates is

$$dV = (dr)(r\,d\phi)(dz),$$

another one of the relationships that you should remember always. Then

$$
\begin{aligned}
\frac{1}{2}U &= \int u\,dV, \\
&= \int_a^c \int_0^{2\pi} \int_0^L \left(\frac{1}{8\pi^2\epsilon_0} \frac{q^2}{L^2 r^2} \right) r\,dr\,d\phi\,dz, \\
&= \frac{q^2}{8\pi^2\epsilon_0 L^2} \int_a^c \int_0^{2\pi} \int_0^L \frac{dr}{r} d\phi\,dz, \\
&= \frac{q^2}{4\pi\epsilon_0 L} \int_a^c \frac{dr}{r}, \\
&= \frac{q^2}{4\pi\epsilon_0 L} \ln\frac{c}{a}.
\end{aligned}
$$

Not too much exciting in that derivation; in the third line we pulled out the constants, and in the fourth line we integrated over ϕ and z, which was trivial. The fifth line was the r integration.

Now we equate this to the value for U that we found above, and we solve for c.

$$
\begin{aligned}
\frac{1}{2}\frac{q^2}{2}\frac{\ln(b/a)}{2\pi\epsilon_0 L} &= \frac{q^2}{4\pi\epsilon_0 L}\ln\frac{c}{a}, \\
\ln(b/a) &= 2\ln(c/a), \\
(b/a) &= (c/a)^2, \\
\sqrt{ab} &= c.
\end{aligned}
$$

Look, we got the right answer!

E30-31 Capacitance with dielectric media is given by Eq. 30-31,

$$C = \frac{\kappa_e \epsilon_0 A}{d}.$$

The various sheets have different dielectric constants and different thicknesses, and we want to maximize C, which means maximizing κ_e/d. For mica this ratio is 54 mm^{-1}, for glass this ratio is 35 mm^{-1}, and for paraffin this ratio is 0.20 mm^{-1}. Mica wins.

E30-35 (a) The capacitance of a cylindrical capacitor is given by Eq. 30-11,

$$C = 2\pi\epsilon_0 \kappa_e \frac{L}{\ln(b/a)}.$$

The factor of κ_e is introduced because there is now a dielectric (the Pyrex drinking glass) between the plates. We can look back to Table 29-2 to get the dielectric properties of Pyrex. The capacitance of our "glass" is then

$$C = 2\pi(8.85\times 10^{-12} \text{F/m})(4.7)\frac{(0.15\,\text{m})}{\ln((3.8\,\text{cm})/(3.6\,\text{cm}))} = 7.3\times 10^{-10}\,\text{F}.$$

(b) The breakdown potential is $(14\text{ kV/mm})(2\text{ mm}) = 28$ kV.

E30-37 (a) Insert the slab so that it is a distance a above the lower plate. Then the distance between the slab and the upper plate is $d - a - b$. Inserting the slab has the same effect as having two capacitors wired in series; the separation of the bottom capacitor is a, while that of the top capacitor is $d - a - b$.

The bottom capacitor has a capacitance of

$$C_1 = \frac{\epsilon_0 A}{a},$$

while the top capacitor has a capacitance of

$$C_2 = \frac{\epsilon_0 A}{d - a - b}.$$

Adding these in series,

$$
\begin{aligned}
\frac{1}{C_{\text{eq}}} &= \frac{1}{C_1} + \frac{1}{C_2}, \\
&= \frac{a}{\epsilon_0 A} + \frac{d - a - b}{\epsilon_0 A}, \\
&= \frac{d - b}{\epsilon_0 A}.
\end{aligned}
$$

So the capacitance of the system after putting the copper slab in is

$$C = \frac{\epsilon_0 A}{d - b}.$$

(b) The energy stored in the system before the slab is inserted is

$$U_i = \frac{q^2}{2C_i} = \frac{q^2}{2} \frac{d}{\epsilon_0 A}$$

while the energy stored after the slab is inserted is

$$U_f = \frac{q^2}{2C_f} = \frac{q^2}{2} \frac{d-b}{\epsilon_0 A}$$

The ratio is

$$\frac{U_i}{U_f} = \frac{d}{d-b}.$$

(c) Since there was more energy *before* the slab was inserted, then the slab must have gone in willingly, *it was pulled in!*. To get the slab back out we will need to do work on the slab equal to the energy difference.

$$U_i - U_f = \frac{q^2}{2} \frac{d}{\epsilon_0 A} - \frac{q^2}{2} \frac{d-b}{\epsilon_0 A} = \frac{q^2}{2} \frac{b}{\epsilon_0 A}.$$

P30-1 The capacitance of the cylindrical capacitor is from Eq. 30-11,

$$C = \frac{2\pi\epsilon_0 L}{\ln(b/a)}.$$

If the cylinders are very close together we can write $b = a+d$, where d, the separation between the cylinders, is a small number, so

$$C = \frac{2\pi\epsilon_0 L}{\ln\left((a+d)/a\right)} = \frac{2\pi\epsilon_0 L}{\ln\left(1+d/a\right)}.$$

Expanding according to the hint,

$$C \approx \frac{2\pi\epsilon_0 L}{d/a} = \frac{2\pi a\epsilon_0 L}{d}.$$

Now $2\pi a$ is the circumference of the cylinder, and L is the length, so $2\pi a L$ is the area of a cylindrical plate. Hence, for small separation between the cylinders we have

$$C \approx \frac{\epsilon_0 A}{d},$$

which is the expression for the parallel plates.

Problems like this are good to be able to do; you might want to verify that the closely spaced concentric spherical shells are also well approximated by parallel plates.

P30-3 See the first part of Exercise 30-37 above.

P30-5 We first need to find the charge on each capacitor. After finding this we can figure out how the charge will redistribute when the switches are closed. Because we want to be able to use our results for *any* problem of this sort, we will avoid numbers until the very end. Let $\Delta V_0 = 96.6\,\text{V}$.

As far as point e is concerned point a looks like it is originally positively charged, and point d is originally negatively charged. It is then convenient to define the charges on the capacitors in terms of the charges on the top sides, so the original charge on C_1 is $q_{1,i} = C_1 \Delta V_0$ while the original charge on C_2 is $q_{2,i} = -C_2 \Delta V_0$. Note the negative sign reflecting the opposite polarity of C_2.

(a) Conservation of charge requires

$$q_{1,i} + q_{2,i} = q_{1,f} + q_{2,f},$$

but since $q = C\Delta V$ and the two capacitors will be at the same potential after the switches are closed we can write

$$C_1 \Delta V_0 - C_2 \Delta V_0 = C_1 \Delta V + C_2 \Delta V,$$
$$\left(C_1 - C_2\right)\Delta V_0 = \left(C_1 + C_2\right)\Delta V,$$
$$\frac{C_1 - C_2}{C_1 + C_2}\Delta V_0 = \Delta V.$$

With numbers,

$$\Delta V = (96.6\,\text{V})\frac{(1.16\,\mu\text{F}) - (3.22\,\mu\text{F})}{(1.16\,\mu\text{F}) + (3.22\,\mu\text{F})} = -45.4\,\text{V}.$$

The negative sign means that the top sides of *both* capacitor will be negatively charged after the switches are closed.

(b) The charge on C_1 is $C_1 \Delta V = (1.16\,\mu\text{F})(45.4\,\text{V}) = 52.7\mu\text{C}$.

(2) The charge on C_2 is $C_2 \Delta V = (3.22\,\mu\text{F})(45.4\,\text{V}) = 146\mu\text{C}$.

P30-7 (a) If terminal a is more positive than terminal b then current can flow that will charge the capacitor on the left, the current can flow through the diode on the top, and the current can charge the capacitor on the right. Current will not flow through the diode on the left. The capacitors are effectively in series.

Since the capacitors are identical and series capacitors have the same charge, we expect the capacitors to have the same potential difference across them. But the total potential difference across both capacitors is equal to 100 V, so the potential difference across either capacitor is 50 V.

The output pins are connected to the capacitor on the right, so the potential difference across the output is 50 V.

(b) If terminal b is more positive than terminal a the current can flow through the diode on the left. If we assume the diode is resistanceless in this configuration then the potential difference across it will be zero. The net result is that the potential difference across the output pins is 0 V.

In real life the potential difference across the diode would not be zero, even if forward biased. It will be somewhere around 0.5 Volts.

P30-11 (a) The charge on the capacitor with stored energy $U_0 = 4.0\,\text{J}$ is q_0, where

$$U_0 = \frac{q_0^2}{2C}.$$

When this capacitor is connected to an identical uncharged capacitor the charge is shared equally, so that the charge on either capacitor is now $q = q_0/2$. The stored energy in *one* capacitor is then

$$U = \frac{q^2}{2C} = \frac{q_0^2/4}{2C} = \frac{1}{4}U_0.$$

But there are two capacitors, so the total energy stored is $2U = U_0/2 = 2.0\,\text{J}$.

(b) Good question. Current had to flow through the connecting wires to get the charge from one capacitor to the other. Originally the second capacitor was uncharged, so the potential difference across that capacitor would have been zero, which means the potential difference across the connecting wires would have been equal to that of the first capacitor, and there would then have been energy dissipation in the wires according to

$$P = i^2 R.$$

That's where the missing energy went.

If you are a true glutton for punishment, you would do the math to verify this. If you are not a true glutton for punishment then you could patiently wait until you get to Problem 31-18, which is similar, but not identical. Note that we never need to know the value for the resistance!

P30-15 According to Problem 14, the force on a plate of a parallel plate capacitor is

$$F = \frac{q^2}{2\epsilon_0 A}.$$

The force per unit area is then

$$\frac{F}{A} = \frac{q^2}{2\epsilon_0 A^2} = \frac{\sigma^2}{2\epsilon_0},$$

where $\sigma = q/A$ is the surface charge density. But we know that the electric field near the surface of a conductor is given by $E = \sigma/\epsilon_0$, so

$$\frac{F}{A} = \frac{1}{2}\epsilon_0 E^2.$$

P30-19   We will treat the system as two capacitors in series by pretending there is an infinitesimally thin conductor between them. The slabs are (I assume) the same thickness. The capacitance of one of the slabs is then given by Eq. 30-31,

$$C_1 = \frac{\kappa_{e1}\epsilon_0 A}{d/2},$$

where $d/2$ is the thickness of the slab. There would be a similar expression for the other slab. The equivalent series capacitance would be given by Eq. 30-21,

$$\frac{1}{C_{eq}} = \frac{1}{C_1} + \frac{1}{C_2},$$
$$= \frac{d/2}{\kappa_{e1}\epsilon_0 A} + \frac{d/2}{\kappa_{e2}\epsilon_0 A},$$
$$= \frac{d}{2\epsilon_0 A}\frac{\kappa_{e2} + \kappa_{e1}}{\kappa_{e1}\kappa_{e2}},$$
$$C_{eq} = \frac{2\epsilon_0 A}{d}\frac{\kappa_{e1}\kappa_{e2}}{\kappa_{e2} + \kappa_{e1}}.$$

The most obvious limiting case would be $\kappa_{e2} = \kappa_{e1}$, in which case the capacitance would simplify to

$$C_{eq} = \frac{2\epsilon_0 A}{d}\frac{\kappa_{e1}\kappa_{e1}}{\kappa_{e1} + \kappa_{e1}} = \frac{2\epsilon_0 A}{d}\frac{\kappa_{e1}}{2},$$

which is what we would expect.

Chapter 31

DC Circuits

E31-3 If the energy is delivered at a rate of 110 W, then the current through the battery is

$$i = \frac{P}{\Delta V} = \frac{(110\,\text{W})}{(12\,\text{V})} = 9.17\,\text{A}.$$

Current is the flow of charge in some period of time, so

$$\Delta t = \frac{\Delta q}{i} = \frac{(125\,\text{A} \cdot \text{h})}{(9.2\,\text{A})} = 13.6\,\text{h},$$

which is the same as 13 hours and 36 minutes.

E31-5 Go all of the way around the circuit. It is a simple one loop circuit, and although it does not matter which way we go around, we will follow the direction of the larger emf. Then

$$(150\,\text{V}) - i(2.0\,\Omega) - (50\,\text{V}) - i(3.0\,\Omega) = 0,$$

where i is positive if it is counterclockwise. Rearranging,

$$100\,\text{V} = i(5.0\,\Omega),$$

or $i = 20\,\text{A}$.

Assuming the potential at P is $V_P = 100\,\text{V}$, then the potential at Q will be given by

$$V_Q = V_P - (50\,\text{V}) - i(3.0\,\Omega) = (100\,\text{V}) - (50\,\text{V}) - (20\,\text{A})(3.0\,\Omega) = -10\,\text{V}.$$

Just a reminder: we went around the loop in a counterclockwise direction.

E31-11 We assign directions to the currents through the four resistors as shown in the figure.

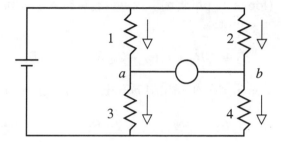

Since the ammeter has no resistance the potential at a is the same as the potential at b. Consequently the potential difference (ΔV_b) across both of the bottom resistors is the same, and the potential difference (ΔV_t) across the two top resistors is also the same (but different from the bottom). We then have the following relationships:

$$\Delta V_t + \Delta V_b = \mathcal{E},$$
$$i_1 + i_2 = i_3 + i_4,$$
$$\Delta V_j = i_j R_j,$$

where the j subscript in the last line refers to resistor 1, 2, 3, or 4.

For the top resistors,

$$\Delta V_1 = \Delta V_2 \text{ implies } 2i_1 = i_2;$$

while for the bottom resistors,

$$\Delta V_3 = \Delta V_4 \text{ implies } i_3 = i_4.$$

Then the junction rule requires $i_4 = 3i_1/2$, and the loop rule requires

$$(i_1)(2R) + (3i_1/2)(R) = \mathcal{E} \text{ or } i_1 = 2\mathcal{E}/(7R).$$

The current that flows through the ammeter is the difference between i_2 and i_4, or $4\mathcal{E}/(7R) - 3\mathcal{E}/(7R) = \mathcal{E}/(7R)$.

E31-13 (a) Assume that the current flows through each source of emf in the same direction as the emf. The the loop rule will give us three equations

$$\mathcal{E}_1 - i_1 R_1 + i_2 R_2 - \mathcal{E}_2 - i_1 R_1 = 0,$$
$$\mathcal{E}_2 - i_2 R_2 + i_3 R_1 - \mathcal{E}_3 + i_3 R_1 = 0,$$
$$\mathcal{E}_1 - i_1 R_1 + i_3 R_1 - \mathcal{E}_3 + i_3 R_1 - i_1 R_1 = 0.$$

These are not independent equations; the third equation can be found by adding the first and the second. Note that we sometimes have positive signs and sometimes negative. The sign will be positive if you go "forward" through a source of emf, or positive is you go against the assigned current direction through a resistance.

The junction rule (looks at point a) gives us $i_1 + i_2 + i_3 = 0$. Use this to eliminate i_2 from the second loop equation,

$$\mathcal{E}_2 + i_1 R_2 + i_3 R_2 + 2i_3 R_1 - \mathcal{E}_3 = 0,$$

and then combine this with the the third equation to eliminate i_3,

$$\mathcal{E}_1 R_2 - \mathcal{E}_3 R_2 + 2i_3 R_1 R_2 + 2\mathcal{E}_2 R_1 + 2i_3 R_1 R_2 + 4i_3 R_1^2 - 2\mathcal{E}_3 R_1 = 0,$$

or

$$i_3 = \frac{2\mathcal{E}_3 R_1 + \mathcal{E}_3 R_2 - \mathcal{E}_1 R_2 - 2\mathcal{E}_2 R_1}{4R_1 R_2 + 4R_1^2} = 0.582\,\text{A}.$$

Then we can find i_1 from

$$i_1 = \frac{\mathcal{E}_3 - \mathcal{E}_2 - i_3 R_2 - 2i_3 R_1}{R_2} = -0.668\,\text{A},$$

where the negative sign indicates the current is *down*.

Finally, we can find i_2 from

$$i_2 = -(i_1 + i_3) = 0.0854\,\text{A}.$$

(b) Start at a and go to b (final minus initial!),

$$+i_2 R_2 - \mathcal{E}_2 = -3.60\,\text{V}.$$

E31-15 (a) We first use $P = Fv$ to find the power output by the electric motor. Then $P = (2.0\,\text{N})(0.50\,\text{m/s}) = 1.0\,\text{W}$.

The potential difference across the motor is

$$\Delta V_{\text{motor}} = \mathcal{E} - ir.$$

The power output from the motor is the rate of energy dissipation, so

$$P_{\text{motor}} = \Delta V_{\text{motor}} i.$$

Combining these two expressions,

$$\begin{aligned} P_{\text{motor}} &= (\mathcal{E} - ir)\,i, \\ &= \mathcal{E}i - i^2 r, \\ 0 &= -i^2 r + \mathcal{E}i - P_{\text{motor}}, \\ 0 &= (0.50\,\Omega)i^2 - (2.0\,\text{V})i + (1.0\,\text{W}). \end{aligned}$$

Rearrange and solve for i,

$$i = \frac{(2.0\,\text{V}) \pm \sqrt{(2.0\,\text{V})^2 - 4(0.50\,\Omega)(1.0\,\text{W})}}{2(0.50\,\Omega)},$$

which has solutions $i = 3.4\,\text{A}$ and $i = 0.59\,\text{A}$.

But which is right? There is not any reason at this point to toss out either answer, so we'll go on to part (b), and then see what happens.

(b) The potential difference across the terminals of the motor is

$$\Delta V_{\text{motor}} = \mathcal{E} - ir$$

which if $i = 3.4\,\text{A}$ yields $\Delta V_{\text{motor}} = 0.3\,\text{V}$, but if $i = 0.59\,\text{A}$ yields $\Delta V_{\text{motor}} = 1.7\,\text{V}$. The battery provides an emf of 2.0 V; it isn't possible for the potential difference across the motor to be larger than this, but both solutions seem to satisfy this constraint, so we will move to the next part and see what happens.

(c) So what is the significance of the two possible solutions? It is a consequence of the fact that power is related to the current squared, and with any quadratics we expect two solutions. Both are possible, but it might be that only one is stable, or even that neither is stable, and a small perturbation to the friction involved in turning the motor will cause the system to break down. We will learn in a later chapter that the effective resistance of an electric motor depends on the speed at which it is spinning, and although that won't affect the problem here as worded, it will affect the physical problem that provided the numbers in this problem!

E31-17 Refer back to the multiple choice questions 8 and 9. There are two very important relations to remember. In parallel connections of two resistors the effective resistance is less than the smaller resistance but larger than half the smaller resistance. In series connections of two resistors the effective resistance is greater than the larger resistance but less than twice the larger resistance.

What does that do for us here? Since the effective resistance of the parallel combination is less than either single resistance and the effective resistance of the series combinations is larger than either single resistance we can conclude that $3.0\,\Omega$ must have been the parallel combination and $16\,\Omega$ must have been the series combination.

The resistors are then $4.0\,\Omega$ and $12\,\Omega$ resistors.

E31-23 (a) Work through the circuit one step at a time. We first "add" R_2, R_3, and R_4 in parallel:

$$\frac{1}{R_{\text{eff}}} = \frac{1}{42.0\,\Omega} + \frac{1}{61.6\,\Omega} + \frac{1}{75.0\,\Omega} = \frac{1}{18.7\,\Omega}$$

We then "add" this resistance in series with R_1,

$$R_{\text{eff}} = (112\,\Omega) + (18.7\,\Omega) = 131\,\Omega.$$

I understand that it is confusing to keep track of the various effective resistances. Drawing circuit diagrams over and over again like in Sample Problem 31-6 is the recommended method for keeping it straight.

(b) The current through the battery is $i = \mathcal{E}/R = (6.22\,\text{V})/(131\,\Omega) = 47.5$ mA. This is also the current through R_1, since all the current through the battery must also go through R_1.

The potential difference across R_1 is $\Delta V_1 = (47.5\,\text{mA})(112\,\Omega) = 5.32\,\text{V}$. The potential difference across each of the three remaining resistors is $6.22\,\text{V} - 5.32\,\text{V} = 0.90\,\text{V}$.

The current through each resistor is then

$$
\begin{aligned}
i_2 &= (0.90\,\text{V})/(42.0\,\Omega) = 21.4\text{ mA}, \\
i_3 &= (0.90\,\text{V})/(61.6\,\Omega) = 14.6\text{ mA}, \\
i_4 &= (0.90\,\text{V})/(75.0\,\Omega) = 12.0\text{ mA}.
\end{aligned}
$$

I realize that it does not add up to 47.5 mA like it should. That's significant figures for you!

E31-25 (a) Work through the circuit one step at a time. First "add" the left two resistors in series, the effective resistance of that branch is $2R$. Then "add" the right two resistors in series, the effective resistance of that branch is also $2R$.

Now we combine the three parallel branches and find the effective resistance to be

$$\frac{1}{R_{\text{eff}}} = \frac{1}{2R} + \frac{1}{R} + \frac{1}{2R} = \frac{4}{2R},$$

or $R_{\text{eff}} = R/2$.

(b) Again, we work through the problem one step at a time. First we "add" the right two resistors in series, the effective resistance of that branch is $2R$. We then combine this branch with the resistor which connects points F and H. This is a parallel connection, so the effective resistance is

$$\frac{1}{R_{\text{eff}}} = \frac{1}{2R} + \frac{1}{R} = \frac{3}{2R},$$

or $2R/3$.

This value is effectively in series with the resistor which connects G and H, so the "total" is $5R/3$.

Finally, we can combine this value in parallel with the resistor that directly connects F and G according to

$$\frac{1}{R_{\text{eff}}} = \frac{1}{R} + \frac{3}{5R} = \frac{8}{5R},$$

or $R_{\text{eff}} = 5R/8$.

E31-28 I promised the publishers that I wouldn't provide solutions for any even numbered exercises. But you might want to look at the solution to *odd* Exercise 30-19!

E31-29 The current through the radio is $i = P/\Delta V = (7.5\,\text{W})/(9.0\,\text{V}) = 0.83\,\text{A}$. The radio was left one for 6 hours, or $2.16 \times 10^4\,\text{s}$. The total charge to flow through the radio in that time is $(0.83\,\text{A})(2.16 \times 10^4\,\text{s}) = 1.8 \times 10^4\,\text{C}$.

E31-33 We want to apply either Eq. 31-21,

$$P_R = i^2 R,$$

or Eq. 31-22,

$$P_R = (\Delta V_R)^2/R,$$

depending on whether we are in series (the current is the same through each bulb), or in parallel (the potential difference across each bulb is the same. The brightness of a bulb will be measured by P, even though P is not necessarily a measure of the rate radiant energy is emitted from the bulb.

(b) If the bulbs are in parallel then $P_R = (\Delta V_R)^2/R$ is how we want to compare the brightness. The potential difference across each bulb is the same, so the bulb with the smaller resistance is brighter.

(b) If the bulbs are in series then $P_R = i^2 R$ is how we want to compare the brightness. Both bulbs have the same current, so the larger value of R results in the brighter bulb.

One direct consequence of this can be tried at home. Wire up a 60 W, 120 V bulb and a 100 W, 120 V bulb in series. Which is brighter? You should observe that the 60 W bulb will be brighter.

E31-39 (a) The current through the wire is

$$i = P/\Delta V = (4800\,\text{W})/(75\,\text{V}) = 64\,\text{A},$$

The resistance of the wire is

$$R = \Delta V / i = (75 \, \text{V})/(64 \, \text{A}) = 1.17 \, \Omega.$$

The length of the wire is then found from

$$L = \frac{RA}{\rho} = \frac{(1.17 \, \Omega)(2.6 \times 10^{-6} \, \text{m}^2)}{(5.0 \times 10^{-7} \, \Omega \text{m})} = 6.1 \, \text{m}.$$

One could easily wind this much nichrome to make a toaster oven. Of course allowing 64 Amps to be drawn through household wiring will likely blow a fuse.

(b) We want to combine the above calculations into one formula, so

$$L = \frac{RA}{\rho} = \frac{A \Delta V / i}{\rho} = \frac{A (\Delta V)^2}{P \rho},$$

then

$$L = \frac{(2.6 \times 10^{-6} \, \text{m}^2)(110 \, \text{V})^2}{(4800 \, \text{W})(5.0 \times 10^{-7} \, \Omega \text{m})} = 13 \, \text{m}.$$

Hmm. We need more wire if the potential difference is increased? Does this make sense? Yes, it does. We need more wire because we need more resistance to *decrease* the current so that the same power output occurs.

E31-41 (a) Integrate both sides of Eq. 31-26;

$$\int_0^q \frac{dq}{q - \mathcal{E}C} = -\int_0^t \frac{dt}{RC},$$

$$\ln(q - \mathcal{E}C)\big|_0^q = -\frac{t}{RC}\Big|_0^t,$$

$$\ln\left(\frac{q - \mathcal{E}C}{-\mathcal{E}C}\right) = -\frac{t}{RC},$$

$$\frac{q - \mathcal{E}C}{-\mathcal{E}C} = e^{-t/RC},$$

$$q = \mathcal{E}C\left(1 - e^{-t/RC}\right).$$

That wasn't so bad, was it?

(b) Rearrange Eq. 31-26 in order to get q terms on the left and t terms on the right, then integrate;

$$\int_{q_0}^q \frac{dq}{q} = -\int_0^t \frac{dt}{RC},$$

$$\ln q\big|_{q_0}^{q} = -\frac{t}{RC}\bigg|_0^t,$$

$$\ln\left(\frac{q}{q_0}\right) = -\frac{t}{RC},$$

$$\frac{q}{q_0} = e^{-t/RC},$$

$$q = q_0 e^{-t/RC}.$$

That wasn't so bad either, was it?

E31-47 The charge on the capacitor needs to build up to a point where the potential across the capacitor is $V_L = 72\,\text{V}$, and this needs to happen within 0.5 seconds. This means that we want to solve

$$C\Delta V_L = C\mathcal{E}\left(1 - e^{T/RC}\right)$$

for R knowing that $T = 0.5\,\text{s}$. This expression can be written as

$$R = -\frac{T}{C\ln(1 - V_L/\mathcal{E})} = -\frac{(0.5\,\text{s})}{(0.15\,\mu\text{C})\ln(1 - (72\,\text{V})/(95\,\text{V}))} = 2.35 \times 10^6\,\Omega.$$

P31-1 The terminal voltage of the battery is given by

$$V = \mathcal{E} - ir,$$

so the internal resistance is

$$r = \frac{\mathcal{E} - V}{i} = \frac{(12.0\,\text{V}) - (11.4\,\text{V})}{(50\,\text{A})} = 0.012\,\Omega,$$

so the battery appears within specs.

The resistance of the wire is given by

$$R = \frac{\Delta V}{i} = \frac{(3.0\,\text{V})}{(50\,\text{A})} = 0.06\,\Omega,$$

so the cable appears to be bad.

What about the motor? Trying it,

$$R = \frac{\Delta V}{i} = \frac{(11.4\,\text{V}) - (3.0\,\text{V})}{(50\,\text{A})} = 0.168\,\Omega,$$

so it appears to be within spec.

P31-9 (a) The three way light-bulb has two filaments (or so we are told in the question). There are four ways for these two filaments to be wired: either one alone, both in series, or both in parallel. Wiring the filaments in series will have the largest total resistance, and since $P = V^2/R$ this arrangement would result in the dimmest light. But we are told the light still operates at the lowest setting, and if a filament burned out in a series arrangement the light would go out.

We then conclude that the lowest setting is one filament, the middle setting is another filament, and the brightest setting is both filaments in parallel.

(b) The beauty of parallel settings is that then power is additive (it is also addictive, but that's a different field.) One filament dissipates 100 W at 120 V; the other filament (the one that burns out) dissipates 200 W at 120 V, and both together dissipate 300 W at 120 V.

The resistance of one filament is then

$$R = \frac{(\Delta V)^2}{P} = \frac{(120\,\text{V})^2}{(100\,\text{W})} = 144\,\Omega.$$

The resistance of the other filament is

$$R = \frac{(\Delta V)^2}{P} = \frac{(120\,\text{V})^2}{(200\,\text{W})} = 72\,\Omega.$$

P31-17 (a) We have $P = 30P_0$ and $i = 4i_0$. Then

$$R = \frac{P}{i^2} = \frac{30P_0}{(4i_0)^2} = \frac{30}{16}R_0.$$

We don't really care what happened with the potential difference, since knowing the change in resistance of the wire should give all the information we need.

The volume of the wire is a constant, even upon drawing the wire out, so $LA = L_0A_0$; the product of the length and the cross sectional area must be a constant.

Resistance is given by $R = \rho L/A$, but $A = L_0A_0/L$, so the length of the wire is

$$L = \sqrt{\frac{A_0 L_0 R}{\rho}} = \sqrt{\frac{30}{16}\frac{A_0 L_0 R_0}{\rho}} = 1.37L_0.$$

(b) We know that $A = L_0A_0/L$, so

$$A = \frac{L}{L_0}A_0 = \frac{A_0}{1.37} = 0.73A_0.$$

P31-19 The capacitor charge as a function of time is given by Eq. 31-27,

$$q = C\mathcal{E}\left(1 - e^{-t/RC}\right),$$

while the current through the circuit (and the resistor) is given by Eq. 31-28,

$$i = \frac{\mathcal{E}}{R}e^{-t/RC}.$$

The energy stored in the capacitor is given by

$$U = \frac{q^2}{2C},$$

so the rate that energy is being stored in the capacitor is

$$P_C = \frac{dU}{dt} = \frac{q}{C}\frac{dq}{dt} = \frac{q}{C}i.$$

The rate of energy dissipation in the resistor is

$$P_R = i^2 R,$$

so the time at which the rate of energy dissipation in the resistor is equal to the rate of energy storage in the capacitor can be found by solving

$$
\begin{aligned}
P_C &= P_R, \\
i^2 R &= \frac{q}{C}i, \\
iRC &= q, \\
\mathcal{E}Ce^{-t/RC} &= C\mathcal{E}\left(1 - e^{-t/RC}\right), \\
e^{-t/RC} &= 1/2, \\
t &= RC\ln 2.
\end{aligned}
$$

Chapter 32

The Magnetic Field

E32-1 Assuming there are no other forces on any of the four particles, a positively charged particle moving in the plane of a page where the magnetic field is into the page will experience a force to the left-hand side of the velocity vector (point your face in the direction of the velocity vector to identify left and right). You can verify this by applying Eq. 32-3, $\vec{F} = q\vec{v} \times \vec{B}$.

All of the paths which involve left hand turns are positive particles (path 1); those paths which involve right hand turns are negative particle (path 2 and path 4); and those paths which don't turn involve neutral particles (path 3).

E32-5 The magnetic force on the proton is given by Eq. 32-3,

$$\vec{F}_B = q\vec{v} \times \vec{B},$$

and since $\vec{v}$ and $\vec{B}$ are perpendicular we can write

$$F_B = qvB = (1.6 \times 10^{-19}\,\mathrm{C})(2.8 \times 10^{7}\,\mathrm{m/s})(30eex - 6\,\mathrm{T}) = 1.3 \times 10^{-16}\mathrm{N}.$$

The gravitational force on the proton is

$$mg = (1.7 \times 10^{-27}\mathrm{kg})(9.8\,\mathrm{m/s}^2) = 1.7 \times 10^{-26}\,\mathrm{N}.$$

The ratio is then 7.6×10^{9}. If, however, you carry the number of significant digits for the intermediate answers farther you will get the answer which is in the back of the book.

E32-9 (a) For a charged particle moving in a circle in a magnetic field we apply Eq. 32-10;

$$r = \frac{mv}{|q|B} = \frac{(9.11 \times 10^{-31}\mathrm{kg})(0.1)(3.00 \times 10^{8}\,\mathrm{m/s})}{(1.6 \times 10^{-19}\,\mathrm{C})(0.50\,\mathrm{T})} = 3.4 \times 10^{-4}\,\mathrm{m}.$$

(b) The (non-relativistic) kinetic energy of the electron is

$$K = \frac{1}{2}mv^2 = \frac{1}{2}m(0.10c)^2 = 0.005mc^2.$$

Now why did we do this this way? If we think back to Chapter 20 on special relativity we might remember that mc^2 is the rest mass energy of the electron, and is an easy to remember number, 0.511 MeV. Consequently, $K = 2.6 \times 10^{-3}$ MeV.

E32-13 (a) Apply Eq. 32-10, but rearrange it as

$$v = \frac{|q|rB}{m} = \frac{2(1.6 \times 10^{-19}\,\text{C})(0.045\,\text{m})(1.2\,\text{T})}{4.0(1.66 \times 10^{-27}\text{kg})} = 2.6 \times 10^6\,\text{m/s}.$$

(b) The speed is equal to the circumference divided by the period, so

$$T = \frac{2\pi r}{v} = \frac{2\pi m}{|q|B} = \frac{2\pi 4.0(1.66 \times 10^{-27}\text{kg})}{2(1.6 \times 10^{-19}\,\text{C})(1.2\,\text{T})} = 1.1 \times 10^{-7}\,\text{s}.$$

We probably could have just used the first expression, but it is worthwhile to be reminded that the period of revolution is independent of the speed or radius of the circle.

(c) The (non-relativistic) kinetic energy is

$$K = \frac{1}{2}mv^2 = \frac{1}{2}m\left(\frac{|q|rB}{m}\right)^2 = \frac{|q|^2r^2B^2}{2m}$$

This does depend on the radius and is in the text as Eq. 32-14. Putting in the given numbers,

$$K = \frac{|q|^2r^2B}{2m} = \frac{(2 \times 1.6 \times 10^{-19}\,\text{C})^2(0.045\,\text{m})^2(1.2\,\text{T})^2}{2(4.0 \times 1.66 \times 10^{-27}\text{kg}))} = 2.24 \times 10^{-14}\text{J}.$$

I see the answers in the back of the book are in keV. To change to electron volts we need merely divide this answer by the charge on one electron, so

$$K = \frac{(2.24 \times 10^{-14}\text{J})}{(1.6 \times 10^{-19}\,\text{C})} = 140\text{ keV}.$$

(d) We need to go back several chapters to answer this. We want to apply $W = q\Delta V$, and we will assume that the work done on the alpha particle is used to give it the final kinetic energy. Then

$$\Delta V = \frac{K}{q} = \frac{(140\text{ keV})}{(2e)} = 70\,\text{V}.$$

Yes, we can cancel the "e" in the numerator with the e in the denominator, because both refer to the charge on one electron.

E32-17 The radius of the circle of a charged particle in a magnetic field is given by Eq. 32-10,

$$r = \frac{mv}{|q|B}.$$

The velocity is related to the kinetic energy by

$$v = \sqrt{\frac{2K}{m}},$$

so

$$r = \frac{\sqrt{2mK}}{|q|B} = \frac{\sqrt{m}}{|q|}\frac{\sqrt{2K}}{B}$$

All three particles are traveling with the same kinetic energy in the same magnetic field. The relevant factors are in front; we just need top compare the mass and charge of each of the three particles. In terms of the proton mass and charge the deuteron mass and charge is 2 and 1; in terms of the proton the alpha particle mass and charge are 4 and 2.

(a) The radius of the deuteron path is $\frac{\sqrt{2}}{1}r_{\text{p}}$.

(b) The radius of the alpha particle path is $\frac{\sqrt{4}}{2}r_{\text{p}} = r_{\text{p}}$.

E32-21 Once again, we want to use Eq. 32-10, except we rearrange for the mass,

$$m = \frac{|q|rB}{v} = \frac{2(1.60\times10^{-19}\,\text{C})(4.72\,\text{m})(1.33\,\text{T})}{0.710(3.00\times10^8\,\text{m/s})} = 9.43\times10^{-27}\,\text{kg}$$

Not a very heavy particle. However, if it is moving at this velocity then the "mass" which we have here is not the true mass, but a relativistic correction. For a particle moving at $0.710c$ we have

$$\gamma = \frac{1}{\sqrt{1 - v^2/c^2}} = \frac{1}{\sqrt{1 - (0.710)^2}} = 1.42,$$

so the true mass of the particle is

$$\frac{(9.43\times10^{-27}\,\text{kg})}{(1.42)} = 6.64\times10^{-27}\text{kg}.$$

The number of nucleons present in this particle is then

$$(6.64\times10^{-27}\text{kg})/(1.67\times10^{-27}\,\text{kg}) = 3.97 \approx 4.$$

The charge was +2, which implies two protons, the other two nucleons would be neutrons, so this must be an alpha particle.

E32-25 We will use Eq. 32-20, $E_H = v_d B$, except we will not take the derivation through to Eq. 32-21. Instead, we will set the drift velocity equal to the speed of the strip. We will, however, set $E_H = \Delta V_H / w$. Then

$$v = \frac{E_H}{B} = \frac{\Delta V_H / w}{B} = \frac{(3.9 \times 10^{-6}\,\text{V})/(0.88 \times 10^{-2}\,\text{m})}{(1.2 \times 10^{-3}\,\text{T})} = 3.7 \times 10^{-1}\,\text{m/s}.$$

E32-29 If there is no tension in the wire then the force of gravity on the rod (the weight) is exactly balanced by the magnetic force on the wire.

The weight is easy enough to find,

$$W = mg.$$

The magnetic force on the wire is given by Eq. 32-26,

$$\vec{F}_B = i\vec{L} \times \vec{B}.$$

Since $\vec{L}$ is perpendicular to $\vec{B}$ we can simplify this as

$$F_B = iLB.$$

Equating these two forces,

$$iLB = mg,$$
$$i = \frac{mg}{LB} = \frac{(0.0130\,\text{kg})(9.81\,\text{m/s}^2)}{(0.620\,\text{m})(0.440\,\text{T})} = 0.467\,\text{A}.$$

Use of an appropriate right hand rule will indicate that the current must be directed to the right in order to have a magnetic force directed upward.

E32-31 (a) We use Eq. 32-26 again, and since the (horizontal) axle is perpendicular to the vertical component of the magnetic field,

$$i = \frac{F}{BL} = \frac{(10,000\,\text{N})}{(10\,\mu\text{T})(3.0\,\text{m})} = 3.3 \times 10^8\,\text{A}.$$

That's quite the current!

(b) The power lost per ohm of resistance in the rails is given by

$$P/r = i^2 = (3.3 \times 10^8\,\text{A})^2 = 1.1 \times 10^{17}\,\text{W}.$$

That's some power loss!

(c) If such a train were to be developed the rails would melt well before the train left the station.

E32-35 The 130 cm side is the hypotenuse of the right triangle. We are not told which way the magnetic field points along the hypotenuse, so I suppose that means we get to choose. We choose that the field points from the shorter side to the longer side.

(a) The magnetic field is parallel to the 130 cm side so there is no magnetic force on that side.

The magnetic force on the 50 cm side is given by Eq. 32-26,

$$\vec{F}_B = i\vec{L} \times \vec{B},$$

and has magnitude

$$F_B = iLB\sin\theta,$$

where θ is the angle between the 50 cm side and the magnetic field. This angle is larger than 90°, but the sine can be found directly from the triangle without ever knowing the value of the angle.

$$\sin\theta = \frac{(120 \text{ cm})}{(130 \text{ cm})} = 0.923,$$

and then the force on the 50 cm side can be found by

$$F_B = (4.00 \text{ A})(0.50 \text{ m})(75.0 \times 10^{-3} \text{ T})\frac{(120 \text{ cm})}{(130 \text{ cm})} = 0.138 \text{ N},$$

and is directed out of the plane of the triangle.

The magnetic force on the 120 cm side is given by Eq. 32-26,

$$\vec{F}_B = i\vec{L} \times \vec{B},$$

and has magnitude

$$F_B = iLB\sin\theta,$$

where θ is the angle between the 1200 cm side and the magnetic field. This angle is larger than 180°, but the sine can be found directly from the triangle without ever knowing the value of the angle.

$$\sin\theta = \frac{(-50 \text{ cm})}{(130 \text{ cm})} = -0.385,$$

and then the force on the 50 cm side can be found by

$$F_B = (4.00 \text{ A})(1.20 \text{ m})(75.0 \times 10^{-3} \text{ T})\frac{(-50 \text{ cm})}{(130 \text{ cm})} = -0.138 \text{ N},$$

and is directed into the plane of the triangle.

(b) Look at the three numbers above.

P32-3 (a) Consider first the cross product, $\vec{v} \times \vec{B}$. The electron moves horizontally, there is a component of the $\vec{B}$ which is down, so the cross product results in a vector which points to the left of the electron's path.

But the force on the electron is given by $\vec{F} = q\vec{v} \times \vec{B}$, and since the electron has a negative charge the force on the electron would be directed to the *right* of the electron's path.

(b) The kinetic energy of the electrons is much less than the rest mass energy, so this is non-relativistic motion. The speed of the electron is then $v = \sqrt{2K/m}$, and the magnetic force on the electron is $F_B = qvB$, where we are assuming $\sin\theta = 1$ because the electron moves horizontally through a magnetic field with a vertical component. We can ignore the effect of the magnetic field's horizontal component because the electron is moving parallel to this component.

The acceleration of the electron because of the magnetic force is then

$$a = \frac{qvB}{m} = \frac{qB}{m}\sqrt{\frac{2K}{m}},$$

$$= \frac{(1.60\times10^{-19}\text{C})(55.0\times10^{-6}\text{T})}{(9.11\times10^{-31}\text{kg})}\sqrt{\frac{2(1.92\times10^{-15}\text{J})}{(9.11\times10^{-31}\text{kg})}} = 6.27\times10^{14}\,\text{m/s}^2.$$

(b) Despite the large acceleration we don't expect the electron to be displaced far because of the large speed. The electron travels a horizontal distance of 20.0 cm in a time of

$$t = \frac{(20.0 \text{ cm})}{\sqrt{2K/m}} = \frac{(20.0 \text{ cm})}{\sqrt{2(1.92\times10^{-15}\text{J})/(9.11\times10^{-31}\text{kg})}} = 3.08\times10^{-9}\,\text{s}.$$

In this time the electron is accelerated to the side through a distance of

$$d = \frac{1}{2}at^2 = \frac{1}{2}(6.27\times10^{14}\,\text{m/s}^2)(3.08\times10^{-9}\,\text{s})^2 = 2.98 \text{ mm}.$$

P32-7 We could take the derivative of the equation in Problem 6, but we want to rearrange the equation first because we *don't* want the variable x to appear in the final expression. Taking the square root of both side we get

$$\sqrt{m} = \left(\frac{B^2q}{8\Delta V}\right)^{\frac{1}{2}} x,$$

and then take the derivative of x with respect to m,

$$\frac{1}{2}\frac{dm}{\sqrt{m}} = \left(\frac{B^2 q}{8\Delta V}\right)^{\frac{1}{2}} dx,$$

and then we want to consider finite differences instead of differential quantities,

$$\Delta m = \left(\frac{mB^2 q}{2\Delta V}\right)^{\frac{1}{2}} \Delta x,$$

(b) Invert the above expression,

$$\Delta x = \left(\frac{2\Delta V}{mB^2 q}\right)^{\frac{1}{2}} \Delta m,$$

and then put in the given values,

$$\begin{aligned}
\Delta x &= \left(\frac{2(7.33\times 10^3 \text{V})}{(35.0)(1.66\times 10^{-27}\text{kg})(0.520\,\text{T})^2(1.60\times 10^{-19}\text{C})}\right)^{\frac{1}{2}} (2.0)(1.66\times 10^{-27}\text{kg}), \\
&= 8.02 \text{ mm.}
\end{aligned}$$

Note that we used 35.0 u for the mass; if we had used 37.0 u the result would have been closer to the answer in the back of the book.

P32-11 (a) The period of motion can be found from the reciprocal of Eq. 32-12,

$$T = \frac{2\pi m}{|q|B} = \frac{2\pi(9.11\times 10^{-31}\text{kg})}{(1.60\times 10^{-19}\text{C})(455\times 10^{-6}\text{T})} = 7.86\times 10^{-8}\text{s.}$$

(b) We need to find the velocity of the electron from the kinetic energy,

$$v = \sqrt{2K/m} = \sqrt{2(22.5 \text{ eV})(1.60\times 10^{-19}\text{ J/eV})/(9.11\times 10^{-31}\text{kg})} = 2.81\times 10^6 \text{ m/s.}$$

This is 1% of the speed of light.

The velocity can written in terms of components which are parallel and perpendicular to the magnetic field. Then

$$v_{\|} = v\cos\theta \text{ and } v_{\perp} = v\sin\theta.$$

The pitch is the parallel distance traveled by the electron in one revolution, so

$$p = v_{\|}T = (2.81\times 10^6 \text{m/s})\cos(65.5^\circ)(7.86\times 10^{-8}\text{s}) = 9.16 \text{ cm.}$$

(c)   The radius of the helical path is given by Eq. 32-10, except that we use the perpendicular velocity component, so

$$R = \frac{mv_\perp}{|q|B} = \frac{(9.11 \times 10^{-31} \text{kg})(2.81 \times 10^6 \text{m/s}) \sin(65.5°)}{(1.60 \times 10^{-19}\text{C})(455 \times 10^{-6}\text{T})} = 3.20 \text{ cm}$$

P32-17   The torque on a current carrying loop depends on the orientation of the loop; the maximum torque occurs when the plane of the loop is parallel to the magnetic field. In this case the magnitude of the torque is from Eq. 32-34 with $\sin \theta = 1$—

$$\tau = NiAB.$$

The area of a circular loop is $A = \pi r^2$ where r is the radius, but since the circumference is $C = 2\pi r$, we can write

$$A = \frac{C^2}{4\pi}.$$

The circumference is *not* the length of the wire, because there may be more than one turn. Instead, $C = L/N$, where N is the number of turns.

Finally, we can write the torque as

$$\tau = Ni\frac{L^2}{4\pi N^2}B = \frac{iL^2B}{4\pi N},$$

which is a maximum when N is a minimum, or $N = 1$.

Chapter 33

The Magnetic Field of a Current

E33-1 (a) The magnetic field from a moving charge is given by Eq. 33-5. If the protons are moving side by side then the angle is $\phi = \pi/2$, so

$$B = \frac{\mu_0}{4\pi} \frac{qv}{r^2}$$

and we are interested is a distance $r = d$. The electric field at that distance is

$$E = \frac{1}{4\pi\epsilon_0} \frac{q}{r^2},$$

where in both of the above expressions q is the charge of the source proton.

On the receiving end is the other proton, and the force on that proton is given by

$$\vec{F} = q(\vec{E} + \vec{v} \times \vec{B}).$$

The velocity is the same as that of the first proton (otherwise they wouldn't be moving side by side.) This velocity is then perpendicular to the magnetic field, and the resulting direction for the cross product will be opposite to the direction of $\vec{E}$. Then for balance,

$$
\begin{aligned}
E &= vB, \\
\frac{1}{4\pi\epsilon_0} \frac{q}{r^2} &= v \frac{\mu_0}{4\pi} \frac{qv}{r^2}, \\
\frac{1}{\epsilon_0 \mu_0} &= v^2.
\end{aligned}
$$

We can solve this easily enough, and we find $v \approx 3 \times 10^8\,\mathrm{m/s}$.

(b) This is clearly a relativistic speed!

E33-5 The magnetic field produced by a long straight wire is given by Eq. 33-13, and this expression is appropriate to use here. Then

$$B = \frac{\mu_0 i}{2\pi d} = \frac{(4\pi \times 10^{-7} \text{N/A}^2)(1.6 \times 10^{-19}\,\text{C})(5.6 \times 10^{14}\,\text{s}^{-1})}{2\,.(0.0015\,\text{m})} = 1.2 \times 10^{-8}\text{T}.$$

E33-9 The field halfway between the wires will be zero if the currents are in the same direction. So they must be in opposite direction. Halfway between the wires is equidistant from either wire, so the magnetic field from one wire is equal in magnitude to the other. Consequently, we really only need to worry about one wire.

Then, for a single long straight wire,

$$B = \frac{\mu_0 i}{2\pi d} \text{ or } i = \frac{\pi dB}{\mu_0}.$$

But where did the "2" go? There are two wires, each one contributes to the magnetic field, so the current in any one wire needs to be only half as large. Finally

$$i = \frac{\pi dB}{\mu_0} = \frac{\pi(0.0405\,\text{m})(296,\mu\text{T})}{(4\pi \times 10^{-7}\text{N/A}^2)} = 30\,\text{A}$$

E33-13 There are four current segments that could contribute to the magnetic field. The straight segments, however, contribute nothing; because of the factor of $\sin\phi$ in Eq. 33-9 (the Biot-Savart law), there must be some *perpendicular* distance between the current segment and the point in question, and the straight segments carry currents either directly toward or directly away from the point P.

That leaves the two rounded segments. We will start with the closer rounded segment. Starting with the Biot-Savart Law (Eq. 33-8)

$$d\vec{\mathbf{B}} = \frac{\mu_0}{4\pi} \frac{i\,d\vec{\mathbf{s}} \times \vec{\mathbf{r}}}{r^3},$$

we can first replace the vector problem with a scalar problem because $d\vec{\mathbf{s}}$ and $\vec{\mathbf{r}}$ are always in the plane of the page and perpendicular to each other so $\vec{\mathbf{B}}$ will be perpendicular to the plane of the page and the cross product can be simplified to

$$dB = \frac{\mu_0}{4\pi} \frac{i\,ds}{r^2}.$$

Since $r = b$ is a constant along the curved path we can immediately integrate this expression and get

$$B_1 = \frac{\mu_0 i}{4\pi} \int_0^\theta \frac{r\,d\theta}{r^2} = \frac{\mu_0 i\theta}{4\pi b}.$$

Note that we used $ds = r\,d\theta$ to integrate along the arc. According to the right hand rule this field is pointing out of the page.

271

There is also a contribution from the top arc; the calculations are almost identical except that this i pointing into the page and $r = a$, so

$$B_2 = \frac{\mu_0 i}{4\pi} \int_\theta^0 \frac{r\, d\theta}{r^2} = -\frac{\mu_0 i \theta}{4\pi a},$$

The negative sign provides direction. The net magnetic field at P is then

$$B = B_1 + B_2 = \frac{\mu_0 i \theta}{4\pi} \left(\frac{1}{b} - \frac{1}{a}\right).$$

E33-15 We imagine the ribbon conductor to be a collection of thin wires, each of thickness dx and carrying a current di. di and dx are related by $di/dx = i/w$. The contribution of one of these thin wires to the magnetic field at P is

$$dB = \frac{\mu_0\, di}{2\pi x},$$

where x is the distance from this thin wire to the point P. We want to change variables to x and integrate, so

$$B = \int dB = \int \frac{\mu_0 i\, dx}{2\pi w x} = \frac{\mu_0 i}{2\pi w} \int \frac{dx}{x}.$$

The limits of integration are from d to $d + w$, so

$$B = \frac{\mu_0 i}{2\pi w} \ln\left(\frac{d + w}{d}\right).$$

E33-19 (a) We can use Eq. 33-21 (an equation you should be able to derive on a test) to find the magnetic field strength at the center of the large loop,

$$B = \frac{\mu_0 i}{2R} = \frac{(4\pi \times 10^{-7}\text{T} \cdot \text{m/A})(13\,\text{A})}{2(0.12\,\text{m})} = 6.8 \times 10^{-5}\text{T}.$$

(b) The torque on the smaller loop in the center is given by Eq. 32-34,

$$\vec{\tau} = N i \vec{\mathbf{A}} \times \vec{\mathbf{B}},$$

but since the magnetic field from the large loop is perpendicular to the plane of the large loop, and the plane of the small loop is also perpendicular to the plane of the large loop, the magnetic field is in the plane of the small loop. This means that $|\vec{\mathbf{A}} \times \vec{\mathbf{B}}| = AB$. Consequently, the magnitude of the torque on the small loop is

$$\tau = NiAB = (50)(1.3\,\text{A})(\pi)(8.2 \times 10^{-3}\text{m})^2(6.8 \times 10^{-5}\text{T}) = 9.3 \times 10^{-7}\text{N} \cdot \text{m}.$$

E33-23 The force on the projectile is given by the integral of

$$d\vec{\mathbf{F}} = i\, d\vec{\mathbf{l}} \times \vec{\mathbf{B}}$$

over the length of the projectile (which is w). The magnetic field strength is *not* a constant between the rails, but the current is, and the magnetic field is will always be perpendicular to $d\vec{\mathbf{l}}$ on the projectile. But we can't go any further until we know the spatial behavior of B.

The magnetic field strength can be found from adding together the contributions from each rail. If the rails are circular and the distance between them is small compared to the length of the wire we can use Eq. 33-13,

$$B = \frac{\mu_0 i}{2\pi x},$$

where x is the distance from the center of the rail. There is one problem, however, because these are not wires of infinite length. Since the current *stops* traveling along the rail when it reaches the projectile we have a rod that is only half of an infinite rod, so we need to multiply by a factor of $1/2$. But there are two rails, and each will contribute to the field, so the net magnetic field strength between the rails is

$$B = \frac{\mu_0 i}{4\pi x} + \frac{\mu_0 i}{4\pi(2r + w - x)}.$$

In that last term we have an expression that is a measure of the distance from the center of the lower rail in terms of the distance x from the center of the upper rail.

The magnitude of the force on the projectile is then

$$
\begin{aligned}
F &= i\int_r^{r+w} B\,dx, \\
&= \frac{\mu_0 i^2}{4\pi}\int_r^{r+w}\left(\frac{1}{x} + \frac{1}{2r + w - x}\right) dx, \\
&= \frac{\mu_0 i^2}{4\pi}2\ln\left(\frac{r+w}{r}\right)
\end{aligned}
$$

We need to verify the direction of the force on the projectile. Between the rails the magnetic field is directed into the page. The easiest way to verify that is to consider one of the many variations of the right hand rule: if you grip a long straight wire with your thumb extended in the direction of the current then your fingers will loop around the wire in the direction of the magnetic field.

The current thought the projectile is down the page; the magnetic field through the projectile is into the page; so the force on the projectile, according to

$$\vec{\mathbf{F}} = i\vec{\mathbf{l}} \times \vec{\mathbf{B}},$$

is to the right.

(b) Numerically the magnitude of the force on the rail is

$$F = \frac{(450 \times 10^3 \text{A})^2 (4\pi \times 10^{-7} \text{N/A}^2)}{2\pi} \ln\left(\frac{(0.067\,\text{m}) + (0.012\,\text{m})}{(0.067\,\text{m})}\right) = 6.65 \times 10^3 \text{ N}$$

The speed of the rail can be found from either energy conservation or basic kinematics. Energy conservation is somewhat easier, so we first find the work done on the projectile,

$$W = Fd = (6.65 \times 10^3 \text{ N})(4.0\,\text{m}) = 2.66 \times 10^4 \text{ J}.$$

This work results in a change in the kinetic energy, so the final speed is

$$v = \sqrt{2K/m} = \sqrt{2(2.66 \times 10^4 \text{ J})/(0.010\,\text{kg})} = 2.31 \times 10^3 \text{ m/s}.$$

E33-27 The magnetic field inside an ideal solenoid is given by Eq. 33-28

$$B = \mu_0 i n,$$

where n is the turns per unit length. Solving for n,

$$n = \frac{B}{\mu_0 i} = \frac{(0.0224\,\text{T})}{(4\pi \times 10^{-7} \text{N/A}^2)(17.8\,\text{A})} = 1.00 \times 10^3 / \text{m}^{-1}.$$

The solenoid has a length of 1.33 m, so the total number of turns is

$$N = nL = (1.00 \times 10^3 / \text{m}^{-1})(1.33\,\text{m}) = 1330,$$

and since each turn has a length of one circumference, then the total length of the wire which makes up the solenoid is

$$(1330)\pi(0.026\,\text{m}) = 109\,\text{m}.$$

E33-33 (a) We don't want to reinvent the wheel. The answer is found from Eq. 33-34, except it looks like

$$B = \frac{\mu_0 i r}{2\pi c^2}.$$

Despite the fact that we didn't do the work here, you should be prepared to do this starting from Ampere's law on a test.

(b) In the region between the wires the magnetic field looks like Eq. 33-13,

$$B = \frac{\mu_0 i}{2\pi r}.$$

This is derived on the right hand side of page 761.

(c)   Now we have some work to do. Ampere's law (Eq. 33-29) is

$$\oint \vec{B} \cdot d\vec{s} = \mu_0 i,$$

where i is the current enclosed. Our Amperian loop will still be a circle centered on the axis of the problem, so the left hand side of the above equation will reduce to $2\pi r B$, just like in Eq. 33-32. The right hand side, however, depends on the *net* current enclosed which is the current i in the center wire minus the fraction of the current enclosed in the outer conductor. The cross sectional area of the outer conductor is

$$\pi(a^2 - b^2),$$

so the fraction of the outer current enclosed in the Amperian loop is

$$i\frac{\pi(r^2 - b^2)}{\pi(a^2 - b^2)} = i\frac{r^2 - b^2}{a^2 - b^2}.$$

The net current in the loop is then

$$i - i\frac{r^2 - b^2}{a^2 - b^2} = i\frac{a^2 - r^2}{a^2 - b^2},$$

so the magnetic field in this region is

$$B = \frac{\mu_0 i}{2\pi}\frac{a^2 - r^2}{a^2 - b^2}.$$

(d)   This part is easy since the net current is zero; consequently $B = 0$.

E33-37 (a)   A positive particle would experience a magnetic force directed to the right for a magnetic field out of the page. This particle is going the other way, so it must be negative.

(b)   The magnetic field of a toroid is given by Eq. 33-36,

$$B = \frac{\mu_0 i N}{2\pi r},$$

while the radius of curvature of a charged particle in a magnetic field is given by Eq. 32-10

$$R = \frac{mv}{|q|B}.$$

We use the R to distinguish it from r. Combining,

$$R = \frac{2\pi mv}{\mu_0 i N |q|}r,$$

so the two radii are directly proportional. This means

$$R/(11 \text{ cm}) = (110 \text{ cm})/(125 \text{ cm}),$$

so $R = 9.7$ cm.

P33-1 A great problem, and of extreme importance. The field from one coil is given by Eq. 33-19

$$B = \frac{\mu_0 i R^2}{2(R^2 + z^2)^{3/2}}.$$

There are N turns in the coil, so we need a factor of N. There are two coils and we are interested in the magnetic field at P, a distance $R/2$ from each coil. The magnetic field strength will be twice the above expression but with $z = R/2$, so

$$B = \frac{2\mu_0 N i R^2}{2(R^2 + (R/2)^2)^{3/2}} = \frac{8\mu_0 N i}{(5)^{3/2} R}.$$

P33-3 This problem is the all important derivation of the Helmholtz coil properties.

(a) The magnetic field from one coil is

$$B_1 = \frac{\mu_0 N i R^2}{2(R^2 + z^2)^{3/2}}.$$

The magnetic field from the other coil, located a distance s away, but for points measured from the first coil, is

$$B_2 = \frac{\mu_0 N i R^2}{2(R^2 + (z - s)^2)^{3/2}}.$$

The magnetic field on the axis between the coils is the sum,

$$B = \frac{\mu_0 N i R^2}{2(R^2 + z^2)^{3/2}} + \frac{\mu_0 N i R^2}{2(R^2 + (z - s)^2)^{3/2}}.$$

This is a *nasty* expression, and we need to take the derivative with respect to z, we can do this in our heads and get

$$\frac{dB}{dz} = -\frac{3\mu_0 N i R^2}{2(R^2 + z^2)^{5/2}} z - \frac{3\mu_0 N i R^2}{2(R^2 + (z - s)^2)^{5/2}}(z - s).$$

At $z = s/2$ this expression vanishes! we expect this by symmetry, because the magnetic field will be strongest in the plane of either coil, so the mid-point should be a local minimum.

(b) Take the derivative again and

$$\frac{d^2 B}{dz^2} = -\frac{3\mu_0 N i R^2}{2(R^2 + z^2)^{5/2}} + \frac{15\mu_0 N i R^2}{2(R^2 + z^2)^{5/2}} z^2$$
$$-\frac{3\mu_0 N i R^2}{2(R^2 + (z - s)^2)^{5/2}} + \frac{15\mu_0 N i R^2}{2(R^2 + (z - s)^2)^{5/2}}(z - s)^2.$$

We could try and simplify this, but we don't really want to; we instead want to set it equal to zero, then let $z = s/2$, and then solve for s. The second derivative will equal zero when

$$-3(R^2 + z^2) + 15z^2 - 3(R^2 + (z - s)^2) + 15(z - s)^2 = 0,$$

and is $z = s/2$ this expression will simplify to

$$
\begin{aligned}
30(s/2)^2 &= 6(R^2 + (s/2)^2), \\
4(s/2)^2 &= R^2, \\
s &= R.
\end{aligned}
$$

P33-7 We want to use the differential expression in Eq. 33-11, except that the limits of integration are going to be different. We have four wire segments. From the top segment,

$$
\begin{aligned}
B_1 &= \left. \frac{\mu_0 i}{4\pi} \frac{d}{\sqrt{z^2 + d^2}} \right|_{-L/4}^{3L/4}, \\
&= \frac{\mu_0 i}{4\pi d} \left(\frac{3L/4}{\sqrt{(3L/4)^2 + d^2}} - \frac{-L/4}{\sqrt{(-L/4)^2 + d^2}} \right).
\end{aligned}
$$

For the top segment $d = L/4$, so this simplifies even further to

$$B_1 = \frac{\mu_0 i}{10\pi L} \left(\sqrt{2}(3\sqrt{5} + 5) \right).$$

The bottom segment has the same integral, but $d = 3L/4$, so

$$B_3 = \frac{\mu_0 i}{30\pi L} \left(\sqrt{2}(\sqrt{5} + 5) \right).$$

By symmetry, the contribution from the right hand side is the same as the bottom, so $B_2 = B_3$, and the contribution from the left hand side is the same as that from the top, so $B_4 = B_1$. Adding all four terms,

$$
\begin{aligned}
B &= \frac{2\mu_0 i}{30\pi L} \left(3\sqrt{2}(3\sqrt{5} + 5) + \sqrt{2}(\sqrt{5} + 5) \right), \\
&= \frac{2\mu_0 i}{3\pi L} (2\sqrt{2} + \sqrt{10}).
\end{aligned}
$$

P33-13 Apply Ampere's law with an Amperian loop that is a circle centered on the center of the wire. Then

$$\oint \vec{\mathbf{B}} \cdot d\vec{s} = \oint B \, ds = B \oint ds = 2\pi r B,$$

because $\vec{B}$ is tangent to the path and B is uniform along the path by symmetry. The current enclosed is

$$i_{\text{enc}} = \int j \, dA.$$

This integral is best done in polar coordinates, so $dA = (dr)(r \, d\theta)$, and then

$$\begin{aligned}
i_{\text{enc}} &= \int_0^r \int_0^{2\pi} (j_0 r/a) \, r \, dr \, d\theta, \\
&= 2\pi j_0/a \int_0^r r^2 dr, \\
&= \frac{2\pi j_0}{3a} r^3.
\end{aligned}$$

When $r = a$ the current enclosed is i, so

$$i = \frac{2\pi j_0 a^2}{3} \text{ or } j_0 = \frac{3i}{2\pi a^2}.$$

The magnetic field strength inside the wire is found by gluing together the two parts of Ampere's law,

$$\begin{aligned}
2\pi r B &= \mu_0 \frac{2\pi j_0}{3a} r^3, \\
B &= \frac{\mu_0 j_0 r^2}{3a}, \\
&= \frac{\mu_0 i r^2}{2\pi a^3}.
\end{aligned}$$

Chapter 34

Faraday's Law of Induction

E34-3 (a) The magnitude emf induced in a loop is given by Eq. 34-4,

$$|\mathcal{E}| = N \left| \frac{d\Phi_B}{dt} \right|,$$

$$= N \left| (12\,\text{mWb/s}^2)t + (7\,\text{mWb/s}) \right|$$

There is only one loop, and we want to evaluate this expression for $t = 2.0\,\text{s}$, so

$$|\mathcal{E}| = (1) \left| (12\,\text{mWb/s}^2)(2.0\,\text{s}) + (7\,\text{mWb/s}) \right| = 31\,\text{mV}.$$

The units are rather thick here. We should verify that a Wb/s is the same a V.

$$\frac{\text{Wb}}{\text{s}} = \frac{\text{Tm}^2}{\text{s}} = \frac{\text{Nm}}{\text{As}} = \frac{\text{J}}{\text{C}} = \text{V}$$

(b) This part isn't harder. The magnetic flux through the loop is increasing when $t = 2.0\,\text{s}$. The induced current needs to flow in such a direction to create a *second* magnetic field to oppose this increase. The original magnetic field is out of the page and we oppose the increase by pointing the other way, so the *second* field will point into the page (inside the loop).

By the right hand rule this means the induced current is clockwise through the loop, or to the left through the resistor.

E34-7 We could re-derive the steps in the sample problem, or we could start with the end result. We'll start with the result,

$$\mathcal{E} = NA\mu_0 n \left| \frac{di}{dt} \right|,$$

except that we have gone ahead and used the derivative instead of the Δ.

The rate of change in the current is

$$\frac{di}{dt} = (3.0\,\text{A/s}) + (1.0\,\text{A/s}^2)t,$$

so the induced emf is

$$\begin{aligned}
\mathcal{E} &= (130)(3.46\times10^{-4}\text{m}^2)(4\pi\times10^{-7}\text{Tm/A})(2.2\times10^4/\text{m})\left((3.0\text{A/s}) + (2.0\text{A/s}^2)t\right), \\
&= (3.73\times10^{-3}\,\text{V}) + (2.48\times10^{-3}\,\text{V/s})t.
\end{aligned}$$

We can plot this to get the emf as a function of time.

(b) When $t = 2.0\,\text{s}$ the induced emf is $8.69\times10^{-3}\,\text{V}$, so the induced current is

$$i = (8.69\times10^{-3}\,\text{V})/(0.15\,\Omega) = 5.8\times10^{-2}\,\text{A}.$$

E34-11 (a) The induced emf, as a function of time, is given by Eq. 34-5,

$$\mathcal{E}(t) = -\frac{d\Phi_B(t)}{dt}$$

This emf drives a current through the loop which obeys

$$\mathcal{E}(t) = i(t)R$$

Combining,

$$i(t) = -\frac{1}{R}\frac{d\Phi_B(t)}{dt}.$$

Don't worry about the negative sign. It matters, but we never need to sort out which direction is positive in this problem. We only need to know that one direction is positive and the other is negative.

Since the current is defined by $i = dq/dt$ we can write

$$\frac{dq(t)}{dt} = -\frac{1}{R}\frac{d\Phi_B(t)}{dt}.$$

Now we do some valid (for physicists) but sloppy (for mathematicians) calculus.

Factor out the dt from both sides, and then integrate:

$$\begin{aligned}
dq(t) &= -\frac{1}{R}\,d\Phi_B(t), \\
\int dq(t) &= -\int \frac{1}{R}\,d\Phi_B(t), \\
q(t) - q(0) &= \frac{1}{R}\left(\Phi_B(0) - \Phi_B(t)\right)
\end{aligned}$$

(b) No. The induced current could have increased from zero to some positive value, then decreased to zero and became negative, so that the net charge to flow through the resistor was zero. This would be like sloshing the charge back and forth through the loop.

E34-17 The magnetic field is out of the page, and the current through the rod is down. Then Eq. 32-26

$$\vec{F} = i\vec{L} \times \vec{B}$$

shows that the direction of the magnetic force is to the right; furthermore, since everything is perpendicular to everything else, we can get rid of the vector nature of the problem and write

$$F = iLB.$$

In this exercise we are guaranteed a constant current; when you do Exercise 18 you have a constant emf from the generator, but there will be a back emf from the motion of the rod (BvL), and this back emf will cause the net current to decrease with increasing speed. Fortunately we don't have to worry about that here.

Newton's second law gives $F = ma$, and the acceleration of an object from rest results in a velocity given by $v = at$. Combining,

$$v(t) = \frac{iLB}{m}t.$$

A similar answer for Exercise 18 would involve exponential functions.

E34-21 We will use the results of Exercise 11 that were worked out above. All we need to do is find the initial flux; flipping the coil up-side-down will simply change the sign of the flux.

So

$$\Phi_B(0) = \vec{B} \cdot \vec{A} = (59\,\mu\text{T})(\pi)(0.13\,\text{m})^2 \sin(20°) = 1.1 \times 10^{-6}\,\text{Wb}.$$

Then using the results of Exercise 11 we have

$$\begin{aligned} q &= \frac{N}{R}(\Phi_B(0) - \Phi_B(t)), \\ &= \frac{950}{85\,\Omega}((1.1 \times 10^{-6}\,\text{Wb}) - (-1.1 \times 10^{-6}\,\text{Wb})), \\ &= 2.5 \times 10^{-5}\,\text{C}. \end{aligned}$$

E34-23 (a) This is an exercise that actually requires integration. We want to know the current in the loop, we get that current by knowing the induced emf, so we must find the magnetic flux, which means we need to know the magnetic field caused by the long, straight wire.

Starting from the beginning, Eq. 33-13 gives

$$B = \frac{\mu_0 i}{2\pi y},$$

where I have used y in place of d because I like y better.

The flux through the loop is given by

$$\Phi_B = \int \vec{B} \cdot d\vec{A},$$

but since the magnetic field from the long straight wire goes through the loop perpendicular to the plane of the loop this expression simplifies to a scalar integral. The loop is a rectangular thing, so we'll use $dA = dx\, dy$, and let x be parallel to the long straight wire.

Combining the above,

$$\begin{aligned}
\Phi_B &= \int_D^{D+b} \int_0^a \left(\frac{\mu_0 i}{2\pi y} \right) dx\, dy, \\
&= \frac{\mu_0 i}{2\pi} a \int_D^{D+b} \frac{dy}{y}, \\
&= \frac{\mu_0 i}{2\pi} a \ln \left(\frac{D+b}{D} \right)
\end{aligned}$$

(b) The flux through the loop is a function of the distance D from the wire. If the loop moves away from the wire at a constant speed v, then the distance D varies as vt. The induced emf is then

$$\begin{aligned}
\mathcal{E} &= -\frac{d\Phi_B}{dt}, \\
&= \frac{\mu_0 i}{2\pi} a \frac{b}{t(vt+b)}.
\end{aligned}$$

The current will be this emf divided by the resistance R. The "back-of-the-book" answer is somewhat different; the answer is expressed in terms of D instead if t. The two answers are otherwise identical.

E34-27 We can use Eq. 34-10; the emf is

$$\mathcal{E} = BA\omega \sin \omega t,$$

This will be a maximum when $\sin \omega t = 1$. The angular frequency, ω is equal to

$$\omega = (1000)(2\pi)/(60) \text{ rad/s} = 105 \text{ rad/s}$$

The maximum emf is then

$$\mathcal{E} = (3.5\,\text{T})\,[(100)(0.5\,\text{m})(0.3\,\text{m})]\,(105\,\text{rad/s}) = 5.5\,\text{kV}.$$

E34-31 The induced electric field can be found from applying Eq. 34-13,

$$\oint \vec{\mathbf{E}} \cdot d\vec{\mathbf{s}} = -\frac{d\Phi_B}{dt}.$$

We start with the left hand side of this expression. The problem has cylindrical symmetry, so the induced electric field lines should be circles centered on the axis of the cylindrical volume. If we choose the path of integration to lie along an electric field line, then the electric field $\vec{\mathbf{E}}$ will be parallel to $d\vec{\mathbf{s}}$, and E will be uniform along this path, so

$$\oint \vec{\mathbf{E}} \cdot d\vec{\mathbf{s}} = \oint E \, ds = E \oint ds = 2\pi r E,$$

where r is the radius of the circular path.

Now for the right hand side. The flux is contained in the path of integration, so $\Phi_B = B\pi r^2$. All of the time dependence of the flux is contained in B, so we can immediately write

$$2\pi r E = -\pi r^2 \frac{dB}{dt} \text{ or } E = -\frac{r}{2}\frac{dB}{dt}.$$

What does the negative sign mean? The path of integration is chosen so that if our right hand fingers curl around the path our thumb gives the direction of the magnetic field which cuts through the path. Since the field points into the page a positive electric field would have a clockwise orientation. Since B is decreasing the derivative is negative, but we get another negative from the equation above, so the electric field has a positive direction.

Now for the magnitude.

$$E = (4.82 \times 10^{-2}\,\text{m})(10.7 \times 10^{-3}\,T/\text{s})/2 = 2.58 \times 10^{-4}\,\text{N/C}.$$

The acceleration of the electron at either a or c then has magnitude

$$a = Eq/m = (2.58 \times 10^{-4}\,\text{N/C})(1.60 \times 10^{-19}\,\text{C})/(9.11 \times 10^{-31}\,\text{kg}) = 4.53 \times 10^{7}\text{m/s}^2.$$

P34-1 The induced current is given by $i = \mathcal{E}/R$. The resistance of the loop is given by $R = \rho L/A$, where A is the cross sectional area. Combining, and writing in terms of the radius of the wire, we have

$$i = \frac{\pi r^2 \mathcal{E}}{\rho L}.$$

The length of the wire is related to the radius of the wire because we have a fixed mass. The total volume of the wire is $\pi r^2 L$, and this is related to the mass and density by $m = \delta \pi r^2 L$. Eliminating r we have

$$i = \frac{m\mathcal{E}}{\rho \delta L^2}.$$

The length of the wire loop is the same as the circumference, which is related to the radius R of the loop by $L = 2\pi R$. The emf is related to the changing flux by $\mathcal{E} = -d\Phi_B/dt$, but if the shape of the loop is fixed this becomes $\mathcal{E} = -A\,dB/dt$. Combining all of this,

$$i = \frac{mA}{\rho\delta(2\pi R)^2}\frac{dB}{dt}.$$

We dropped the negative sign because we are only interested in absolute values here. Really.

Now $A = \pi R^2$, so this expression can also be written as

$$i = \frac{m\pi R^2}{\rho\delta(2\pi R)^2}\frac{dB}{dt} = \frac{m}{4\pi\rho\delta}\frac{dB}{dt}.$$

It is rather amazing that the answer is independent of the radius of the wire loop. It might be interesting to see if this result would hold true for other shapes as well.

P34-5 This is a integral best performed in rectangular coordinates, then $dA = (dx)(dy)$. The magnetic field is perpendicular to the surface area, so $\vec{\mathbf{B}} \cdot d\vec{\mathbf{A}} = B\,dA$. The flux is then

$$
\begin{aligned}
\Phi_B &= \int \vec{\mathbf{B}} \cdot d\vec{\mathbf{A}} = \int B\,dA, \\
&= \int_0^a \int_0^a (4\,\mathrm{T/m \cdot s^2})t^2 y\,dy\,dx, \\
&= (4\,\mathrm{T/m \cdot s^2})t^2 \left(\frac{1}{2}a^2\right)a, \\
&= (2\,\mathrm{T/m \cdot s^2})a^3 t^2.
\end{aligned}
$$

But $a = 2.0$ cm, so this becomes

$$\Phi_B = (2\,\mathrm{T/m \cdot s^2})(0.02\,\mathrm{m})^3 t^2 = (1.6 \times 10^{-5}\mathrm{Wb/s^2})t^2.$$

The emf around the square is given by

$$\mathcal{E} = -\frac{d\Phi_B}{dt} = -(3.2 \times 10^{-5}\mathrm{Wb/s^2})t,$$

and at $t = 2.5$ s this is -8.0×10^{-5}V. Since the magnetic field is directed out of the page, a positive emf would be counterclockwise (hold your right thumb in the direction of the magnetic field and your fingers will give a counter clockwise sense around the loop). But the answer was negative, so the emf must be clockwise.

P34-7 The magnetic field is perpendicular to the surface area, so $\vec{\mathbf{B}} \cdot d\vec{\mathbf{A}} = B\, dA$. The flux is then

$$\Phi_B = \int \vec{\mathbf{B}} \cdot d\vec{\mathbf{A}} = \int B\, dA = BA,$$

since the magnetic field is uniform. The area is $A = \pi r^2$, where r is the radius of the loop. The induced emf is

$$\mathcal{E} = -\frac{d\Phi_B}{dt} = -2\pi r B \frac{dr}{dt}.$$

It is given that $B = 0.785\,\mathrm{T}$, $r = 1.23\,\mathrm{m}$, and $dr/dt = -7.50\times10^{-2}\mathrm{m/s}$. The negative sign indicate a decreasing radius. Then

$$\mathcal{E} = -2\pi(1.23\,\mathrm{m})(0.785\,\mathrm{T})(-7.50\times10^{-2}\mathrm{m/s}) = 0.455\,\mathrm{V}.$$

P34-11 It does say approximate, so we will be making some rather bold assumptions here. First we will find an expression for the emf. Since B is constant, the emf must be caused by a change in the area; in this case a shift in position. The small square where $B \neq 0$ has a width a and sweeps around the disk with a speed $r\omega$. An approximation for the emf is then

$$\mathcal{E} = Bar\omega.$$

This emf causes a current. We don't know exactly where the current flows, but we can reasonably assume that it occurs near the location of the magnetic field. Let us assume that it is constrained to that region of the disk. The resistance of this portion of the disk is the approximately

$$R = \frac{1}{\sigma}\frac{L}{A} = \frac{1}{\sigma}\frac{a}{at} = \frac{1}{\sigma t},$$

where we have assumed that the current is flowing radially when defining the cross sectional area of the "resistor". The induced current is then (on the order of)

$$\frac{\mathcal{E}}{R} = \frac{Bar\omega}{1/(\sigma t)} = Bar\omega\sigma t.$$

This current experiences a breaking force according to $F = BIl$, so

$$F = B^2 a^2 r\omega\sigma t,$$

where l is the length through which the current flows, which is a.

Finally we can find the torque from $\tau = rF$, and

$$\tau = B^2 a^2 r^2 \omega\sigma t.$$

Was this legitimate? I hope so, it appears to have worked.

Chapter 35

Magnetic Properties of Materials

E35-1 The magnetic dipole moment of a simple current loop is given by Eq. 35-3,

$$\mu = iA.$$

If the Earth's magnetic dipole moment were produced by a single current around the core, then that current would be

$$i = \frac{\mu}{A} = \frac{(8.0 \times 10^{22}\,\text{J/T})}{\pi(3.5 \times 10^6\text{m})^2} = 2.1 \times 10^9\,\text{A}$$

E35-5 (a) The result from Problem 33-4 for a square loop of wire was

$$B(z) = \frac{4\mu_0 i a^2}{\pi(4z^2 + a^2)(4z^2 + 2a^2)^{1/2}}.$$

For z much, much larger than a we can ignore any a terms which are added to or subtracted from z terms. This means that

$$4z^2 + a^2 \to 4z^2 \text{ and } (4z^2 + 2a^2)^{1/2} \to 2z,$$

but we can't ignore the a^2 in the numerator.

The expression for B then simplifies to

$$B(z) = \frac{\mu_0 i a^2}{2\pi z^3},$$

which certainly looks like Eq. 35-4.

(b) we can rearrange this expression and get

$$B(z) = \frac{\mu_0}{2\pi z^3} i a^2,$$

where it is rather evident that ia^2 must correspond to $\vec{\mu}$, the dipole moment, in Eq. 35-4. So that must be the answer.

E35-9 (a)   The electric field at this distance from the proton is

$$E = \frac{1}{4\pi(8.85\times10^{-12}\mathrm{C}^2/\mathrm{N}\cdot\mathrm{m}^2)}\frac{(1.60\times10^{-19}\mathrm{C})}{(5.29\times10^{-11}\mathrm{m})^2} = 5.14\times10^{11}\mathrm{N/C}.$$

That's huge!

(b)   The magnetic field at this from the proton is given by the dipole approximation,

$$
\begin{aligned}
B(z) &= \frac{\mu_0\mu}{2\pi z^3}, \\
&= \frac{(4\pi\times10^{-7}\mathrm{N/A}^2)(1.41\times10^{-26}\mathrm{A/m}^2)}{2\pi(5.29\times10^{-11}\mathrm{m})^3}, \\
&= 1.90\times10^{-2}\,\mathrm{T}
\end{aligned}
$$

That's small.

E35-13   The magnetic dipole moment is given by $\mu = MV$, Eq. 35-13. Then

$$\mu = (5,300\,\mathrm{A/m})(0.048\,\mathrm{m})\pi(0.0055\,\mathrm{m})^2 = 0.024\,\mathrm{A/m}^2.$$

Note that I prefer $\mathrm{A/m}^2$ for the units of the dipole moment. It is the same as J/T; but this is something you can verify on your own.

E35-17 (a)   Look at the figure. At 50% (which is 0.5 on the vertical axis), the curve is at $B_0/T \approx 0.55\,\mathrm{T/K}$. Since $T = 300\,\mathrm{K}$, we have $B_0 \approx 165\,\mathrm{T}$. That's quite the magnetic field!

(b)   Same figure, but now look at the 90% mark. $B_0/T \approx 1.60\,\mathrm{T/K}$, so $B_0 \approx$ 480 T.

(c)   Good question. I think both field are far beyond our current abilities.

E35-23 (a)   According to the text in section 35-5 the magnetization M depends on the history of the ferromagnet. We'll assume, however, that all of the iron atoms are perfectly aligned. Then the dipole moment of the earth will be related to the dipole moment of one atom by

$$\mu_{\mathrm{Earth}} = N\mu_{\mathrm{Fe}},$$

where N is the number of iron atoms in the magnetized sphere. If m_A is the relative atomic mass of iron, then the total mass is

$$m = \frac{Nm_A}{A} = \frac{m_A}{A}\frac{\mu_{\mathrm{Earth}}}{\mu_{\mathrm{Fe}}},$$

where A is Avogadro's number. Next, the volume of a sphere of mass m is

$$V = \frac{m}{\rho} = \frac{m_A}{\rho A} \frac{\mu_{\text{Earth}}}{\mu_{\text{Fe}}},$$

where ρ is the density.

And finally, the radius of a sphere with this volume would be

$$r = \left(\frac{3V}{4\pi}\right)^{1/3} = \left(\frac{3\mu_{\text{Earth}} m_A}{4\pi\rho\mu_{\text{Fe}} A}\right)^{1/3}.$$

I think that I did those substitutions correctly. Now we find the radius by substituting in the known values,

$$r = \left(\frac{3(8.0\times 10^{22}\,\text{J/T})(56\text{ g/mol})}{4\pi(14\times 10^6 \text{g/m}^3)(2.1\times 10^{-23}\,\text{J/T})(6.0\times 10^{23}/\text{mol})}\right)^{1/3} = 1.8\times 10^5 \text{m}.$$

(b)   The fractional volume is the cube of the fractional radius, so the answer is

$$(1.8\times 10^5\,\text{m}/6.4\times 10^6)^3 = 2.2\times 10^{-5}.$$

E35-29   The total magnetic flux through a closed surface is zero. There is inward flux on faces one, three, and five for a total of -9 Wb. There is outward flux on faces two and four for a total of +6 Wb. The difference is +3 Wb; consequently the outward flux on the sixth face must be +3 Wb.

P35-1   We can imagine the rotating disk as being composed of a number of rotating rings of radius r, width dr, and circumference $2\pi r$. The surface charge density on the disk is $\sigma = q/\pi R^2$, and consequently the (differential) charge on any ring is

$$dq = \sigma(2\pi r)(dr) = \frac{2qr}{R^2}dr$$

The rings "rotate" with angular frequency ω, or period $T = 2\pi/\omega$. The effective (differential) current for each ring is then

$$di = \frac{dq}{T} = \frac{qr\omega}{\pi R^2}dr.$$

Each ring contributes to the magnetic moment, and we can glue all of this together as

$$\begin{aligned}
\mu &= \int d\mu, \\
&= \int \pi r^2\, di, \\
&= \int_0^R \frac{qr^3\omega}{R^2}dr, \\
&= \frac{qR^2\omega}{4}.
\end{aligned}$$

P35-7 (a) Centripetal acceleration is given by $a = r\omega^2$. Then

$$
\begin{aligned}
a - a_0 &= r(\omega_0 + \Delta\omega)^2 - r\omega_0^2, \\
&= 2r\omega_0\Delta\omega + r(\Delta\omega_0)^2, \\
&\approx 2r\omega_0\Delta\omega.
\end{aligned}
$$

(b) The change in centripetal acceleration is caused by the **additional magnetic** force, which has magnitude

$$
F_B = qvB = er\omega B.
$$

Then

$$
\Delta\omega = \frac{a - a_0}{2r\omega_0} = \frac{eB}{2m}.
$$

Note that we boldly canceled ω against ω_0 in this last expression; we are assuming that $\Delta\omega$ is small, and for these problems it is.

Chapter 36

Inductance

E36-1 The important relationship is Eq. 36-4, written as

$$\Phi_B = \frac{iL}{N} = \frac{(5.0\,\text{mA})(8.0\,\text{mH})}{(400)} = 1.0 \times 10^{-7}\text{Wb}$$

E36-5 (a) Eq. 36-1 can be used to find the inductance of the coil.

$$L = \frac{\mathcal{E}_L}{di/dt} = \frac{(3.0\,\text{mV})}{(5.0\,\text{A/s})} = 6.0 \times 10^{-4}\text{H}.$$

(b) Eq. 36-4 can then be used to find the number of turns in the coil.

$$N = \frac{iL}{\Phi_B} = \frac{(8.0\,\text{A})(6.0 \times 10^{-4}\text{H})}{(40\mu\text{Wb})} = 120$$

E36-9 (a) If two inductors are connected in parallel then the current through each inductor will add to the total current through the circuit,

$$i = i_1 + i_2,$$

We can take the derivative of the current with respect to time and then

$$di/dt = di_1/dt + di_2/dt,$$

The potential difference across each inductor is the same, so if we divide by $\mathcal{E}$ and apply we get

$$\frac{di/dt}{\mathcal{E}} = \frac{di_1/dt}{\mathcal{E}} + \frac{di_2/dt}{\mathcal{E}},$$

But

$$\frac{di/dt}{\mathcal{E}} = \frac{1}{L},$$

so the previous expression can also be written as

$$\frac{1}{L_{\text{eq}}} = \frac{1}{L_1} + \frac{1}{L_2}.$$

(b) If the inductors are close enough together then the magnetic field from one coil will induce currents in the other coil. Then we will need to consider mutual induction effects, but that is a topic not covered in this text.

E36-13 (a) From Eq. 36-4 we find the inductance to be

$$L = \frac{N\Phi_B}{i} = \frac{(26.2 \times 10^{-3} \text{Wb})}{(5.48 \text{ A})} = 4.78 \times 10^{-3} \text{H}.$$

Note that Φ_B is the *flux*, while the quantity $N\Phi_B$ is the *number of flux linkages.*

(b) We can find the time constant from Eq. 36-14,

$$\tau_L = L/R = (4.78 \times 10^{-3} \text{H})/(0.745 \,\Omega) = 6.42 \times 10^{-3} \text{ s}.$$

The we can invert Eq. 36-13 to get

$$
\begin{aligned}
t &= -\tau_L \ln\left(1 - \frac{Ri(t)}{\mathcal{E}}\right), \\
&= -(6.42 \times 10^{-3} \text{ s}) \ln\left(1 - \frac{(0.745 \text{ A})(2.53 \text{ A})}{(6.00 \text{ V})}\right) = 2.42 \times 10^{-3} \text{ s}.
\end{aligned}
$$

E36-17 We want to take the derivative of the current in Eq. 36-13 with respect to time,

$$\frac{di}{dt} = \frac{\mathcal{E}}{R} \frac{1}{\tau_L} e^{-t/\tau_L} = \frac{\mathcal{E}}{L} e^{-t/\tau_L}.$$

Then $\tau_L = (5.0 \times 10^{-2} \text{H})/(180 \,\Omega) = 2.78 \times 10^{-4} \text{s}$. Using this we find the rate of change in the current when $t = 1.2$ ms to be

$$\frac{di}{dt} = \frac{(45 \text{ V})}{((5.0 \times 10^{-2} \text{H})} e^{-(1.2 \times 10^{-3} \text{S})/(2.78 \times 10^{-4} \text{S})} = 12 \text{ A/s}.$$

E36-21 (I) When the switch is just closed there is *no* current through the inductor or R_2, so the potential difference across the inductor must be 10 V. The potential difference across R_1 is always 10 V when the switch is closed, regardless of the amount of time elapsed since closing.

(a) $i_1 = (10 \text{ V})/(5.0 \,\Omega) = 2.0 \text{ A}.$

(b) Zero; read the above paragraph.

(c) The current through the switch is the sum of the above two currents, or 2.0 A.

(d) Zero, since the current through R_2 is zero.

(e) 10 V, since the potential across R_2 is zero.

(f) Look at the results of Exercise 36-17. When $t = 0$ the rate of change of the current is $di/dt = \mathcal{E}/L$. Then

$$di/dt = (10\,\text{V})/(5.0\,\text{H}) = 2.0\,\text{A/s}.$$

(II) After the switch has been closed for a long period of time the currents are stable and the inductor no longer has an effect on the circuit. Then the circuit is a simple two resistor parallel network, each resistor has a potential difference of 10 V across it.

(a) Still 2.0 A; nothing has changed.

(b) $i_2 = (10\,\text{V})/(10\,\Omega) = 1.0\,\text{A}.$

(c) Add the two currents and the current through the switch will be 3.0 A.

(d) 10 V; see the above discussion.

(e) Zero, since the current is no longer changing.

(f) Zero, since the current is no longer changing.

E36-27 The energy density of an electric field is given by Eq. 36-23; that of a magnetic field is given by Eq. 36-22. Equating,

$$\frac{\epsilon_0}{2} E^2 = \frac{1}{2\mu_0} B^2,$$

$$E = \frac{B}{\sqrt{\epsilon_0 \mu_0}}.$$

The quantity

$$\frac{1}{\sqrt{\epsilon_0 \mu_0}}$$

occurs frequently enough to merit additional attention. It is

$$\frac{1}{\sqrt{(8.85 \times 10^{-12} \text{C}^2/\text{N} \cdot \text{m}^2)(4\pi \times 10^{-7} \text{N/A}^2)}} = 3.00 \times 10^8 \,\text{m/s}.$$

This should look like a familiar constant; it is the speed of light in a vacuum.
Now back to the problem at hand.

$$E = (3.00 \times 10^8 \text{m/s})(0.50\,\text{T}) = 1.5 \times 10^8\,\text{V/m}.$$

E36-31 This shell has a volume of

$$V = \frac{4\pi}{3}\left((R_E + a)^3 - R_E{}^3\right).$$

Since $a << R_E$ we can expand the polynomials but keep only the terms which are linear in a. Then

$$V \approx 4\pi R_E{}^2 a = 4\pi(6.37 \times 10^6\text{m})^2(1.6 \times 10^4\text{m}) = 8.2 \times 10^{18}\text{m}^3.$$

The magnetic energy density is found from Eq. 36-22,

$$u_B = \frac{1}{2\mu_0}B^2 = \frac{(60 \times 10^{-6}\,\text{T})^2}{2(4\pi \times 10^{-7}\text{N/A}^2)} = 1.43 \times 10^{-3}\text{J/m}^3.$$

The total energy is then $(1.43 \times 10^{-3}\text{J/m}^3)(8.2eex18\text{m}^3) = 1.2 \times 10^{16}\text{J}.$

E36-37 Closing a switch has the effect of "shorting" out the relevant circuit element, which effectively removes it from the circuit. If S_1 is closed we have

$$\tau_C = RC \text{ or } C = \tau_C/R,$$

if instead S_2 is closed we have

$$\tau_L = L/R \text{ or } L = R\tau_L,$$

but if instead S_3 is closed we have a LC circuit which will oscillate with period

$$T = \frac{2\pi}{\omega} = 2\pi\sqrt{LC}.$$

Substituting from the expressions above,

$$T = \frac{2\pi}{\omega} = 2\pi\sqrt{\tau_L \tau_C}.$$

Just for fun you should consider finding the behavior of the circuit if all three switched are open.

E36-41 (a) An LC circuit oscillates so that the energy is converted from all magnetic to all electrical *twice* each cycle. It occurs twice because once the energy is magnetic with the current flowing in one direction through the inductor, and later the energy is magnetic with the current flowing the other direction through the inductor.
The period is then *four* times 1.52μs, or 6.08μs.

(b) The frequency is the reciprocal of the period, or 164000 Hz.

(c) Since it occurs twice during each oscillation it is equal to half a period, or 3.04μs.

E36-49 The frequency of such a system is given by Eq. 36-26,

$$f = \frac{1}{2\pi\sqrt{LC}}.$$

Note that maximum frequency occurs with minimum capacitance. Then

$$\frac{f_1}{f_2} = \sqrt{\frac{C_2}{C_1}} = \sqrt{\frac{(365 \text{ pF})}{(10 \text{ pF})}} = 6.04.$$

(b) The desired ratio is $1.60/0.54 = 2.96$ Adding a capacitor in parallel will result in an effective capacitance given by

$$C_{1,\text{eff}} = C_1 + C_{\text{add}},$$

with a similar expression for C_2. We want to choose C_{add} so that

$$\frac{f_1}{f_2} = \sqrt{\frac{C_{2,\text{eff}}}{C_{1,\text{eff}}}} = 2.96.$$

Solving,

$$\begin{aligned}
C_{2,\text{eff}} &= C_{1,\text{eff}}(2.96)^2, \\
C_2 + C_{\text{add}} &= (C_1 + C_{\text{add}})8.76, \\
C_{\text{add}} &= \frac{C_2 - 8.76C_1}{7.76}, \\
&= \frac{(365 \text{ pF}) - 8.76(10 \text{ pF})}{7.76} = 36 \text{ pF}.
\end{aligned}$$

The necessary inductance is then

$$L = \frac{1}{4\pi^2 f^2 C} = \frac{1}{4\pi^2(0.54\times10^6\text{Hz})^2(401\times10^{-12}\text{F})} = 2.2\times10^{-4}\text{H}.$$

Note that we used the new capacitance in this calculation, and not the original capacitance of the variable capacitor.

E36-55 The damped (angular) frequency is given by Eq. 36-41; the fractional change would then be

$$\frac{\omega - \omega'}{\omega} = 1 - \sqrt{1 - (R/2L\omega)^2} = 1 - \sqrt{1 - (R^2C/4L)}.$$

Setting this equal to 0.01% and then solving for R,

$$R = \sqrt{\frac{4L}{C}\left(1 - (1 - 0.0001)^2\right)} = \sqrt{\frac{4(12.6 \times 10^{-3}\text{H})}{(1.15 \times 10^{-6}\text{F})}(1.9999 \times 10^{-4})} = 2.96\,\Omega.$$

P36-1 The inductance of a toroid is

$$L = \frac{\mu_0 N^2 h}{2\pi}\ln\frac{b}{a}.$$

If the toroid is very large and thin then we can write $b = a + \delta$, where $\delta << a$. The natural log then can be approximated as

$$\ln\frac{b}{a} = \ln\left(1 + \frac{\delta}{a}\right) \approx \frac{\delta}{a}.$$

The product of δ and h is the cross sectional area of the toroid, while $2\pi a$ is the circumference, which we will call l. The inductance of the toroid then reduces to

$$L \approx \frac{\mu_0 N^2}{2\pi}\frac{\delta}{a} = \frac{\mu_0 N^2 A}{l}.$$

But N is the number of turns, which can also be written as $N = nl$, where n is the turns per unit length. Substitute this in and we arrive at Eq. 36-7.

P36-5 This is a classic problem, always good for an exam question. The magnetic field in the region between the conductors of a coaxial cable is given by

$$B = \frac{\mu_0 i}{2\pi r},$$

so the flux through an area of length l, width $b - a$, and perpendicular to $\vec{B}$ is

$$\begin{aligned}\Phi_B &= \int \vec{B} \cdot d\vec{A} = \int B\,dA, \\ &= \int_a^b \int_0^l \frac{\mu_0 i}{2\pi r}\,dz\,dr, \\ &= \frac{\mu_0 i l}{2\pi}\ln\frac{b}{a}.\end{aligned}$$

We evaluated this integral is cylindrical coordinates: $dA = (dr)(dz)$. As you have been warned so many times before, learn these differentials!

The inductance is then

$$L = \frac{\Phi_B}{i} = \frac{\mu_0 l}{2\pi}\ln\frac{b}{a}.$$

P36-11 (a) In Chapter 33 we found the magnetic field inside a wire carrying a uniform current density is

$$B = \frac{\mu_0 i r}{2\pi R^2} \text{ (Eq. 33-34)}.$$

The magnetic energy density in this wire is

$$u_B = \frac{1}{2\mu_0} B^2 = \frac{\mu_0 i^2 r^2}{8\pi^2 R^4}.$$

We want to integrate in cylindrical coordinates over the volume of the wire. Then $dV = (dr)(r\,d\theta)(dz)$, so

$$
\begin{aligned}
U_B &= \int u_B dV, \\
&= \int_0^R \int_0^l \int_0^{2\pi} \frac{\mu_0 i^2 r^2}{8\pi^2 R^4} d\theta\, dz\, r dr, \\
&= \frac{\mu_0 i^2 l}{4\pi R^4} \int_0^R r^3\, dr, \\
&= \frac{\mu_0 i^2 l}{16\pi}.
\end{aligned}
$$

(b) Solve

$$U_B = \frac{L}{2} i^2$$

for L, and

$$L = \frac{2U_B}{i^2} = \frac{\mu_0 l}{8\pi}.$$

P36-15 We start by focusing on the charge on the capacitor, given by Eq. 36-40 as

$$q = q_{\mathrm{m}} e^{-Rt/2L} \cos(\omega' t + \phi).$$

After one oscillation the cosine term has returned to the original value but the exponential term has attenuated the charge on the capacitor according to

$$q = q_{\mathrm{m}} e^{-RT/2L},$$

where T is the period. The fractional energy loss on the capacitor is then

$$\frac{U_0 - U}{U_0} = 1 - \frac{q^2}{q_m^2} = 1 - e^{-RT/L}.$$

For small enough damping we can expand the exponent. Not only that, but $T = 2\pi/\omega$, so

$$\frac{\Delta U}{U} \approx 2\pi R/\omega L.$$

Chapter 37

Alternating Current Circuits

E37-1 The frequency, f, is related to the angular frequency ω by

$$\omega = 2\pi f = 2\pi(60 \text{ Hz}) = 377 \text{ rad/s}$$

The current is alternating because that is what the generator is designed to produce. It does this through the configuration of the magnets and coils of wire. One complete turn of the generator will (could?) produce one "cycle"; hence, the generator is turning 60 times per second. Not only does this set the frequency, it also sets the emf, since the emf is proportional to the speed at which the coils move through the magnetic field.

This is an important point. AC motors and transformers are designed to operate at specific frequencies. Changing the rotational speed of the generator affects both the frequency and the emf.

But changing the load on the circuit will change the current through the generator coils, and this will change the magnetic force on the coils, which will result in the coils rotating at a different rate, and so on.

It is therefore crucial that the generators turn at the same rate, regardless of the load. This can be done with speed governors, but at least in some small systems it is done by allowing the "unneeded" power to heat large containers of water.

E37-5 (a) The reactance of the capacitor is from Eq. 37-11, $X_C = 1/\omega C$. The AC generator from Exercise 4 has $\mathcal{E} = (25.0 \text{ V})\sin(377 \text{ rad/s})t$. So the reactance is

$$X_C = \frac{1}{\omega C} = \frac{1}{(377 \text{ rad/s})(4.1 \mu\text{F})} = 647\,\Omega.$$

The maximum value of the current is found from Eq. 37-13,

$$i_m = \frac{(\Delta V_C)_{\text{max}}}{X_C} = \frac{(25.0 \text{ V})}{(647\,\Omega)} = 3.86 \times 10^{-2}\text{A}.$$

(b) The generator emf is 90° out of phase with the current, so when the current is a maximum the emf is zero.

(c) The emf is -13.8 V when

$$\omega t = \arcsin \frac{(-13.8\,\text{V})}{(25.0\,\text{V})} = 0.585 \text{ rad}.$$

The current leads the voltage by $90° = \pi/2$, so

$$i = i_{\text{m}}\sin(\omega t - \phi) = (3.86\times 10^{-2}\text{A})\sin(0.585 - \pi/2) = -3.22\times 10^{-2}\text{A}.$$

(d) Since both the current and the emf are negative the product is positive and the generator is supplying energy to the circuit.

E37-9 A circuit is considered inductive if $X_L > X_C$, this happens when i_{m} lags $\mathcal{E}_{\text{m}}$. If, on the other hand, $X_L < X_C$, and i_{m} leads $\mathcal{E}_{\text{m}}$, we refer to the circuit as capacitive. This is discussed on page 850, although it is slightly hidden in the text of column one.

(a) At resonance, $X_L = X_C$. Since $X_L = \omega L$ and $X_C = 1/\omega C$ we expect that X_L grows with increasing frequency, while X_C decreases with increasing frequency.
Consequently, at frequencies above the resonant frequency $X_L > X_C$ and the circuit is predominantly inductive. But what does this really mean? It means that the inductor plays a major role in the current through the circuit while the capacitor plays a minor role. The more inductive a circuit is, the less significant any capacitance is on the behavior of the circuit.
For frequencies below the resonant frequency the reverse is true.

(b) Right at the resonant frequency the inductive effects are exactly canceled by the capacitive effects. The impedance is equal to the resistance, and it is (almost) as if neither the capacitor or inductor are even in the circuit.

E37-13 (a) The voltage across the generator is the generator emf, so when it is a maximum from Sample Problem 37-3, it is 36 V. This corresponds to $\omega t = \pi/2$.

(b) The current through the circuit is given by $i = i_{\text{m}}\sin(\omega t - \phi)$. We found in Sample Problem 37-3 that $i_{\text{m}} = 0.196\,\text{A}$ and $\phi = -29.4° = 0.513$ rad.
For a resistive load we apply Eq. 37-3,

$$\Delta V_R = i_{\text{m}}R\sin(\omega t - \phi) = (0.196\,\text{A})(160\Omega)\sin((\pi/2) - (-0.513)) = 27.3\,\text{V}.$$

(c) For a capacitive load we apply Eq. 37-12,

$$\Delta V_C = i_\mathrm{m} X_C \sin(\omega t - \phi - \pi/2) = (0.196\,\mathrm{A})(177\Omega)\sin(-(-0.513)) = 17.0\,\mathrm{V}.$$

(d) For an inductive load we apply Eq. 37-7,

$$\Delta V_L = i_\mathrm{m} X_L \sin(\omega t - \phi + \pi/2) = (0.196\,\mathrm{A})(87\Omega)\sin(\pi - (-0.513)) = -8.4\,\mathrm{V}.$$

(e) $(27.3\,\mathrm{V}) + (17.0\,\mathrm{V}) + (-8.4\,\mathrm{V}) = 35.9\,\mathrm{V}$, which is close enough to 36 for me.

E37-17 The resistance would be given by Eq. 37-32,

$$R = \frac{P_\mathrm{av}}{i_\mathrm{rms}{}^2} = \frac{(0.10)(746\,\mathrm{W})}{(0.650\,\mathrm{A})^2} = 177\,\Omega.$$

This would not be the same as the direct current resistance of the coils of a stopped motor, because there would be no inductive effects.

E37-21 The rms value of any sinusoidal quantity is related to the maximum value by $\sqrt{2}\,v_\mathrm{rms} = v_\mathrm{max}$. Since this factor of $\sqrt{2}$ appears in all of the expressions, we can conclude that if the rms values are equal then so are the maximum values. This means that

$$(\Delta V_R)_\mathrm{max} = (\Delta V_C)_\mathrm{max} = (\Delta V_L)_\mathrm{max}$$

or

$$i_m R = i_m X_C = i_m X_L$$

or, with one last simplification,

$$R = X_L = X_C.$$

Focus on the right hand side of the last equality. If $X_C = X_L$ then we have a resonance condition, and the impedance (see Eq. 37-20) is a minimum, and is equal to R. Then, according to Eq. 37-21,

$$i_m = \frac{\mathcal{E}_\mathrm{m}}{R},$$

which has the immediate consequence that the rms voltage across the resistor is the same as the rms voltage across the generator. So everything is 100 V.

E37-25 (a) We need to find the impedance before doing anything else. This is defined as in Eq. 37-20,

$$Z = \sqrt{R^2 + (X_L - X_C)^2},$$

The resistance is $R = 15.0\,\Omega$. The inductive reactance is

$$X_C = \frac{1}{\omega C} = \frac{1}{2\pi(550\,\mathrm{s}^{-1})(4.72\,\mu\mathrm{F})} = 61.3\,\Omega.$$

The inductive reactance is given by

$$X_L = \omega L = 2\pi(550\,\mathrm{s}^{-1})(25.3\,\mathrm{mH}) = 87.4\,\Omega.$$

The impedance is then

$$Z = \sqrt{(15.0\,\Omega)^2 + ((87.4\,\Omega) - (61.3\,\Omega))^2} = 30.1\,\Omega.$$

Finally, the rms current is

$$i_{\mathrm{rms}} = \frac{\mathcal{E}_{\mathrm{rms}}}{Z} = \frac{(75.0\,\mathrm{V})}{(30.1\,\Omega)} = 2.49\,\mathrm{A}.$$

(b) The rms voltages between any two points is given by

$$(\Delta V)_{\mathrm{rms}} = i_{\mathrm{rms}}Z,$$

where Z is *not* the impedance of the circuit but instead the impedance between the two points in question. When only one device is between the two points the impedance is equal to the reactance (or resistance) of that device.

We're not going to show *all* of the work here, but we will put together a nice table for you

Points	Impedance Expression	Impedance Value	$(\Delta V)_{\mathrm{rms}}$,
ab	$Z = R$	$Z = 15.0\,\Omega$	37.4 V,
bc	$Z = X_C$	$Z = 61.3\,\Omega$	153 V,
cd	$Z = X_L$	$Z = 87.4\,\Omega$	218 V,
bd	$Z = \lvert X_L - X_C \rvert$	$Z = 26.1\,\Omega$	65 V,
ac	$Z = \sqrt{R^2 + X_C^2}$	$Z = 63.1\,\Omega$	157 V,

Note that this last one was ΔV_{ac}, and not ΔV_{ad}, because it is more entertaining. You probably should use ΔV_{ad} for your homework.

(c) The average power dissipated from a capacitor or inductor is zero; that of the resistor is

$$P_R = [(\Delta V_R)_{\mathrm{rms}}]^2/R = (37.4\,\mathrm{V})^2/(15.0\,\Omega) = 93.3\,\mathrm{W}.$$

E37-27 Apply Eq. 37-41,

$$\Delta V_s = \Delta V_p \frac{N_s}{N_p} = (150\,\text{V})\frac{(780)}{(65)} = 1.8 \times 10^3\,\text{V}.$$

E37-29 The autotransformer could have a primary connected between taps T_1 and T_2 (200 turns), T_1 and T_3 (1000 turns), and T_2 and T_3 (800 turns).

The same possibilities are true for the secondary connections. Ignoring the one-to-one connections there are 6 choices— three are step up, and three are step down. The step up ratios are $1000/200 = 5$, $800/200 = 4$, and $1000/800 = 1.25$. The step down ratios are the reciprocals of these three values.

P37-3 (a) Since the maximum values for the voltages across the individual devices are proportional to the reactances (or resistances) for devices in series (the constant of proportionality is the maximum current), we have $X_L = 2R$ and $X_C = R$.

From Eq. 37-18,

$$\tan\phi = \frac{X_L - X_C}{R} = \frac{2R - R}{R} = 1,$$

or $\phi = 45°$.

(b) The impedance of the circuit, in terms of the resistive element, is

$$Z = \sqrt{R^2 + (X_L - X_C)^2} = \sqrt{R^2 + (2R - R)^2} = \sqrt{2}\,R.$$

But $\mathcal{E}_m = i_m Z$, so $Z = (34.4\,\text{V})/(0.320\,\text{A}) = 108\Omega$. Then we can use our previous work to finds that $R = 76\Omega$.

P37-5 All three wires have emfs which vary sinusoidally in time; if we choose *any* two wires the phase difference will have an absolute value of 120°. We can then choose any two wires and expect (by symmetry) to get the same result. We choose 1 and 2. The potential difference is then

$$V_1 - V_2 = V_m \left(\sin \omega t - \sin(\omega t - 120°) \right).$$

We need to add these two sine functions to get just one. We use

$$\sin\alpha - \sin\beta = 2\sin\frac{1}{2}(\alpha - \beta)\cos\frac{1}{2}(\alpha + \beta).$$

Then

$$
\begin{aligned}
V_1 - V_2 &= 2V_m \sin\frac{1}{2}(120°)\cos\frac{1}{2}(2\omega t - 120°), \\
&= 2V_m\left(\frac{\sqrt{3}}{2}\right)\cos(\omega t - 60°), \\
&= \sqrt{3}V_m \sin(\omega t + 30°).
\end{aligned}
$$

P37-11 (a) The resistance of this bulb is

$$R = \frac{(\Delta V)^2}{P} = \frac{(120\,\text{V})^2}{(1000\,\text{W})} = 14.4\,\Omega.$$

The power is directly related to the brightness; if the bulb is to be varied in brightness by a factor of 5 then it would have a minimum power of 200 W. The rms current through the bulb at this power would be

$$i_{\text{rms}} = \sqrt{P/R} = \sqrt{(200\,\text{W})/(14.4\,\Omega)} = 3.73\,\text{A}.$$

The impedance of the circuit must have been

$$Z = \frac{\mathcal{E}_{\text{rms}}}{i_{\text{rms}}} = \frac{(120\,\text{V})}{(3.73\,\text{A})} = 32.2\,\Omega.$$

The inductive reactance would then be

$$X_L = \sqrt{Z^2 - R^2} = \sqrt{(32.2\,\Omega)^2 - (14.4\,\Omega)^2} = 28.8\,\Omega.$$

Finally, the inductance would be

$$L = X_L/\omega = (28.8\,\Omega)/(2\pi(60.0\,\text{s}^{-1})) = 7.64\,\text{H}.$$

(b) One could use a variable resistor, and since it would be in series with the lamp a value of

$$32.2\,\Omega - 14.4\,\Omega = 17.8\,\Omega$$

would work. But the resistor would get hot, while on average there is no power radiated from a pure inductor.

Chapter 38

Maxwell's Equations and Electromagnetic Waves

E38-3 We could use the results of Exercise 2, and change the potential difference across the plates of the capacitor at a rate

$$\frac{dV}{dt} = \frac{i_\mathrm{d}}{C} = \frac{(1.0\,\mathrm{mA})}{(1.0\mu\mathrm{F})} = 1.0\ \mathrm{kV/s}.$$

But how would we do that? We would need to provide a constant current to the capacitor, and that constant current would be

$$i = \frac{dQ}{dt} = \frac{d}{dt}CV = C\frac{dV}{dt} = i_\mathrm{d}.$$

E38-7 The displacement current is defined in Eq. 38-8,

$$i_\mathrm{d} = \epsilon_0 \frac{d\Phi_E}{dt}.$$

Since the electric field is uniform in the area and perpendicular to the surface area we have

$$\Phi_E = \int \vec{\mathbf{E}} \cdot d\vec{\mathbf{A}} = \int E\, dA = E \int dA = EA.$$

The displacement current is then

$$i_\mathrm{d} = \epsilon_0 A \frac{dE}{dt} = (8.85\times 10^{-12}\mathrm{F/m})(1.9\,\mathrm{m}^2)\frac{dE}{dt}.$$

(a) In the first region the electric field decreases by 0.2 MV/m in 4μs, so

$$i_\mathrm{d} = (8.85\times 10^{-12}\mathrm{F/m})(1.9\,\mathrm{m}^2)\frac{(-0.2\times 10^6\mathrm{V/m})}{(4\times 10^{-6}\mathrm{s})} = -0.84\,\mathrm{A}.$$

(b) The electric field is constant so there is no change in the electric flux, and hence there is no displacement current.

(c) In the last region the electric field decreases by 0.4 MV/m in 5μs, so

$$i_{\rm d} = (8.85 \times 10^{-12}{\rm F/m})(1.9\,{\rm m}^2)\frac{(-0.4 \times 10^6 {\rm V/m})}{(5 \times 10^{-6}{\rm s})} = -1.3\,{\rm A}.$$

The negative signs aren't very important here until we try to establish a direction to either the electric field or the current.

E38-11 (a) Consider the path $abefa$. The closed line integral consists of *two* parts: $b \rightarrow e$ and $e \rightarrow f \rightarrow a \rightarrow b$. Then

$$\oint \vec{\bf E} \cdot d\vec{s} = -\frac{d\Phi}{dt}$$

can be written as

$$\int_{b \rightarrow e} \vec{\bf E} \cdot d\vec{s} + \int_{e \rightarrow f \rightarrow a \rightarrow b} \vec{\bf E} \cdot d\vec{s} = -\frac{d}{dt}\Phi_{abef}.$$

Now consider the path $bcdeb$. The closed line integral consists of *two* parts: $b \rightarrow c \rightarrow d \rightarrow e$ and $e \rightarrow b$. Then

$$\oint \vec{\bf E} \cdot d\vec{s} = -\frac{d\Phi}{dt}$$

can be written as

$$\int_{b \rightarrow c \rightarrow d \rightarrow e} \vec{\bf E} \cdot d\vec{s} + \int_{e \rightarrow b} \vec{\bf E} \cdot d\vec{s} = -\frac{d}{dt}\Phi_{bcde}.$$

These two expressions can be added together, and since

$$\int_{e \rightarrow b} \vec{\bf E} \cdot d\vec{s} = -\int_{b \rightarrow e} \vec{\bf E} \cdot d\vec{s}$$

we get

$$\int_{e \rightarrow f \rightarrow a \rightarrow b} \vec{\bf E} \cdot d\vec{s} + \int_{b \rightarrow c \rightarrow d \rightarrow e} \vec{\bf E} \cdot d\vec{s} = -\frac{d}{dt}\left(\Phi_{abef} + \Phi_{bcde}\right).$$

The left hand side of this is just the line integral over the closed path $efadcde$; the right hand side is the net change in flux through the two surfaces. Then we can simplify this expression as

$$\oint \vec{\bf E} \cdot d\vec{s} = -\frac{d\Phi}{dt}.$$

Looks pretty consistent to me.

(b) Do everything above again, except substitute B for E.

(c) If the equations were not self consistent we would arrive at different values of E and B depending on how we defined our surfaces. This multi-valued result would be quite unphysical.

E38-15 A series LC circuit will oscillate naturally at a frequency

$$f = \frac{\omega}{2\pi} = \frac{1}{2\pi\sqrt{LC}}$$

We will need to combine this with $v = f\lambda$, where $v = c$ is the speed of EM waves.

We want to know the inductance required to produce an EM wave of wavelength $\lambda = 550 \times 10^{-9}$m, so

$$L = \frac{\lambda^2}{4\pi^2 c^2 C} = \frac{(550 \times 10^{-9}\text{m})^2}{4\pi^2 (3.00 \times 10^8 \text{m/s})^2 (17 \times 10^{-12}\text{ F})} = 5.01 \times 10^{-21} \text{ H}.$$

This is a small inductance! But how small is it really?

Assume this inductor is made from a single loop of wire. What would be the size? Since inductance is given by $L = N\Phi_B/i$, then a estimate of the inductance would be found from a minimum calculation of the flux. The magnetic field at the very center of a loop of wire is at least a not unreasonable estimate of the average magnetic field in the loop, so an estimate for the flux is

$$\Phi_B \approx BA = \frac{\mu_0 i}{2R}\pi R^2 = \frac{\pi \mu_0 i R}{2},$$

where R is the radius of the loop. The inductance of this loop is then about

$$L \approx \frac{\pi \mu_0 R}{2},$$

which means the radius of the loop would be

$$R \approx \frac{2L}{\pi \mu_0} = 2.54 \times 10^{-15}\text{m}.$$

This is about the same size as a proton.

E38-17 The electric and magnetic field of an electromagnetic wave are related by Eqs. 38-15 and 38-16,

$$B = \frac{E}{c} = \frac{(321\mu\text{V/m})}{(3.00 \times 10^8\text{m/s})} = 1.07 \text{ pT},$$

and that's a small field.

E38-23 Intensity is given by Eq. 38-28, which is simply an expression of power divided by surface area. To find the intensity of the TV signal at α-Centauri we need to find the distance in meters;

$$r = (4.30 \text{ light-years})(3.00 \times 10^8 \text{ m/s})(3.15 \times 10^7 \text{ s/year}) = 4.06 \times 10^{16} \text{ m}.$$

The intensity of the signal when it has arrived at out nearest neighbor is then

$$I = \frac{P}{4\pi r^2} = \frac{(960 \text{ kW})}{4\pi(4.06 \times 10^{16} \text{ m})^2} = 4.63 \times 10^{-29} \text{ W/m}^2$$

Without a doubt this is a small number. But is it too small to be detected? Consider Exercise 38-30, where the radio telescope at Arecibo (did you read Contact by Carl Sagan or see the movie starring Jodie Foster?) is capable of detecting a signal with an intensity on the order of 10^{-27} W/m^2. The α-Centarians would need a telescope with a radius 10 times that of Arecibo, assuming the electronic reception equipment is the same.

E38-27 (a) Intensity is related to distance by Eq. 38-28. If r_1 is the original distance from the street lamp and I_1 the intensity at that distance, then

$$I_1 = \frac{P}{4\pi r_1^2}.$$

There is a similar expression for the closer distance $r_2 = r_1 - 162 \text{ m}$ and the intensity at that distance $I_2 = 1.50 I_1$. We can combine the two expression for intensity,

$$
\begin{aligned}
I_2 &= 1.50 I_1, \\
\frac{P}{4\pi r_2^2} &= 1.50 \frac{P}{4\pi r_1^2}, \\
r_1^2 &= 1.50 r_2^2, \\
r_1 &= \sqrt{1.50}\,(r_1 - 162 \text{ m}).
\end{aligned}
$$

The last line is easy enough to solve and we find $r_1 = 883 \text{ m}$.

(b) No, we can't find the power output from the lamp, because we were never provided with an absolute intensity reference.

E38-31 (a) The electric field amplitude is related to the intensity by Eq. 38-26,

$$I = \frac{E_{\mathrm{m}}^2}{2\mu_0 c},$$

or

$$E_{\mathrm{m}} = \sqrt{2\mu_0 c I} = \sqrt{2(4\pi \times 10^{-7} \text{H/m})(3.00 \times 10^8 \text{m/s})(7.83\mu\text{W/m}^2)} = 7.68 \times 10^{-2} \text{ V/m}.$$

(b) The magnetic field amplitude is given by

$$B_{\mathrm{m}} = \frac{E_{\mathrm{m}}}{c} = \frac{(7.68 \times 10^{-2}\,\mathrm{V/m})}{(3.00 \times 10^{8}\mathrm{m/s})} = 2.56 \times 10^{-10}\,\mathrm{T}$$

(c) The power radiated by the transmitter can be found from Eq. 38-28,

$$P = 4\pi r^2 I = 4\pi (11.3\ \mathrm{km})^2 (7.83 \mu\mathrm{W/m}^2) = 12.6\ \mathrm{kW}.$$

E38-33 Radiation pressure for absorption is given by Eq. 38-34, but we need to find the energy absorbed before we can apply that. We are given an intensity, a surface area, and a time, so

$$\Delta U = (1.1 \times 10^3 \mathrm{W/m}^2)(1.3\,\mathrm{m}^2)(9.0 \times 10^3 \mathrm{s}) = 1.3 \times 10^7 \mathrm{J}.$$

The momentum delivered is

$$p = (\Delta U)/c = (1.3 \times 10^7 \mathrm{J})/(3.00 \times 10^8 \mathrm{m/s}) = 4.3 \times 10^{-2} \mathrm{kg} \cdot \mathrm{m/s}.$$

E38-39 We can treat the object as having two surfaces, one completely reflecting and the other completely absorbing. If the entire surface has an area A then the absorbing part has an area fA while the reflecting part has area $(1 - f)A$. The average force is then the sum of the force on each part,

$$F_{\mathrm{av}} = \frac{I}{c} f A + \frac{2I}{c}(1 - f)A,$$

which can be written in terms of pressure as

$$\frac{F_{\mathrm{av}}}{A} = \frac{I}{c}(2 - f).$$

P38-7 Look back to Chapter 14 for a discussion on the elliptic orbit. On page 312 it is pointed out that the closest distance to the sun is $R_{\mathrm{p}} = a(1 - e)$ while the farthest distance is $R_{\mathrm{a}} = a(1 + e)$, where a is the semimajor axis and e the eccentricity.

The fractional variation in intensity is

$$\frac{\Delta I}{I} \approx \frac{I_{\mathrm{p}} - I_{\mathrm{a}}}{I_{\mathrm{a}}},$$

$$= \frac{I_{\mathrm{p}}}{I_{\mathrm{a}}} - 1,$$

$$= \frac{R_{\mathrm{a}}^2}{R_{\mathrm{p}}^2} - 1,$$

$$= \frac{(1 + e)^2}{(1 - e)^2} - 1.$$

We need to expand this expression for small e using

$$(1 + e)^2 \approx 1 + 2e,$$

and

$$(1 - e)^{-2} \approx 1 + 2e,$$

and finally

$$(1 + 2e)^2 \approx 1 + 4e.$$

Combining,

$$\frac{\Delta I}{I} \approx (1 + 2e)^2 - 1 \approx 4e.$$

P38-9   Eq. 38-14 requires

$$
\begin{aligned}
\frac{\partial E}{\partial x} &= -\frac{\partial B}{\partial t}, \\
E_{\mathrm{m}} k \cos kx \sin \omega t &= B_{\mathrm{m}} \omega \cos kx \sin \omega t, \\
E_{\mathrm{m}} k &= B_{\mathrm{m}} \omega.
\end{aligned}
$$

Eq. 38-17 requires

$$
\begin{aligned}
\mu_0 \epsilon_0 \frac{\partial E}{\partial t} &= -\frac{\partial B}{\partial x}, \\
\mu_0 \epsilon_0 E_{\mathrm{m}} \omega \sin kx \cos \omega t &= B_{\mathrm{m}} k \sin kx \cos \omega t, \\
\mu_0 \epsilon_0 E_{\mathrm{m}} \omega &= B_{\mathrm{m}} k.
\end{aligned}
$$

Dividing one expression by the other,

$$\mu_0 \epsilon_0 k^2 = \omega^2,$$

or

$$\frac{\omega}{k} = c = \frac{1}{\sqrt{\mu_0 \epsilon_0}}$$

Not only that, but $E_{\mathrm{m}} = cB_{\mathrm{m}}$. You've seen an expression similar to this before, and you'll see expressions similar to it again.

(b)   We'll assume that Eq. 38-21 is applicable here. Then

$$
\begin{aligned}
S &= \frac{1}{\mu_0} = \frac{E_{\mathrm{m}} B_{\mathrm{m}}}{\mu_0} \sin kx \sin \omega t \cos kx \cos \omega t, \\
&= \frac{E_{\mathrm{m}}^2}{4\mu_0 c} \sin 2kx \sin 2\omega t
\end{aligned}
$$

is the magnitude of the instantaneous Poynting vector.

(c) The time averaged power flow across any surface is the value of

$$\frac{1}{T} \int_0^T \int \vec{S} \cdot d\vec{A}\, dt,$$

where T is the period of the oscillation. We'll just gloss over any concerns about direction, and assume that the $\vec{S}$ will be constant in direction so that we will, at most, need to concern ourselves about a constant factor $\cos\theta$. We can then deal with a scalar, instead of vector, integral, and we can integrate it in any order we want. We want to do the t integration first, because an integral over $\sin\omega t$ for a period $T = 2\pi/\omega$ is zero. Then we are done!

(d) There is no energy flow; the energy remains inside the container.

P38-11 (a) We've already calculated B previously. It is

$$B = \frac{\mu_0 i}{2\pi r} \text{ where } i = \frac{\mathcal{E}}{R}.$$

It is, of course, a vector field, where the field lines are concentric circles around the axis of the cable. The sense of the circles can be determined from the right hand rule: point your right thumb in the direction of the current on the inner conductor and then your fingers will grip the wire in the direction of the magnetic field lines. In this convention if $\mathcal{E}$ is positive the current on the inner conductor flows away from the battery.

The electric field is only slightly harder. The electric field of a long straight wire has the form $E = k/r$, where k is some constant. But

$$\Delta V = -\int \vec{E} \cdot d\vec{s} = -\int_a^b E\, dr = -k\ln(b/a).$$

In this problem the inner conductor is at the higher potential, so

$$k = \frac{-\Delta V}{\ln(b/a)} = \frac{\mathcal{E}}{\ln(b/a)},$$

and then the electric field is

$$E = \frac{\mathcal{E}}{r\ln(b/a)}.$$

This is also a vector field, and if $\mathcal{E}$ is positive the electric field points radially out from the central conductor.

(b) The Poynting vector is

$$\vec{S} = \frac{1}{\mu_0}\vec{E} \times \vec{B};$$

$\vec{E}$ is radial while $\vec{B}$ is circular, so they are perpendicular. Assuming that $\mathcal{E}$ is positive the direction of $\vec{S}$ is away from the battery. Switching the sign of $\mathcal{E}$ (connecting the battery in reverse) will flip the direction of both $\vec{E}$ and $\vec{B}$, so $\vec{S}$ will pick up *two* negative signs and therefore *still* point away from the battery.

The magnitude is

$$S = \frac{EB}{\mu_0} = \frac{\mathcal{E}^2}{2\pi R \ln(b/a)r^2}$$

(c) We want to evaluate a surface integral in polar coordinates (we do have cylindrical symmetry, eh?) and so $dA = (dr)(rd\theta)$. We have already established that $\vec{S}$ is pointing away from the battery parallel to the central axis. Then we can integrate

$$
\begin{aligned}
P &= \int \vec{S} \cdot d\vec{A} = \int S\, dA, \\
&= \int_a^b \int_0^{2\pi} \frac{\mathcal{E}^2}{2\pi R \ln(b/a)r^2}\, d\theta\, r\, dr, \\
&= \int_a^b \frac{\mathcal{E}^2}{R \ln(b/a)r}\, dr, \\
&= \frac{\mathcal{E}^2}{R}.
\end{aligned}
$$

(d) Read part (b) above.

Chapter 39

Light Waves

E39-3 (a) Apply $v = f\lambda$. Then

$$f = (3.0 \times 10^8 \text{m/s})/(0.067 \times 10^{-15}\text{m}) = 4.5 \times 10^{24} \text{ Hz}.$$

(b) $\lambda = (3.0 \times 10^8 \text{m/s})/(30 \text{ Hz}) = 1.0 \times 10^7 \text{m}.$

E39-5 (a) We refer to Fig. 39-6 to answer this question. The limits are approximately 520 nm and 620 nm.

(b) The wavelength for which the eye is most sensitive is 550 nm. This corresponds to to a frequency of

$$f = c/\lambda = (3.00 \times 10^8 \text{ m/s})/(550 \times 10^{-9}\text{m}) = 5.45 \times 10^{14} \text{ Hz}.$$

This frequency corresponds to a period of $T = 1/f = 1.83 \times 10^{-15}\text{s}.$

E39-9 This is a question of how much time it takes light to travel 4 cm, because the light traveled from the Earth to the moon, bounced off of the reflector, and then traveled back. The time to travel 4 cm is $\Delta t = (0.04 \text{ m})/(3 \times 10^8 \text{ m/s}) = 0.13$ ns. Note that I interpreted the question differently than the answer in the back of the book.

E39-11 We will choose the mirror to lie in the xy plane at $z = 0$. There is no loss of generality in doing so; we had to define our coordinate system somehow. The choice is convenient in that any normal is then parallel to the z axis. Furthermore, we can arbitrarily define the incident ray to originate at $(0, 0, z_1)$. Lastly, we can rotate the coordinate system about the z axis so that the reflected ray passes through the point $(0, y_3, z_3)$.

311

The point of reflection for this ray is somewhere on the surface of the mirror, say $(x_2, y_2, 0)$. This distance traveled from the point 1 to the reflection point 2 is

$$d_{12} = \sqrt{(0 - x_2)^2 + (0 - y_2)^2 + (z_1 - 0)^2} = \sqrt{x_2^2 + y_2^2 + z_1^2}$$

and the distance traveled from the reflection point 2 to the final point 3 is

$$d_{23} = \sqrt{(x_2 - 0)^2 + (y_2 - y_3)^2 + (0 - z_3)^2} = \sqrt{x_2^2 + (y_2 - y_3)^2 + z_3^2}.$$

The only point which is free to move is the reflection point, $(x_2, y_2, 0)$, and that point can only move in the xy plane. Fermat's principle states that the reflection point will be such to minimize the total distance,

$$d_{12} + d_{23} = \sqrt{x_2^2 + y_2^2 + z_1^2} + \sqrt{x_2^2 + (y_2 - y_3)^2 + z_3^2}.$$

We do this minimization by taking the partial derivative with respect to both x_2 and y_2. But we can do part by inspection alone. Any non-zero value of x_2 can only *add* to the total distance, regardless of the value of any of the other quantities. Consequently, $x_2 = 0$ is one of the conditions for minimization.

We are done! Although you are invited to finish the minimization process, once we know that $x_2 = 0$ we have that point 1, point 2, and point 3 all lie in the yz plane. The normal is parallel to the z axis, so it also lies in the yz plane. Everything is then in the yz plane.

E39-17 The speed of light in a substance with index of refraction n is given by $v = c/n$, where c is the speed of light in a vacuum. An electron will then emit Cerenkov radiation in this particular liquid if the speed exceeds

$$v = c/n = (3.00 \times 10^8 \,\text{m/s})/(1.54) = 1.95 \times 10^8 \,\text{m/s}.$$

We are done, but it is nice to go a little farther. A quick review of relativity (Chapter 20) shows that this speed corresponds to a Lorentz factor of

$$\gamma = \frac{1}{\sqrt{1 - v^2/c^2}} = \frac{1}{\sqrt{1 - 1/(1.54)^2}} = 1.31;$$

notice how we used $v/c = 1/n$, which was rather sneaky. The *relativistic* energy of the electron is then

$$E = \gamma mc^2 = (1.31)(0.511 \,\text{MeV}) = 0.672 \,\text{MeV}.$$

Much of this we could have done in our heads.

E39-19 The angle of the refracted ray is $\theta_2 = 90°$, the angle of the incident ray can be found by trigonometry,

$$\tan\theta_1 = \frac{(1.14\,\text{m})}{(0.85\,\text{m})} = 1.34,$$

or $\theta_1 = 53.3°$.

We can use these two angles, along with the index of refraction of air, to find that the index of refraction of the liquid from Eq. 39-4,

$$n_1 = n_2 \frac{\sin\theta_2}{\sin\theta_1} = (1.00)\frac{(\sin 90°)}{(\sin 53.3°)} = 1.25.$$

There are no units attached to this quantity.

E39-25 We'll rely heavily on the figure for our arguments. Let x be the distance between the points on the surface where the vertical ray crosses and the bent ray crosses.

Maybe we should include our own figure.

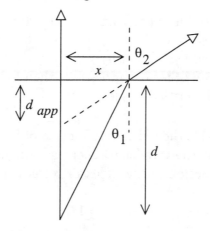

In this exercise we will take advantage of the fact that, for small angles θ,

$$\sin\theta \approx \tan\theta \approx \theta$$

In this approximation Snell's law takes on the particularly simple form

$$n_1\theta_1 = n_2\theta_2$$

The two angles here are conveniently found from the figure,

$$\theta_1 \approx \tan\theta_1 = \frac{x}{d},$$

and

$$\theta_2 \approx \tan\theta_2 = \frac{x}{d_{\text{app}}}.$$

Inserting these two angles into the simplified Snell's law, as well as substituting $n_1 = n$ and $n_2 = 1.0$,

$$n_1\theta_1 = n_2\theta_2,$$
$$n\frac{x}{d} = \frac{x}{d_{\text{app}}},$$
$$d_{\text{app}} = \frac{d}{n}.$$

E39-33 (a) The critical angle is given by Eq. 39-17,

$$\theta_{\text{c}} = \sin^{-1}\frac{n_2}{n_1} = \sin^{-1}\frac{(1.586)}{(1.667)} = 72.07°.$$

(b) Critical angles only exist when "attempting" to travel from a medium of higher index of refraction to a medium of lower index of refraction; in this case from A to B.

E39-37 The light strikes the quartz-air interface from the inside; it is originally "white", so if the reflected ray is to appear "bluish" (reddish) then the refracted ray should have been "reddish" (bluish). Since part of the light undergoes total internal reflection while the other part does not, then the angle of incidence must be approximately equal to the critical angle.

(a) Look at Fig. 39-11, the index of refraction of fused quartz is given as a function of the wavelength. As the wavelength increases the index of refraction decreases. The critical angle is a function of the index of refraction; for a substance in air the critical angle is given by

$$\sin\theta_{\text{c}} = 1/n.$$

As n decreases $1/n$ increases so θ_{c} increases. For fused quartz, then, as wavelength increases θ_{c} also increases.

In short, red light has a larger critical angle than blue light. If the angle of incidence is midway between the critical angle of red and the critical angle of blue, then the blue component of the light will experience total internal reflection while the red component will pass through as a refracted ray.

So yes, the light can be made to appear bluish.

(b) No, the light can't be made to appear reddish. See above.

(c) Choose an angle of incidence between the two critical angles as described in part (a). Using a value of $n = 1.46$ from Fig. 39-11,

$$\theta_c = \sin^{-1}(1/1.46) = 43.2°.$$

Getting the effect to work will require considerable sensitivity.

E39-41 The Doppler theory for light gives

$$f = f_0 \frac{1 - u/c}{\sqrt{1 - u^2/c^2}} = f_0 \frac{1 - (0.2)}{\sqrt{1 - (0.2)^2}} = 0.82\, f_0.$$

The frequency is shifted down to about 80%, which means the wavelength is shifted up by an additional 25%. Blue light (480 nm) would appear yellow/orange (585 nm).

E39-45 The sun rotates once every 26 days at the equator, while the radius is 7.0×10^8 m. The speed of a point on the equator is then

$$v = \frac{2\pi R}{T} = \frac{2\pi(7.0 \times 10^8 \text{m})}{(2.2 \times 10^6 \text{s})} = 2.0 \times 10^3 \text{ m/s}.$$

This corresponds to a velocity parameter of

$$\beta = u/c = (2.0 \times 10^3 \text{ m/s})/(3.0 \times 10^8 \text{ m/s}) = 6.7 \times 10^{-6}.$$

This is a case of small numbers, so we'll use the formula that you derived in Exercise 39-40:

$$\Delta\lambda = \lambda\beta = (553 \text{ nm})(6.7 \times 10^{-6}) = 3.7 \times 10^{-3} \text{ nm}.$$

E39-51 The frequency observed by the detector from the first source is (Eq. 39-31)

$$f = f_1 \sqrt{1 - (0.717)^2} = 0.697 f_1.$$

The frequency observed by the detector from the second source is (Eq. 39-30)

$$f = f_2 \frac{\sqrt{1 - (0.717)^2}}{1 + (0.717) \cos\theta} = \frac{0.697 f_2}{1 + (0.717) \cos\theta}.$$

We need to equate these and solve for θ. Then

$$
\begin{aligned}
0.697 f_1 &= \frac{0.697 f_2}{1 + 0.717 \cos\theta}, \\
1 + 0.717 \cos\theta &= f_2/f_1, \\
\cos\theta &= (f_2/f_1 - 1)/0.717, \\
\theta &= 101.1°.
\end{aligned}
$$

Subtract from 180° to find the angle with the line of sight.

P39-5 (a) As was done in Ex. 39-25 above we use the small angle approximation of

$$\sin\theta \approx \theta \approx \tan\theta$$

The incident angle is θ; if the light were to go in a straight line we would expect it to strike a distance y_1 beneath the normal on the right hand side. The various distances are related to the angle by

$$\theta \approx \tan\theta \approx y_1/t.$$

The light, however, does *not* go in a straight line, it is refracted according to (the small angle approximation to) Snell's law,

$$n_1\theta_1 = n_2\theta_2,$$

which we will simplify further by letting $\theta_1 = \theta$, $n_2 = n$, and $n_1 = 1$,

$$\theta = n\theta_2.$$

The point where the refracted ray *does* strike is related to the angle by

$$\theta_2 \approx \tan\theta_2 = y_2/t.$$

Combining the three expressions,

$$y_1 = ny_2.$$

The difference, $y_1 - y_2$ is the vertical distance between the displaced ray and the original ray as measured on the plate glass. A little algebra yields

$$
\begin{aligned}
y_1 - y_2 &= y_1 - y_1/n, \\
&= y_1\left(1 - 1/n\right), \\
&= t\theta\frac{n-1}{n}.
\end{aligned}
$$

The perpendicular distance x is related to this difference by

$$\cos\theta = x/(y_1 - y_2).$$

In the small angle approximation $\cos\theta \approx 1 - \theta^2/2$. If θ is sufficiently small we can ignore the square term, and $x \approx y_2 - y_1$.

(b) Remember to use *radians* and not degrees whenever the small angle approximation is applied. Then

$$x = (1.0\text{ cm})(0.175\text{ rad})\frac{(1.52) - 1}{(1.52)} = 0.060\text{ cm}.$$

P39-7 The "big idea" of Problem 6 is that when light travels through layers the angle that it makes in any layer depends only on the incident angle, the index of refraction where that incident angle occurs, and the index of refraction at the current point.

That means that light which leaves the surface of the runway at 90° to the normal will make an angle

$$n_0 \sin 90° = n_0(1 + ay) \sin \theta$$

at some height y above the runway. It is mildly entertaining to note that the value of n_0 is unimportant, only the value of a!

The expression

$$\sin \theta = \frac{1}{1 + ay} \approx 1 - ay$$

can be used to find the angle made by the curved path against the normal as a function of y. The slope of the curve at any point is given by

$$\frac{dy}{dx} = \tan(90° - \theta) = \cot \theta = \frac{\cos \theta}{\sin \theta}.$$

Now we need to know $\cos \theta$. It is

$$\cos \theta = \sqrt{1 - \sin^2 \theta} \approx \sqrt{2ay}.$$

Combining

$$\frac{dy}{dx} \approx \frac{\sqrt{2ay}}{1 - ay},$$

and now we integrate. We will ignore the ay term in the denominator because it will always be small compared to 1. Then

$$\int_0^d dx = \int_0^h \frac{dy}{\sqrt{2ay}},$$

$$d = \sqrt{\frac{2h}{a}} = \sqrt{\frac{2(1.7\,\mathrm{m})}{(1.5 \times 10^{-6}\mathrm{m}^{-1})}} = 1500\,\mathrm{m}.$$

P39-11 (a) The fraction of light energy which escapes from the water is dependent on the critical angle. Light radiates in all directions from the source, but only that which strikes the surface at an angle less than the critical angle will escape. This critical angle is

$$\sin \theta_{\mathrm{c}} = 1/n.$$

We want to find the solid angle of the light which escapes; this is found by integrating

$$\Omega = \int_0^{2\pi} \int_0^{\theta_c} \sin \theta \, d\theta \, d\phi.$$

317

This is *not* a hard integral to do. The result is

$$\Omega = 2\pi(1 - \cos\theta_c).$$

There are 4π steradians in a spherical surface, so the fraction which escapes is

$$f = \frac{1}{2}(1 - \cos\theta_c) = \frac{1}{2}(1 - \sqrt{1 - \sin^2\theta_c}).$$

The last substitution is easy enough. We never needed to know the depth h.

(b) $f = \frac{1}{2}(1 - \sqrt{1 - (1/(1.3))^2}) = 0.18$.

P39-13 Consider the two possible extremes: a ray of light can propagate in a straight line directly down the axis of the fiber, or it can reflect off of the sides with the minimum possible angle of incidence. Start with the harder option.

The minimum angle of incidence that will still involve reflection is the critical angle, so

$$\sin\theta_c = \frac{n_2}{n_1}.$$

This light ray has farther to travel than the ray down the fiber axis because it is traveling at an angle. The distance traveled by this ray is

$$L' = L/\sin\theta_c = L\frac{n_1}{n_2},$$

The time taken for this bouncing ray to travel a length L down the fiber is then

$$t' = \frac{L'}{v} = \frac{L'n_1}{c} = \frac{L}{c}\frac{n_1^2}{n_2}.$$

Now for the easier ray. It travels straight down the fiber in a time

$$t = \frac{L}{c}n_1.$$

The difference is

$$t' - t = \Delta t = \frac{L}{c}\left(\frac{n_1^2}{n_2} - n_1\right) = \frac{Ln_1}{cn_2}(n_1 - n_2).$$

(b) For the numbers is Problem 12 we have

$$\Delta t = \frac{(350 \times 10^3\,\text{m})(1.58)}{(3.00 \times 10^8\text{m/s})(1.53)}((1.58) - (1.53)) = 6.02 \times 10^{-5}\text{s}.$$

Chapter 40

Mirrors and Lenses

E40-3 Drawing a picture can help, but I'm going to use words here. If the mirror rotates through an angle α then the angle of incidence will increase by an angle α, and so will the angle of reflection. But that means that the angle between the incident angle and the reflected angle has increased by α twice.

E40-7 The apparent depth of the swimming pool is given by the work done for Exercise 39-25,

$$d_{\text{app}} = d/n$$

The water then "appears" to be only 186 cm/1.33 = 140 cm deep. The apparent distance between the light and the mirror is then 250 cm + 140 cm = 390 cm; consequently the image of the light is 390 cm beneath the surface of the mirror.

E40-9 We want to know over what surface area of the mirror are rays of light reflected from the object into the eye. By similar triangles the diameter of the pupil and the diameter of the part of the mirror (d) which reflects light into the eye are related by

$$\frac{d}{(10 \text{ cm})} = \frac{(5.0 \text{ mm})}{(24 \text{ cm}) + (10 \text{ cm})},$$

which has solution $d = 1.47$ mm The area of the circle on the mirror is

$$A = \pi(1.47 \text{ mm})^2/4 = 1.7 \text{ mm}^2.$$

E40-13 The image is magnified by a factor of 2.7, so the image distance is 2.7 times farther from the mirror than the object. An important question to ask is whether or not the image is real or virtual. If it is a virtual image it is behind the mirror and someone looking at the mirror could see it. If it were a real image it would be in front of the mirror, and the man, who serves as the object and is therefore closer to the mirror than the image, would not be able to see it.

So we shall assume that the image is virtual. The image distance is then a negative number. The focal length is half of the radius of curvature, so we want to solve Eq. 40-6, with $f = 17.5$ cm and $i = -2.7o$

$$\frac{1}{(17.5 \text{ cm})} = \frac{1}{o} + \frac{1}{-2.7o} = \frac{0.63}{o},$$

which has solution $o = 11$ cm.

It is real easy to ignore the all important negative signs in these problems. Use *all* of the (relevant) information that is given to set up the equations!

E40-17 (a)   If I am easily convinced does that mean that I don't need to show any work?

Consider the point A. Light from this point travels along the line ABC and will be parallel to the horizontal center line from the center of the cylinder. Since the tangent to a circle defines the outer limit of the intersection with a line, this line must describe the apparent size.

Are we convinced yet?

(b)   The angle of incidence of ray AB is given by

$$\sin \theta_1 = r/R.$$

The angle of refraction of ray BC is given by

$$\sin \theta_2 = r^*/R.$$

Snell's law, and a little algebra, yields

$$
\begin{aligned}
n_1 \sin \theta_1 &= n_2 \sin \theta_2, \\
n_1 \frac{r}{R} &= n_2 \frac{r^*}{R}, \\
nr &= r^*.
\end{aligned}
$$

In the last line we used the fact that $n_2 = 1$, because it is in the air, and $n_1 = n$, the index of refraction of the glass.

E40-21   The image location can be found from Eq. 40-15,

$$\frac{1}{i} = \frac{1}{f} - \frac{1}{o} = \frac{1}{(-30 \text{ cm})} - \frac{1}{(20 \text{ cm})} = \frac{1}{-12 \text{ cm}},$$

so the image is located 12 cm from the thin lens, *on the same side as the object.*

E40-27 This exercise requires just a little bit of a algebra. According to the definitions, $o = f + x$ and $i = f + x'$. Starting with Eq. 40-15,

$$\frac{1}{o} + \frac{1}{i} = \frac{1}{f},$$

$$\frac{i+o}{oi} = \frac{1}{f},$$

$$\frac{2f + x + x'}{(f+x)(f+x')} = \frac{1}{f},$$

$$2f^2 + fx + fx' = f^2 + fx + fx' + xx',$$

$$f^2 = xx'.$$

E40-31 Step through the exercise one lens at a time. The object is 40 cm to the left of a converging lens with a focal length of $+20$ cm. The image from this first lens will be located by solving

$$\frac{1}{i} = \frac{1}{f} - \frac{1}{o} = \frac{1}{(20 \text{ cm})} - \frac{1}{(40 \text{ cm})} = \frac{1}{40 \text{ cm}},$$

so $i = 40$ cm. Since i is positive it is a real image, and it is located to the right of the converging lens. This image becomes the object for the diverging lens.

But wait! The image from the converging lens is located 40 cm - 10 cm from the diverging lens, but it is located on the wrong side: the diverging lens is "in the way" so the rays which would form the image hit the diverging lens before they have a chance to form the image. That means that the real image from the converging lens is a *virtual* object in the diverging lens, so that the object distance for the diverging lens is $o = -30$ cm.

The image formed by the diverging lens is located by solving

$$\frac{1}{i} = \frac{1}{f} - \frac{1}{o} = \frac{1}{(-15 \text{ cm})} - \frac{1}{(-30 \text{ cm})} = \frac{1}{-30 \text{ cm}},$$

or $i = -30$ cm. This would mean the image formed by the diverging lens would be a virtual image, and would be located to the left of the diverging lens.

The image is virtual, so it is upright. The magnification from the first lens is

$$m_1 = -i/o = -(40 \text{ cm})/(40 \text{ cm})) = -1;$$

the magnification from the second lens is

$$m_2 = -i/o = -(-30 \text{ cm})/(-30 \text{ cm})) = -1;$$

which implies an overall magnification of $m_1 m_2 = 1$.

E40-39 Microscope magnification is given by Eq. 40-33. We need to first find the focal length of the objective lens before we can use this formula. We are told in the text, however, that the microscope is constructed so the at the object is placed just beyond the focal point of the objective lens, then $f_{ob} \approx 12.0$ mm. Similarly, the intermediate image is formed at the focal point of the eyepiece, so $f_{ey} \approx 48.0$ mm. The magnification is then

$$m = \frac{-s(250 \text{ mm})}{f_{ob}f_{ey}} = -\frac{(285 \text{ mm})(250 \text{ mm})}{(12.0 \text{ mm})(48.0 \text{ mm})} = 124.$$

A more accurate answer can be found by calculating the *real* focal length of the objective lens, which is 11.4 mm, but since there is a *huge* uncertainty in the near point of the eye, I see no point in trying to be more accurate than this.

P40-3 I think it is easier to answer part (b) first, so...

(b) There are two ends to the object of length L, one of these ends is a distance o_1 from the mirror, and the other is a distance o_2 from the mirror. The images of the two ends will be located at i_1 and i_2.

Since we are told that the object has a short length L we will assume that a differential approach to the problem is in order. Then

$$L = \Delta o = o_1 - o_2 \text{ and } L' = \Delta i = i_1 - i_2,$$

Finding the ratio of L'/L is then reduced to

$$\frac{L'}{L} = \frac{\Delta i}{\Delta o} \approx \frac{di}{do}.$$

We can take the derivative of Eq. 40-15 with respect to changes in o and i,

$$\frac{di}{i^2} + \frac{do}{o^2} = 0,$$

or

$$\frac{L'}{L} \approx \frac{di}{do} = -\frac{i^2}{o^2} = -m^2,$$

where m is the lateral magnification.

What does the negative sign mean? The end of the object which is closer to the mirror will form the end of the image which is farther away.

(a) Since i is given by

$$\frac{1}{i} = \frac{1}{f} - \frac{1}{o} = \frac{o - f}{of},$$

322

the fraction i/o can also be written

$$\frac{i}{o} = \frac{of}{o(o-f)} = \frac{f}{o-f}.$$

Then

$$L \approx -\frac{i^2}{o^2} = -\left(\frac{f}{o-f}\right)^2$$

P40-7 The image (which will appear on the screen) and object are a distance $D = o + i$ apart. We can use this information to eliminate one variable from Eq. 40-15,

$$\frac{1}{o} + \frac{1}{i} = \frac{1}{f},$$

$$\frac{1}{o} + \frac{1}{D-o} = \frac{1}{f},$$

$$\frac{D}{o(D-o)} = \frac{1}{f},$$

$$o^2 - oD + fD = 0.$$

This last expression is a quadratic, and we would expect to get two solutions for o. These solutions will be of the form "something" plus/minus "something else"; the distance between the two locations for o will evidently be twice the "something else", which is then

$$d = o_+ - o_- = \sqrt{(-D)^2 - 4(fD)} = \sqrt{D(D-4f)}.$$

(b) The ratio of the image sizes is m_+/m_-, or i_+o_-/i_-o_+. Now it seems we must find the actual values of o_+ and o_-. From the quadratic in part (a) we have

$$o_\pm = \frac{D \pm \sqrt{D(D-4f)}}{2} = \frac{D \pm d}{2},$$

so the ratio is

$$\frac{o_-}{o_+} = \left(\frac{D-d}{D+d}\right).$$

But $i_- = o_+$, and vice-versa, so the ratio of the image sizes is this quantity squared.

P40-11 We'll solve the problem by finding out what happens if you put an object in front of the combination of lenses.

Let the object distance be o_1. The first lens will create an image at i_1, where

$$\frac{1}{i_1} = \frac{1}{f_1} - \frac{1}{o_1}$$

This image will act as an object for the second lens.

If the first image is real (i_1 positive) then the image will be on the "wrong" side of the second lens, and as such the real image will act like a virtual object. In short, $o_2 = -i_1$ will give the correct sign to the object distance when the image from the first lens acts like an object for the second lens. The image formed by the second lens will then be at

$$\begin{aligned}
\frac{1}{i_2} &= \frac{1}{f_2} - \frac{1}{o_2}, \\
&= \frac{1}{f_2} + \frac{1}{i_2}, \\
&= \frac{1}{f_2} + \frac{1}{f_1} - \frac{1}{o_1}.
\end{aligned}$$

In this case it appears as if the combination

$$\frac{1}{f_2} + \frac{1}{f_1}$$

is equivalent to the reciprocal of a focal length. We will go ahead and make this connection, and

$$\frac{1}{f} = \frac{1}{f_2} + \frac{1}{f_1} = \frac{f_1 + f_2}{f_1 f_2}.$$

The rest is straightforward enough.

P40-15 We want the maximum linear motion of the train to move no more than 0.75 mm on the film; this means we want to find the size of an object on the train that will form a 0.75 mm image. The object distance is much larger than the focal length, so the image distance is approximately equal to the focal length. The magnification is then $m = -i/o = (3.6 \text{ cm})/(44.5 \text{ m}) = -0.00081$.

The size of an object on the train that would produce a 0.75 mm image on the film is then 0.75 mm$/0.00081 = 0.93$ m.

How much time does it take the train to move that far?

$$t = \frac{(0.93 \text{ m})}{(135 \text{ km/hr})(1/3600 \text{ hr/s})} = 25 \text{ ms}.$$

Chapter 41

Interference

E41-1 Double-slit interference patterns follow Eqs. 41-1 and 41-2 *under "normal" circumstances.* In this problem we look for the location of the third-order bright fringe, so

$$\theta = \sin^{-1}\frac{m\lambda}{d} = \sin^{-1}\frac{(3)(554 \times 10^{-9}\text{m})}{(7.7 \times 10^{-6}\text{m})} = 12.5° = 0.22 \text{ rad.}$$

E41-5 Since the angles are *very* small, we can assume $\sin\theta \approx \theta$ for angles measured in radians.

If the interference fringes are $0.23°$ apart, then the angular position of the first bright fringe is $0.23°$ away from the central maximum. Eq. 41-1, written with the small angle approximation in mind, is

$$d\theta = \lambda$$

for this first ($m = 1$) bright fringe. The goal is to find the wavelength which increases θ by 10%. To do this we must increase the right hand side of the equation by 10%, which means increasing λ by 10%. The new wavelength will be

$$\lambda' = 1.1\lambda = 1.1(589 \text{ nm}) = 650 \text{ nm}$$

E41-9 Can we use the small angle approximation? Since the distance on the screen (18 mm) is much smaller than the distance to the screen (50 cm), the answer is yes. Then a variation of Eq. 41-3 is in order:

$$y_m = \left(m + \frac{1}{2}\right)\frac{\lambda D}{d}$$

We are given the distance (on the screen) between the first minima ($m = 0$) and the tenth minima ($m = 9$). Then

$$18 \text{ mm} = y_9 - y_0 = 9\frac{\lambda(50 \text{ cm})}{(0.15 \text{ mm})},$$

or

$$\lambda = 6 \times 10^{-4} \text{ mm} = 600 \text{ nm}.$$

E41-11   This figure should explain it well enough.

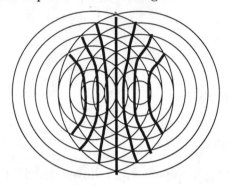

E41-15   Leading by 90° is the same as leading by a quarter wavelength, since there are 360° in a circle. The distance from A to the detector is 100 m longer than the distance from B to the detector. Since the wavelength is 400 m, 100 m corresponds to a quarter wavelength.

So a wave peak starts out from source A and travels to the detector. When it has traveled a quarter wavelength a wave peak leaves source B. But when the wave peak from A has traveled a quarter wavelength it is now located at the same distance from the detector as source B, which means the two wave peaks arrive at the detector at the same time.

They are in phase.

E41-19 (a)   We want to know the path length difference of the two sources to the detector. Assume the detector is at x and the second source is at $y = d$. The distance S_1D is x; the distance S_2D is $\sqrt{x^2 + d^2}$. The difference is

$$\sqrt{x^2 + d^2} - x.$$

If this difference is an integral number of wavelengths then we have a maximum; if instead it is a half integral number of wavelengths we have a minimum. For part (a) we are looking for the maxima, so we set the path length difference equal to $m\lambda$ and solve for x_m.

$$\sqrt{x_m^2 + d^2} - x_m = m\lambda,$$
$$x_m^2 + d^2 = (m\lambda + x_m)^2,$$
$$x_m^2 + d^2 = m^2\lambda^2 + 2m\lambda x_m + x_m^2,$$
$$x_m = \frac{d^2 - m^2\lambda^2}{2m\lambda}$$

The first question we need to ask is what happens when $m = 0$. The right hand side becomes indeterminate, so we need to go back to the first line in the above derivation. If $m = 0$ then $d^2 = 0$; since this is *not* true in this problem, there is no $m = 0$ solution.

In fact, we may have even more troubles. x_m needs to be a positive value, so the maximum allowed value for m will be given by

$$m^2 \lambda^2 < d^2,$$
$$m < d/\lambda = (4.17\,\text{m})/(1.06\,\text{m}) = 3.93;$$

but since m is an integer, $m = 3$ is the maximum value.

The first three maxima occur at $m = 3$, $m = 2$, and $m = 1$. These maxima are located at

$$x_3 = \frac{(4.17\,\text{m})^2 - (3)^2(1.06\,\text{m})^2}{2(3)(1.06\,\text{m})} = 1.14 \text{ m},$$

$$x_2 = \frac{(4.17\,\text{m})^2 - (2)^2(1.06\,\text{m})^2}{2(2)(1.06\,\text{m})} = 3.04 \text{ m},$$

$$x_1 = \frac{(4.17\,\text{m})^2 - (1)^2(1.06\,\text{m})^2}{2(1)(1.06\,\text{m})} = 7.67 \text{ m}.$$

Interestingly enough, as m decreases the maxima get farther away!

(b) The closest maxima to the origin occurs at $x = \pm 6.94$ cm. What then is $x = 0$? It is a local minimum, but the intensity isn't zero. It corresponds to a point where the path length difference is 3.93 wavelengths. It should be half an integer to be a complete minimum.

E41-23 Light from above the oil slick can be reflected back up from the top of the oil layer or from the bottom of the oil layer. For both reflections the light is reflecting off a substance with a higher index of refraction so both reflected rays pick up a phase change of π. Since both waves have this phase the equation for a maxima is

$$2d + \frac{1}{2}\lambda_n + \frac{1}{2}\lambda_n = m\lambda_n.$$

Remember that $\lambda_n = \lambda/n$, where n is the index of refraction of the thin film. Then

$$2nd = (m - 1)\lambda$$

is the condition for a maxima. We know $n = 1.20$ and $d = 460$ nm. We don't know m or λ. It might seem as if there isn't enough information to solve the problem, but we can. We need to find the wavelength in the visible range (400 nm to 700 nm) which has an integer m. Trial and error might work. If $\lambda = 700$ nm, then m is

$$m = \frac{2nd}{\lambda} + 1 = \frac{2(1.20)(460 \text{ nm})}{(700 \text{ nm})} + 1 = 2.58$$

But m needs to be an integer. If we increase m to 3, then

$$\lambda = \frac{2(1.20)(460 \text{ nm})}{(3-1)} = 552 \text{ nm}$$

which is clearly in the visible range. Trying $m = 2$ will result in a value in the infrared range (1100 nm), while $m = 4$ will yield a value in the ultraviolet range (368 nm). So the oil slick will appear green.

(b)   One of the most profound aspects of thin film interference is that wavelengths which are maximally reflected are minimally transmitted, and vice versa. Finding the maximally transmitted wavelengths is the same as finding the minimally reflected wavelengths, or looking for values of m that are half integer.

The most obvious choice is $m = 3.5$, and then

$$\lambda = \frac{2(1.20)(460 \text{ nm})}{(3.5-1)} = 442 \text{ nm}.$$

Using $m = 1.5$ results in a wavelength of 736 nm, which is probably not visible, while $m = 4.5$ would be far in the ultraviolet.

E41-27   As with the oil on the water in Ex. 41-23, both the light which reflects off of the acetone and the light which reflects off of the glass undergoes a phase shift of π. Then the maxima for reflection are given by

$$2nd = (m-1)\lambda$$

We don't know m, but at some integer value of m we have $\lambda = 700$ nm. If m is increased by exactly $\frac{1}{2}$ then we are at a minimum of $\lambda = 600$ nm. Consequently,

$$2(1.25)d = (m-1)(700 \text{ nm}) \text{ and } 2(1.25)d = (m-1/2)(600 \text{ nm}),$$

we can set these two expressions equal to each other to find m,

$$(m-1)(700 \text{ nm}) = (m-1/2)(600 \text{ nm}),$$

so $m = 4$. Then we can find the thickness,

$$d = (4-1)(700 \text{ nm})/2(1.25) = 840 \text{ nm}.$$

E41-33 We can start with the last equation from Sample Problem 41-5,

$$r = \sqrt{(m - \frac{1}{2})\lambda R},$$

and solve for m,

$$m = \frac{r^2}{\lambda R} + \frac{1}{2}$$

In this exercise $R = 5.0\,\text{m}$, $r = 0.01\,\text{m}$, and $\lambda = 589$ nm. Then

$$m = \frac{(0.01\,\text{m})^2}{(589\,\text{nm})(5.0\,\text{m})} = 34$$

is the number of rings observed.

(b) Putting the apparatus in water effectively changes the wavelength to

$$(589\,\text{nm})/(1.33) = 443\,\text{nm},$$

so the number of rings will now be

$$m = \frac{(0.01\,\text{m})^2}{(443\,\text{nm})(5.0\,\text{m})} = 45.$$

E41-39 When M_2 moves through a distance of $\lambda/2$ a fringe has will be produced, destroyed, and then produced again. This is because the light travels twice through any change in distance. The wavelength of light is then

$$\lambda = \frac{2(0.233\,\text{mm})}{792} = 588\,\text{nm}.$$

P41-1 (a) This is a small angle problem, so we use Eq. 41-4. The distance to the screen is $2 \times 20\,\text{m}$, because the light travels to the mirror and back again. Then

$$d = \frac{\lambda D}{\Delta y} = \frac{(632.8\,\text{nm})(40.0\,\text{m})}{(0.1\,\text{m})} = 0.253\,\text{mm}.$$

(b) Placing the cellophane over one slit will cause the interference pattern to shift to the left or right, but not disappear or change size. How does it shift? Since we are picking up 2.5 waves then we are, in effect, swapping bright fringes for dark fringes.

P41-5 The intensity is given by Eq. 41-17, which, in the small angle approximation, can be written as

$$I_\theta = 4I_0 \cos^2\left(\frac{\pi d\theta}{\lambda}\right).$$

The intensity will be half of the maximum when

$$\frac{1}{2} = \cos^2\left(\frac{\pi d\Delta\theta/2}{\lambda}\right)$$

or

$$\frac{\pi}{4} = \frac{\pi d\Delta\theta}{2\lambda},$$

which will happen if $\Delta\theta = \lambda/2d$.

P41-7 We actually did this problem in Exercise 41-27, although slightly differently. One maximum is

$$2(1.32)d = (m - 1/2)(679 \text{ nm}),$$

the other is

$$2(1.32)d = (m + 1/2)(485 \text{ nm}).$$

Set these equations equal to each other,

$$(m - 1/2)(679 \text{ nm}) = (m + 1/2)(485 \text{ nm}),$$

and find $m = 3$. Then the thickness is

$$d = (3 - 1/2)(679 \text{ nm})/2(1.32) = 643 \text{ nm}.$$

P41-11 (a) Look back at the work for Sample Problem 41-5 where it was found

$$r_m = \sqrt{(m - \frac{1}{2})\lambda R},$$

We can write this as

$$r_m = \sqrt{\left(1 - \frac{1}{2m}\right)m\lambda R}$$

and expand the part in parentheses in a binomial expansion,

$$r_m \approx \left(1 - \frac{1}{2}\frac{1}{2m}\right)\sqrt{m\lambda R}.$$

We will do the same with

$$r_{m+1} = \sqrt{(m + 1 - \frac{1}{2})\lambda R},$$

expanding

$$r_{m+1} = \sqrt{\left(1 + \frac{1}{2m}\right) m \lambda R}$$

to get

$$r_{m+1} \approx \left(1 + \frac{1}{2}\frac{1}{2m}\right) \sqrt{m \lambda R}.$$

Then

$$\Delta r \approx \frac{1}{2m}\sqrt{m \lambda R},$$

or

$$\Delta r \approx \frac{1}{2}\sqrt{\lambda R/m}.$$

(b)   The area between adjacent rings is found from the difference,

$$A = \pi \left(r_{m+1}^2 - r_m^2\right),$$

and into this expression we will substitute the *exact* values for r_m and r_{m+1},

$$\begin{aligned} A &= \pi \left((m + 1 - \frac{1}{2})\lambda R - (m - \frac{1}{2})\lambda R\right), \\ &= \pi \lambda R. \end{aligned}$$

Unlike part (a), we did not need to assume $m \gg 1$ in order to arrive at this expression; it is exact for all m.

Chapter 42

Diffraction

E42-3 (a) This is a valid small angle approximation problem: the distance between the points on the screen is much less than the distance to the screen. Then

$$\theta \approx \frac{(0.0162\,\text{m})}{(2.16\,\text{m})} = 7.5 \times 10^{-3} \text{ rad}.$$

There really is no reason to convert this to degrees.

(b) The diffraction minima are described by Eq. 42-3,

$$
\begin{aligned}
a\sin\theta &= m\lambda, \\
a\sin(7.5 \times 10^{-3} \text{ rad}) &= (2)(441 \times 10^{-9}\,\text{m}), \\
a &= 1.18 \times 10^{-4}\,\text{m}.
\end{aligned}
$$

This is about the width of a razor blade; the slit could have been produced by scraping a pair of razor blades across a smoked microscope slide.

E42-5 We again use Eq. 42-3, but we will need to throw in a few extra subscripts to distinguish between which wavelength we are dealing with. If the angles match, then so will the sine of the angles. We then have

$$\sin\theta_{a,1} = \sin\theta_{b,2}$$

or, using Eq. 42-3,

$$\frac{(1)\lambda_a}{a} = \frac{(2)\lambda_b}{a},$$

from which we can deduce $\lambda_a = 2\lambda_b$.

(b) Will any other minima coincide? We want to solve for the values of m_a and m_b that will be integers and have the same angle. Using Eq. 42-3 one more time,

$$\frac{m_a \lambda_a}{a} = \frac{m_b \lambda_b}{a},$$

and then substituting into this the relationship between the wavelengths,

$$m_a = m_b/2.$$

whenever m_b is an even integer m_a is an integer. Then *all* of the diffraction minima from λ_a are overlapped by a minima from λ_b.

E42-11 (a) This is a small angle approximation problem, so

$$\theta = (1.13 \text{ cm})/(3.48 \text{ m}) = 3.25 \times 10^{-3} \text{ rad.}$$

(b) A convenient measure of the phase difference, α is related to θ through Eq. 42-7,

$$\alpha = \frac{\pi a}{\lambda} \sin \theta = \frac{\pi(25.2 \times 10^{-6}\text{m})}{(538 \times 10^{-9}\text{m})} \sin(3.25 \times 10^{-3} \text{ rad}) = 0.478 \text{ rad}$$

(c) The intensity at a point is related to the intensity at the central maximum by Eq. 42-8,

$$\frac{I_\theta}{I_m} = \left(\frac{\sin \alpha}{\alpha}\right)^2 = \left(\frac{\sin(0.478 \text{ rad})}{(0.478 \text{ rad})}\right)^2 = 0.926$$

E42-15 (a) Rayleigh's criterion for resolving images (Eq. 42-11) requires that two objects have an angular separation of at least

$$\theta_R = \sin^{-1}\left(\frac{1.22\lambda}{d}\right) = \sin^{-1}\left(\frac{1.22(562 \times 10^{-9})}{(5.00 \times 10^{-3}\text{m})}\right) = 1.37 \times 10^{-4} \text{ rad}$$

This quantity is so small that we never really needed to press the inverse sine button on the calculator; the small angle approximation is *very* applicable here.

(b) Once again, this is a small angle, so we can use the small angle approximation to find the distance to the car. In that case

$$\theta_R = y/D,$$

where y is the headlight separation and D the distance to the car. Solving,

$$D = y/\theta_R = (1.42 \text{ m})/(1.37 \times 10^{-4} \text{ rad}) = 1.04 \times 10^4 \text{m},$$

or about six or seven miles.

Interestingly enough, as the car gets closer the headlight will cause your pupils to contract, which will decrease the distance of resolution, which will make it harder to tell if it is two headlights or one that is approaching you. But then you shouldn't be looking into the headlights of an oncoming car anyway!

E42-17 The smallest resolvable angular separation will be given by Eq. 42-11,

$$\theta_R = \sin^{-1}\left(\frac{1.22\lambda}{d}\right) = \sin^{-1}\left(\frac{1.22(565 \times 10^{-9}\text{m})}{(5.08\,\text{m})}\right) = 1.36 \times 10^{-7}\text{ rad},$$

a very small angle indeed.

The smallest objects resolvable on the Moon's surface by this telescope have a size y where

$$y = D\theta_R = (3.84 \times 10^8\,\text{m})(1.36 \times 10^{-7}\text{ rad}) = 52.2\,\text{m}$$

This is much bigger than any of the spacecraft that have been left on the moon; consequently, we can't *directly* see from Earth any evidence that we have been on the moon.

E42-21 This is basically the opposite of the approaching headlights exercise. Here we want to know how far we need to stand from the painting so that we can't resolve the individual dots.

Using Eq. 42-11, we find the minimum resolvable angular separation is given by

$$\theta_R = \sin^{-1}\left(\frac{1.22\lambda}{d}\right) = \sin^{-1}\left(\frac{1.22(475 \times 10^{-9}\text{m})}{(4.4 \times 10^{-3}\text{m})}\right) = 1.32 \times 10^{-4}\text{ rad}$$

The dots are 2 mm apart, so we want to stand a distance D away such that

$$D > y/\theta_R = (2 \times 10^{-3}\text{m})/(1.32 \times 10^{-4}\text{ rad}) = 15\,\text{m}.$$

E42-25 The interference fringe pattern depends on λ and d, the slit separation. The intensity "envelope" of this pattern is set by the diffraction characteristics, which are determined by λ and a, the individual slit width.

The linear separation of the fringes is given by

$$\frac{\Delta y}{D} = \Delta\theta = \frac{\lambda}{d} \text{ or } \Delta y = \frac{\lambda D}{d}$$

for sufficiently small d compared to λ.

E42-29 (a) The first diffraction minimum is given at an angle θ such that

$$a \sin \theta = \lambda;$$

the order of the interference maximum at that point is given by

$$d \sin \theta = m\lambda.$$

Dividing one expression by the other we get

$$d/a = m,$$

with solution $m = (0.150)/(0.030) = 5$. The fact that the answer is *exactly* 5 implies that the fifth interference maximum is squelched by the diffraction minimum. Then there are only four complete fringes on either side of the central maximum. Add this to the central maximum and we get nine as the answer.

(b) For the third fringe $m = 3$, so $d \sin \theta = 3\lambda$. Then β is Eq. 42-14 is 3π, while α in Eq. 42-16 is

$$\alpha = \frac{\pi a}{\lambda} \frac{3\lambda}{d} = 3\pi \frac{a}{d},$$

so the relative intensity of the third fringe is, from Eq. 42-17,

$$(\cos 3\pi)^2 \left(\frac{\sin(3\pi a/d)}{(3\pi a/d)} \right)^2 = 0.255.$$

P42-3 We want to take the derivative of Eq. 42-8 with respect to α, so

$$
\begin{aligned}
\frac{dI_\theta}{d\alpha} &= \frac{d}{d\alpha} I_{\mathrm{m}} \left(\frac{\sin \alpha}{\alpha} \right)^2, \\
&= I_{\mathrm{m}} 2 \left(\frac{\sin \alpha}{\alpha} \right) \left(\frac{\cos \alpha}{\alpha} - \frac{\sin \alpha}{\alpha^2} \right), \\
&= I_{\mathrm{m}} 2 \frac{\sin \alpha}{\alpha^3} \left(\alpha \cos \alpha - \sin \alpha \right).
\end{aligned}
$$

This equals zero whenever $\sin \alpha = 0$ or $\alpha \cos \alpha = \sin \alpha$; the former is the case for a minima while the latter is the case for the maxima. The maxima case can also be written as

$$\tan \alpha = \alpha.$$

(b) Note that as the order of the maxima increases the solutions get closer and closer to odd integers times $\pi/2$. The solutions are

$$\alpha = 0, 1.43\pi, 2.46\pi, \text{ etc.}$$

(c) The m values are $m = \alpha/\pi - 1/2$, and correspond to

$$m = 0.5, 0.93, 1.96, \text{ etc.}$$

These values will get closer and closer to integers as the values are increased.

P42-7 (a) The missing fringe at $\theta = 5°$ is a good hint as to what is going on. There should be some sort of *interference* fringe, unless the *diffraction* pattern has a minimum at that point. This would be the first minimum, so

$$a\sin(5°) = (440 \times 10^{-9}\text{m})$$

would be a good measure of the width of each slit. Then $a = 5.05 \times 10^{-6}\text{m}$.

(b) If the diffraction pattern envelope were not present we could expect that the fourth interference maxima beyond the central maximum would occur at this point, and then

$$d\sin(5°) = 4(440 \times 10^{-9}\text{m})$$

yielding

$$d = 2.02 \times 10^{-5}\text{m}.$$

(c) Apply Eq. 42-17, where $\beta = m\pi$ and

$$\alpha = \frac{\pi a}{\lambda}\sin\theta = \frac{\pi a}{\lambda}\frac{m\lambda}{d} = m\frac{\pi a}{d} = m\pi/4.$$

Then for $m = 1$ we have

$$I_1 = (7)\left(\frac{\sin(\pi/4)}{(\pi/4)}\right)^2 = 5.7;$$

while for $m = 2$ we have

$$I_2 = (7)\left(\frac{\sin(2\pi/4)}{(2\pi/4)}\right)^2 = 2.8.$$

These are in good agreement with the figure.

Chapter 43

Gratings and Spectra

E43-3 *These kinds of problems are fun!*
We want to find a relationship between the angle and the order number which is linear. We'll plot the data in this representation, and then use a least squares fit to find the wavelength.

The relationship will be linear if we graph m horizontally against $\sin\theta$ vertically. The m values are plotted on the horizontal axis because they are *exact*; the least squares method works by minimizing the error in the *vertical* direction. The data to be plotted is

m	θ	$\sin\theta$	m	θ	$\sin\theta$
1	17.6°	0.302	-1	-17.6°	-0.302
2	37.3°	0.606	-2	-37.1°	-0.603
3	65.2°	0.908	-3	-65.0°	-0.906

Don't make the mistake of assuming that the best straight line must pass through the origin; there is no reason to believe that the equipment was calibrated so exactly that $\theta = 0°$ is any more accurate than any other measurement.

We won't go into the theory of why a least square fit works, or even how it works. We certainly won't derive it. Most calculators will do a linear least squares without even asking.

On my calculator I get the best straight line fit as

$$0.302m + 8.33 \times 10^{-4} = \sin\theta_m,$$

which means that

$$\lambda = (0.302)(1.73\,\mu\text{m}) = 522\text{ nm}.$$

E43-5 (a) The principle maxima occur at points given by Eq. 43-1,

$$\sin\theta_m = m\frac{\lambda}{d}.$$

The difference of the sine of the angle between any two adjacent orders is

$$\sin\theta_{m+1} - \sin\theta_m = (m+1)\frac{\lambda}{d} - m\frac{\lambda}{d} = \frac{\lambda}{d}.$$

Using the information provided we can find d from

$$d = \frac{\lambda}{\sin\theta_{m+1} - \sin\theta_m} = \frac{(600 \times 10^{-9})}{(0.30) - (0.20)} = 6\,\mu\text{m}.$$

It doesn't take much imagination to recognize that the second and third order maxima were given.

(b) If the fourth order maxima is missing it must be because the diffraction pattern envelope has a minimum at that point. Any fourth order maxima should have occurred at $\sin\theta_4 = 0.4$. If it is a diffraction minima then

$$a\sin\theta_m = m\lambda \text{ where } \sin\theta_m = 0.4$$

Note that we *do not* know what m is; it isn't necessarily 4, because this isn't the same relationship as the one for the principle maxima. It could be that the first diffraction minima occurs at the fourth principle maxima, or it could be that the third (or any other) diffraction minima occurs at the fourth principle maxima. Nonetheless, we can solve this expression and find

$$a = m\frac{\lambda}{\sin\theta_m} = m\frac{(600 \times 10^{-9}\text{m})}{(0.4)} = m1.5\,\mu\text{m}.$$

The minimum width is when $m = 1$, or $a = 1.5\,\mu\text{m}$.

Just for fun, note that $m = 2$ can't happen, because then the second order maxima at $\sin\theta = 0.2$ would also be missing. $m = 4$ won't work either, because then none of the maxima would have been present (not only that, but the slits would be so wide that there would be no "non-slit" area.) The only other possibility is $m = 3$.

(c) The visible orders would be integer values of m *except* for when m is a multiple of four.

E43-9 A grating with 400 rulings/mm has a slit separation of

$$d = \frac{1}{400\text{ mm}^{-1}} = 2.5 \times 10^{-3}\text{ mm}.$$

To find the number of orders of the entire visible spectrum that will be present we need only consider the wavelength which will be on the outside of the maxima. That will be the longer wavelengths, so we only need to look at the 700 nm behavior. Using Eq. 43-1,

$$d \sin \theta = m\lambda,$$

and using the maximum angle 90°, we find

$$m < \frac{d}{\lambda} = \frac{(2.5 \times 10^{-6}\text{m})}{(700 \times 10^{-9}\text{m})} = 3.57,$$

so there can be at most three orders of the entire spectrum.

E43-11 If the second-order spectra overlaps the third-order, it is because the 700 nm second-order line is at a larger angle than the 400 nm third-order line.

Start with the wavelengths multiplied by the appropriate order parameter, then divide both side by d, and finally apply Eq. 43-1.

$$2(700 \text{ nm}) > 3(400 \text{ nm}),$$
$$\frac{2(700 \text{ nm})}{d} > \frac{3(400 \text{ nm})}{d},$$
$$\sin \theta_{2,\lambda=700} > \sin \theta_{3,\lambda=400},$$

regardless of the value of d.

E43-17 The required resolving power of the grating is given by Eq. 43-10

$$R = \frac{\lambda}{\Delta\lambda} = \frac{(589.0 \text{ nm})}{(589.6 \text{ nm}) - (589.0 \text{ nm})} = 982.$$

Just a little note: it might look like we have three or four significant digits, but we don't; there is only one significant digit in the denominator after taking the difference, so there is only one in the final answer. Our resolving power is then $R = 1000$.

Using Eq. 43-11 we can find the number of grating lines required. We are looking at the second-order maxima, so

$$N = \frac{R}{m} = \frac{(1000)}{(2)} = 500.$$

E43-21 (a) We find the ruling spacing by Eq. 43-1,

$$d = \frac{m\lambda}{\sin \theta_m} = \frac{(3)(589 \text{ nm})}{\sin(10.2°)} = 9.98 \,\mu\text{m}.$$

That's so close to $10 \,\mu$m that I wonder if I made a mistake.

(b) The resolving power of the grating needs to be at least $R = 1000$ for the third-order line; see the work for Ex. 43-17 above. The number of lines required is given by Eq. 43-11,

$$N = \frac{R}{m} = \frac{(1000)}{(3)} = 333,$$

so the width of the grating (or at least the part that is being used) is $333(9.98\,\mu m) = 3.3$ mm. We've been fast and dirty with our significant figures. As such, my answers may differ from those in the back of the book.

E43-25 Bragg reflection is given by Eq. 43-12

$$2d\sin\theta = m\lambda,$$

where the angles are measured not against the normal, but against the plane. The value of d depends on the family of planes under consideration, but it is at never larger than a_0, the unit cell dimension.

We are looking for the smallest angle; this will correspond to the largest d and the smallest m. That means $m = 1$ and $d = 0.313$ nm. Then the minimum angle is

$$\theta = \sin^{-1}\frac{(1)(29.3 \times 10^{-12}\,\text{m})}{2(0.313 \times 10^{-9}\,\text{m})} = 2.68°.$$

E43-27 We apply Eq. 43-12 to each of the peaks and find the product

$$m\lambda = 2d\sin\theta.$$

The four values are 26 pm, 39 pm, 52 pm, and 78 pm. The last two values are twice the first two, so the wavelengths are 26 pm and 39 pm.

E43-33 We use Eq. 43-12 to first find d;

$$d = \frac{m\lambda}{2\sin\theta} = \frac{(1)(0.261 \times 10^{-9}\,\text{m})}{2\sin(63.8°)} = 1.45 \times 10^{-10}\,\text{m}.$$

d is the spacing between the planes in Fig. 43-28; it correspond to half of the diagonal distance between two cell centers. Then

$$(2d)^2 = a_0^2 + a_0^2,$$

or

$$a_0 = \sqrt{2}d = \sqrt{2}(1.45 \times 10^{-10}\,\text{m}) = 0.205\,\text{nm}.$$

P43-1 Since the slits are *so* narrow we only need to consider interference effects, not diffraction effects. There are three waves which contribute at any point. The phase angle between adjacent waves is

$$\phi = 2\pi d \sin\theta / \lambda.$$

We can add the electric field vectors as was done in the previous chapters, or we can do it in a different order as is shown in the figure below.

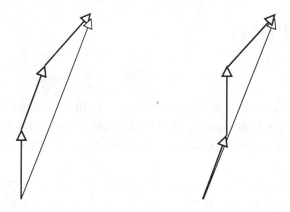

Then the vectors sum to
$$E(1 + 2\cos\phi).$$

We need to square this quantity, and then normalize it so that the central maximum is the maximum. Then

$$I = I_\mathrm{m} \frac{(1 + 4\cos\phi + 4\cos^2\phi)}{9}.$$

P43-3 (a) and (b) A plot of the intensity quickly reveals that there is an alternation of large maximum, then a smaller maximum, etc. The large maxima are at $\phi = 2n\pi$, the smaller maxima are half way between those values.

(c) The intensity at these secondary maxima is then

$$I = I_\mathrm{m} \frac{(1 + 4\cos\pi + 4\cos^2\pi)}{9} = \frac{I_\mathrm{m}}{9}.$$

Note that the minima are *not* located half-way between the maxima!

Chapter 44

Polarization

E44-1 (a) The direction of propagation is determined by considering the argument of the sine function. As t increases y must decrease to keep the sine function "looking" the same, so the wave is propagating in the negative y direction.

(b) The electric field is orthogonal (perpendicular) to the magnetic field (so $E_x = 0$) and the direction of motion (so $E_y = 0$); Consequently, the only non-zero term is E_z. The magnitude of E will be equal to the magnitude of B times c. Since $\vec{S} = \vec{E} \times \vec{B}/\mu_0$, when $\vec{B}$ points in the positive x direction then $\vec{E}$ must point in the negative z direction in order that $\vec{S}$ point in the negative y direction. Then

$$E_z = -cB\sin(ky + \omega t).$$

(c) The polarization is given by the direction of the electric field, so the wave is linearly polarized in the z direction.

E44-5 This is an all time favorite physics question. The classic mistake is to assume that the first sheet polarizes the light and the third, being at 90° to the first, prevents any light from passing. The mistake occurs because light the light which is transmitted through a polarizing sheet is polarized in the direction of the sheet.

The first sheet polarizes the un-polarized light, half of the intensity is transmitted, so $I_1 = \frac{1}{2}I_0$.

The second sheet transmits according to Eq. 44-1,

$$I_2 = I_1\cos^2\theta = \frac{1}{2}I_0\cos^2(45°) = \frac{1}{4}I_0,$$

and the transmitted light is polarized in the direction of the second sheet.

The third sheet is 45° to the second sheet, so the intensity of the light which is transmitted through the third sheet is

$$I_3 = I_2\cos^2\theta = \frac{1}{4}I_0\cos^2(45°) = \frac{1}{8}I_0.$$

E44-9 Since the incident beam is unpolarized the first sheet transmits 1/2 of the original intensity. The transmitted beam then has a polarization set by the first sheet: 58.8° to the vertical. The second sheet is horizontal, which puts it 31.2° to the first sheet. Then the second sheet transmits $\cos^2(31.2°)$ of the intensity incident on the second sheet. The final intensity transmitted by the second sheet can be found from the product of these terms,

$$I = (43.3\,\text{W/m}^2) \left(\frac{1}{2}\right) \left(\cos^2(31.2°)\right) = 15.8\,\text{W/m}^2.$$

E44-11 (a) The angle for complete polarization of the reflected ray is Brewster's angle, and is given by Eq. 44-3 (since the first medium is air)

$$\theta_\text{p} = \tan^{-1} n = \tan^{-1}(1.33) = 53.1°.$$

(b) Since the index of refraction depends (slightly) on frequency, then so does Brewster's angle.

E44-15 (a) The incident wave is at 45° to the optical axis. This means that there are two components; assume they originally point in the $+y$ and $+z$ direction. When they travel through the half wave plate they are now out of phase by 180°; this means that when one component is in the $+y$ direction the other is in the $-z$ direction. In effect the polarization has been rotated by 90°.

(b) Since the half wave plate will delay one component so that it emerges 180° "later" than it should, it will in effect reverse the handedness of the circular polarization.

(c) Pretend that an unpolarized beam can be broken into two orthogonal linearly polarized components. Both are then rotated through 90°; but when recombined it looks like the original beam. As such, there is no apparent change.

P44-1 Intensity is proportional to the electric field *squared*, so the original intensity reaching the eye is I_0, with components $I_\text{h} = (2.3)^2 I_\text{v}$, and then

$$I_0 = I_\text{h} + I_\text{v} = 6.3 I_\text{v} \text{ or } I_\text{v} = 0.16 I_0.$$

Similarly, $I_\text{h} = (2.3)^2 I_\text{v} = 0.84 I_0.$

(a) When the sun-bather is standing only the vertical component passes, while

(b) when the sun-bather is lying down only the horizontal component passes.

P44-3 Each sheet transmits a fraction

$$\cos^2 \alpha = \cos^2 \left(\frac{\theta}{N} \right).$$

There are N sheets, so the fraction transmitted through the stack is

$$\left(\cos^2 \left(\frac{\theta}{N} \right) \right)^N.$$

We want to evaluate this in the limit as $N \to \infty$.

As N gets larger we can use a small angle approximation to the cosine function,

$$\cos x \approx 1 - \frac{1}{2} x^2 \text{ for } x \ll 1$$

The the transmitted intensity is

$$\left(1 - \frac{1}{2} \frac{\theta^2}{N^2} \right)^{2N}.$$

This expression can also be expanded in a binomial expansion to get

$$1 - 2N \frac{1}{2} \frac{\theta^2}{N^2},$$

which in the limit as $N \to \infty$ approaches 1.

The stack then transmits *all* of the light which makes it past the first filter. Assuming the light is originally unpolarized, then the stack transmits half the original intensity.

P44-5 Since passing through a quarter wave plate twice can rotate the polarization of a linearly polarized wave by 90°, then if the light passes through a polarizer, through the plate, reflects off the coin, then through the plate, and through the polarizer, it would be possible that when it passes through the polarizer the second time it is 90° to the polarizer and no light will pass. You won't see the coin.

On the other hand if the light passes first through the plate, then through the polarizer, then is reflected, the passes again through the polarizer, *all* the reflected light will pass through he polarizer and eventually work its way out through the plate. So the coin will be visible.

Hence, side A must be the polarizing sheet, and that sheet must be at 45° to the optical axis.

Chapter 45

The Nature of Light

E45-1 (a) The energy of a photon is given by Eq. 45-1, $E = hf$. Since $c = f\lambda$, where c is the speed of light, we can also write

$$E = hf = \frac{hc}{\lambda}.$$

In fact, h and c almost always occur together in particle physics, so remembering the value of hc (and later $hc/2\pi$) in terms of "appropriate" units is a worthwhile venture.

Putting in "best" numbers

$$hc = \frac{(6.62606876 \times 10^{-34} \text{J} \cdot \text{s})}{(1.602176462 \times 10^{-19} \text{C})} (2.99792458 \times 10^{8} \text{m/s}) = 1.23984 \times 10^{-6} \text{ eV} \cdot \text{m}.$$

This means that $hc = 1240$ eV $\cdot$ nm is accurate to almost one part in 8000!

(b) $E = (1240 \text{ eV} \cdot \text{nm})/(589 \text{ nm}) = 2.11$ eV.

E45-5 When talking about the regions in the sun's spectrum it is more common to refer to wavelengths than frequencies. So we will use the results of Exercise 45-1(a), and solve

$$\lambda = hc/E = (1240 \text{ eV} \cdot \text{nm})/E.$$

The energies are between $E = (1.0 \times 10^{18} \text{J})/(1.6 \times 10^{-19} \text{C}) = 6.25$ eV and $E = (1.0 \times 10^{16} \text{J})/(1.6 \times 10^{-19} \text{C}) = 625$ eV. These energies correspond to wavelengths between 198 nm and 1.98 nm; this is the ultraviolet range.

E45-7 (a) Since the power is the same, the bulb with the larger energy per photon will emits *fewer* photons per second. Since longer wavelengths have lower energies, the bulb emitting 700 nm must be giving off more photons per second.

(b) How many more photons per second? If E_1 is the energy per photon for one of the bulbs, then $N_1 = P/E_1$ is the number of photons per second emitted. The difference is then

$$N_1 - N_2 = \frac{P}{E_1} - \frac{P}{E_2} = \frac{P}{hc}(\lambda_1 - \lambda_2),$$

or

$$N_1 - N_2 = \frac{(130\,\text{W})}{(6.63\times 10^{-34}\,\text{J·s})(3.00\times 10^8\,\text{m/s})}((700\times 10^{-9}\,\text{m}) - (400\times 10^{-9}\,\text{m})) = 1.96\times 10^{20}.$$

E45-15 (a) We want to apply Eq. 45-6,

$$R(\lambda, T) = \frac{2\pi c^2 h}{\lambda^5}\frac{1}{e^{hc/\lambda kT} - 1}.$$

We know the ratio of the spectral radiancies at two different wavelengths. Dividing the above equation at the first wavelength by the same equation at the second wavelength,

$$3.5 = \frac{\lambda_1^5\left(e^{hc/\lambda_1 kT} - 1\right)}{\lambda_2^5\left(e^{hc/\lambda_2 kT} - 1\right)},$$

where $\lambda_1 = 200$ nm and $\lambda_2 = 400$ nm. We can considerably simplify this expression if we let

$$x = e^{hc/\lambda_2 kT},$$

because since $\lambda_2 = 2\lambda_1$ we would have

$$e^{hc/\lambda_1 kT} = e^{2hc/\lambda_2 kT} = x^2.$$

Then we get

$$3.5 = \left(\frac{1}{2}\right)^5\frac{x^2 - 1}{x - 1} = \frac{1}{32}(x + 1).$$

We will use the results of Exercise 45-1 for the exponents and then rearrange to get

$$T = \frac{hc}{\lambda_1 k \ln(111)} = \frac{(3.10\ \text{eV})}{(8.62\times 10^{-5}\ \text{eV/K})\ln(111)} = 7640\,\text{K}.$$

(b) The method is the same, except that instead of 3.5 we have 1/3.5; this means the equation for x is

$$\frac{1}{3.5} = \frac{1}{32}(x + 1),$$

with solution $x = 8.14$, so then

$$T = \frac{hc}{\lambda_1 k \ln(8.14)} = \frac{(3.10\ \text{eV})}{(8.62\times 10^{-5}\ \text{eV/K})\ln(8.14)} = 17200\,\text{K}.$$

E45-19 (a) In order to show a photo-electric effect the energy of the photon must exceed the energy needed to remove an electron from the metal. We are given a wavelength, so we will use the results of Exercise 45-1 to find the energy of the corresponding photon,

$$E = \frac{hc}{\lambda} = \frac{(1240 \text{ eV} \cdot \text{nm})}{(678 \text{ nm})} = 1.83 \text{ eV}.$$

Since this energy is less than than the minimum energy required to remove an electron then the photo-electric effect will not occur.

(b) The cut-off wavelength is the longest possible wavelength of a photon that will still result in the photo-electric effect occurring. That wavelength is

$$\lambda = \frac{(1240 \text{ eV} \cdot \text{nm})}{E} = \frac{(1240 \text{ eV} \cdot \text{nm})}{(2.28 \text{ eV})} = 544 \text{ nm}.$$

This would be visible as green. In the event that the photon wavelength is longer than this cut-off value there will be *no* photo electric effect.

E45-23 (a) The stopping potential is given by Eq. 45-11,

$$V_0 = \frac{h}{e}f - \frac{\phi}{e},$$

which can also be written as

$$V_0 = \frac{hc}{e\lambda} - \frac{\phi}{e},$$

so

$$V_0 = \frac{(1240 \text{ eV} \cdot \text{nm})}{e(410 \text{ nm}} - \frac{(1.85 \text{ eV}}{e} = 1.17 \text{ V}.$$

(b) These are *not* relativistic electrons, so

$$v = \sqrt{2K/m} = c\sqrt{2K/mc^2} = c\sqrt{2(1.17 \text{ eV})/(0.511 \times 10^6 \text{ eV})} = 2.14 \times 10^{-3}c,$$

or $v = 64200$ m/s.

E45-29 If the sodium atom was originally at rest, then a collision with the photon wouldn't slow it down, it would speed it up. So we will interpret this question to mean "by how much does the speed of a sodium atom change on absorbing a photon..."

To make matters worse, the absorption of a photon means that this is *not* an elastic collision, so we can only apply momentum conservation principles.

The initial momentum of the system is the momentum of the photon, $p = h/\lambda$. This momentum is imparted to the sodium atom, so the final speed of the sodium is $v = p/m$, where m is the mass of the sodium. Then

$$v = \frac{h}{\lambda m} = \frac{(6.63 \times 10^{-34} \text{J} \cdot \text{s})}{(589 \times 10^{-9}\text{m})(23)(1.7 \times 10^{-27}\text{kg})} = 2.9 \text{ cm/s}.$$

E45-33 The change in the wavelength of a photon during Compton scattering is given by Eq. 45-17,

$$\lambda' - \lambda = \frac{h}{mc}(1 - \cos\phi).$$

We are *not* using the expression with the form $\Delta\lambda$ because $\Delta\lambda$ and ΔE are *not* simply related.

The wavelength is related to frequency by $c = f\lambda$, while the frequency is related to the energy by Eq. 45-1, $E = hf$. Then

$$\begin{aligned}
\Delta E &= E - E' = hf - hf', \\
&= hc\left(\frac{1}{\lambda} - \frac{1}{\lambda'}\right), \\
&= hc\frac{\lambda' - \lambda}{\lambda\lambda'}.
\end{aligned}$$

Into this last expression we substitute the Compton formula. Then

$$\Delta E = \frac{h^2}{m}\frac{(1 - \cos\phi)}{\lambda\lambda'}.$$

Now $E = hf = hc/\lambda$, and we can divide this on both sides of the above equation. Also, $\lambda' = c/f'$, and we can substitute this into the right hand side of the above equation. Both of these steps result in

$$\frac{\Delta E}{E} = \frac{hf'}{mc^2}(1 - \cos\phi).$$

Note that mc^2 is the rest energy of the scattering particle (usually an electron), while hf' is the energy of the scattered photon.

E45-37 (a) The change in wavelength is independent of the wavelength and is given by Eq. 45-17,

$$\Delta\lambda = \frac{hc}{mc^2}(1 - \cos\phi) = 2\frac{(1240\ \text{eV}\cdot\text{nm})}{(0.511\times10^6\ \text{eV})} = 4.85\times10^{-3}\ \text{nm}.$$

The cosine term didn't just disappear. Since the photon was scattered through $180°$ the cosine was equal to -1, and then $1 - (-1) = 2$. Note also that we took advantage of hc and mc^2 when evaluating this quantity. It really does make the numbers easier.

(b) The change in energy is given by

$$\begin{aligned}
\Delta E &= \frac{hc}{\lambda_\text{f}} - \frac{hc}{\lambda_\text{i}}, \\
&= hc\left(\frac{1}{\lambda_\text{i} + \Delta\lambda} - \frac{1}{\lambda_\text{i}}\right), \\
&= (1240\ \text{eV}\cdot\text{nm})\left(\frac{1}{(9.77\ \text{pm}) + (4.85\ \text{pm})} - \frac{1}{(9.77\ \text{pm})}\right) = -42.1\ \text{keV}
\end{aligned}$$

348

(c) This energy went to the electron, so the final kinetic energy of the electron is 42.1 keV.

P45-1 The radiant intensity is given by Eq. 45-3, $I = \sigma T^4$. The power that is radiated through the opening is $P = IA$, where A is the area of the opening. But energy goes both ways through the opening; it is the *difference* that will give the net power transfer. Then

$$P_{\text{net}} = (I_0 - I_1)A = \sigma A \left(T_0^4 - T_1^4 \right).$$

Put in the numbers, and

$$P_{\text{net}} = (5.67 \times 10^{-8} \text{W/m}^2 \cdot \text{K}^4)(5.20 \times 10^{-4} \text{m}^2) \left((488 \,\text{K})^4 - (299 \,\text{K})^4 \right) = 1.44 \,\text{W}.$$

Did you remember to change to Kelvins?

P45-3 Light from the sun will "heat-up" the thin black screen. As the temperature of the screen increases it will begin to radiate energy. When the rate of energy radiation from the screen is equal to the rate at which the energy from the sun strikes the screen we will have equilibrium. We need first to find an expression for the rate at which energy from the sun strikes the screen.

The temperature of the sun is T_S. The radiant intensity is given by Eq. 45-3, $I_S = \sigma T_S^4$. The total power radiated by the sun is the product of this radiant intensity and the surface area of the sun, so

$$P_S = 4\pi r_S^2 \sigma T_S^4,$$

where r_S is the radius of the sun.

Assuming that the lens is on the surface of the Earth (a reasonable assumption), then we can find the power incident on the lens if we know the intensity of sunlight at the distance of the Earth from the sun. That intensity is

$$I_E = \frac{P_S}{A} = \frac{P_S}{4\pi R_E^2},$$

where R_E is the radius of the Earth's orbit. Combining,

$$I_E = \sigma T_S^4 \left(\frac{r_S}{R_E} \right)^2$$

If you think about it, this expression makes sense; the intensity is proportional to distance squared, and the radiant intensity is the intensity at the surface of the object.

The total power incident on the lens is then

$$P_{\text{lens}} = I_E A_{\text{lens}} = \sigma T_S^4 \left(\frac{r_S}{R_E} \right)^2 \pi r_l^2,$$

where r_l is the radius of the lens. All of the energy that strikes the lens is focused on the image, so the power incident on the lens is also incident on the image.

The screen radiates as the temperature increases. The radiant intensity is $I = \sigma T^4$, where T is the temperature of the screen. The power radiated is this intensity times the surface area, so

$$P = IA = 2\pi r_i^2 \sigma T^4.$$

The factor of "2" is because the screen has two sides, while r_i is the radius of the image. All we need to do is set this equal to P_{lens} and we are done. So

$$2\pi r_i^2 \sigma T^4 = \sigma T_S^4 \left(\frac{r_S}{R_E}\right)^2 \pi r_l^2,$$

or

$$T^4 = \frac{1}{2} T_S^4 \left(\frac{r_S\, r_l}{R_E\, r_i}\right)^2.$$

Maybe we aren't done.

The radius of the image of the sun divided by the radius of the sun is the magnification of the lens. But magnification is also related to image distance divided by object distance, so

$$\frac{r_i}{r_S} = |m| = \frac{i}{o},$$

Distances should be measured from the lens, but since the sun is so far from the Earth, we won't be far off in stating $o \approx R_E$. Since the object is so far from the lens, the image will be very, very close to the focal point, so we can also state $i \approx f$. Then

$$\frac{r_i}{r_S} = \frac{f}{R_E},$$

so the expression for the temperature of the thin black screen is considerably simplified to

$$T^4 = \frac{1}{2} T_S^4 \left(\frac{r_l}{f}\right)^2.$$

Now we can put in some of the numbers.

$$T = \frac{1}{2^{1/4}}(5800\,\text{K}) \sqrt{\frac{(1.9\,\text{cm})}{(26\,\text{cm})}} = 1300\,\text{K}.$$

Interestingly enough, one could do this problem backward to find the temperature of the sun. But it is *much* easier to find the temperature from Wien's displacement law.

P45-5 (a) This problem is much easier than the previous one. If the planet has a temperature T, then the radiant intensity of the planet will be $I\sigma T^4$, and the rate of energy radiation from the planet will be

$$P = 4\pi R^2 \sigma T^4,$$

where R is the radius of the planet.

A steady state planet temperature requires that the energy from the sun arrive at the same rate as the energy is radiated from the planet. The intensity of the energy from the sun a distance r from the sun is

$$P_{\text{sun}}/4\pi r^2,$$

and the total power incident on the planet is then

$$P = \pi R^2 \frac{P_{\text{sun}}}{4\pi r^2}.$$

Equating,

$$4\pi R^2 \sigma T^4 = \pi R^2 \frac{P_{\text{sun}}}{4\pi r^2},$$
$$T^4 = \frac{P_{\text{sun}}}{16\pi\sigma r^2}.$$

We are done, but it is mildly entertaining to take this problem a step farther. The power from the sun is related to the temperature of the sun according to the work done in Problem 3. Combining that work with this we find

$$T^4 = \frac{1}{4}T_{\text{s}}^4\left(\frac{R_{\text{s}}}{r}\right)^2,$$

or

$$T = \frac{1}{\sqrt{2}}T_{\text{s}}\sqrt{\frac{R_{\text{s}}}{r}}.$$

We can use this expression to find the likely distance from a star for habitable planets. Not that we are in a position to go visit those planets if we did find them...

(b) Using the last equation and the numbers from Problem 3,

$$T = \frac{1}{\sqrt{2}}(5800\,\text{K})\sqrt{\frac{(6.96\times10^8\text{m})}{(1.5\times10^{11}\text{m})}} = 279\,\text{K}.$$

That's about 43° F. Cool, but with a sweater, comfortable.

P45-11 (a) The maximum value of $\Delta\lambda$ is $2h/mc$. The maximum energy lost by the photon is then

$$
\begin{aligned}
\Delta E &= \frac{hc}{\lambda_f} - \frac{hc}{\lambda_i}, \\
&= hc\left(\frac{1}{\lambda_i + \Delta\lambda} - \frac{1}{\lambda_i}\right), \\
&= hc\frac{-2h/mc}{\lambda(\lambda + 2h/mc)},
\end{aligned}
$$

where in the last line we wrote λ for λ_i. The energy given to the electron is the negative of this, so

$$
K_{max} = \frac{2h^2}{m\lambda(\lambda + 2h/mc)}.
$$

Multiplying through by $1^2 = (E\lambda/hc)^2$ we get

$$
K_{max} = \frac{2E^2}{mc^2(1 + 2hc/\lambda mc^2)}.
$$

or

$$
K_{max} = \frac{E^2}{mc^2/2 + E}.
$$

(b) The answer is

$$
K_{max} = \frac{(17.5 \text{ keV})^2}{(511 \text{ eV})/2 + (17.5 \text{ keV})} = 1.12 \text{ keV}.
$$

Chapter 46

The Nature of Matter

E46-1 (a) Apply Eq. 46-1, $\lambda = h/p$. The momentum of the bullet is

$$p = mv = (0.041\,\text{kg})(960\,\text{m/s}) = 39\text{kg}\cdot\text{m/s},$$

so the corresponding wavelength is

$$\lambda = h/p = (6.63\times10^{-34}\text{J}\cdot\text{s})/(39\text{kg}\cdot\text{m/s}) = 1.7\times10^{-35}\,\text{m}.$$

(b) This length is much too small to be significant. How much too small? If the radius of the galaxy were one meter, this distance would correspond to the diameter of a proton.

E46-5 (a) Apply Eq. 46-1, $p = h/\lambda$. The proton speed would then be

$$v = \frac{h}{m\lambda} = c\frac{hc}{mc^2\lambda} = c\frac{(1240\text{ MeV}\cdot\text{fm})}{(938\text{ MeV})(113\text{ fm})} = 0.0117c.$$

This is good, because it means we were justified in using the non-relativistic equations. Then $v = 3.51\times10^6\text{m/s}$.

(b) The kinetic energy of this electron would be

$$K = \frac{1}{2}mv^2 = \frac{1}{2}(938\text{ MeV})(0.0117)^2 = 64.2\text{ keV}.$$

The potential through which it would need to be accelerated is 64.2 kV.

E46-9 The relativistic relationship between energy and momentum is

$$E^2 = p^2c^2 + m^2c^4,$$

and if the energy is very large (compared to mc^2), then the contribution of the mass to the above expression is small, and

$$E^2 \approx p^2 c^2.$$

Then from Eq. 46-1,

$$\lambda = \frac{h}{p} = \frac{hc}{pc} = \frac{hc}{E} = \frac{(1240 \text{ MeV} \cdot fm)}{(50 \times 10^3 \text{ MeV})} = 2.5 \times 10^{-2} \text{ fm}.$$

E46-13 (a)   The classical expression for kinetic energy is

$$p = \sqrt{2mK},$$

so

$$\lambda = \frac{h}{p} = \frac{hc}{\sqrt{2mc^2 K}} = \frac{(1240 \text{ keV} \cdot pm)}{\sqrt{2(511 \text{ keV})(25.0 \text{ keV})}} = 7.76 \text{ pm}.$$

(a)   The relativistic expression for momentum is

$$pc = sqrt{E^2} - m^2 c^4 = \sqrt{(mc^2 + K)^2 - m^2 c^4} = \sqrt{K^2 + 2mc^2 K}.$$

Then

$$\lambda = \frac{hc}{pc} = \frac{(1240 \text{ keV} \cdot pm)}{\sqrt{(25.0 \text{ keV})^2 + 2(511 \text{ keV})(25.0 \text{ keV})}} = 7.66 \text{ pm}.$$

E46-17 (a)   Since $\sin 52° = 0.78$, then $2(\lambda/d) = 1.57 > 1$, so there is no diffraction order other than the first.

(b)   For an accelerating potential of 54 volts we have $\lambda/d = 0.78$. Increasing the potential will increase the kinetic energy, increase the momentum, and decrease the wavelength. d won't change. The kinetic energy is increased by a factor of $60/54 = 1.11$, the momentum increases by a factor of $\sqrt{1.11} = 1.05$, so the wavelength changes by a factor of $1/1.05 = 0.952$. The new angle is then

$$\theta = \arcsin(0.952 \times 0.78) = 48°.$$

E46-21   Apply Eq. 46-9,

$$\Delta E \geq \frac{h}{2\pi \Delta t} = \frac{4.14 \times 10^{-15} \text{ eV} \cdot \text{s})}{2\pi (8.7 \times 10^{-12} \text{s})} = 7.6 \times 10^{-5} \text{ eV}.$$

This is *much* smaller than the photon energy.

E46-25 We want $v \approx \Delta v$, which means $p \approx \Delta p$. Apply Eq. 46-8, and

$$\Delta x \geq \frac{h}{2\pi\Delta p} \approx \frac{h}{2\pi p}.$$

According to Eq. 46-1, the de Broglie wavelength is related to the momentum by

$$\lambda = h/p,$$

so

$$\Delta x \geq \frac{\lambda}{2\pi}.$$

E46-27 (a) A particle confined in a (one dimensional) box of size L will have a position uncertainty of no more than $\Delta x \approx L$. The momentum uncertainty will then be no less than

$$\Delta p \geq \frac{h}{2\pi\Delta x} \approx \frac{h}{2\pi L}.$$

so

$$\Delta p \approx \frac{(6.63\times 10^{-34}\,\text{J}\cdot\text{s})}{2\pi(\times 10^{-14}\,\text{m})} = 1\times 10^{-20}\text{kg}\cdot\text{m/s}.$$

(b) Assuming that $p \approx \Delta p$, we have

$$p \geq \frac{h}{2\pi L},$$

and then the electron will have a (minimum) kinetic energy of

$$E \approx \frac{p^2}{2m} \approx \frac{h^2}{8\pi^2 m L^2}.$$

or

$$E \approx \frac{(hc)^2}{8\pi^2 mc^2 L^2} = \frac{(1240\,\text{MeV}\cdot\text{fm})^2}{8\pi^2(0.511\,\text{MeV})(10\,\text{fm})^2} = 381\,\text{MeV}.$$

This is so large compared to the mass energy of the electron that we must consider relativistic effects. It will be very relativistic ($381 \gg 0.5$!), so we can use $E = pc$ as was derived in Exercise 9. Then

$$E = \frac{hc}{2\pi L} = \frac{(1240\,\text{MeV}\cdot\text{fm})}{2\pi(10\,\text{fm})} = 19.7\,\text{MeV}.$$

This is the *total* energy; so we subtract 0.511 MeV to get $K = 19$ MeV.

E46-29 The transmission coefficient is given by Eq. 46-25,

$$T = 16\frac{E}{U_0}\left(1 - \frac{E}{U_0}\right)e^{-2kL},$$

but this approximation is only valid for $E \ll U_0$ and/or $L \gg k$. The wave number, k, is given by Eq. 46-23,

$$k = \frac{2\pi}{h}\sqrt{2m(U_0 - E)}$$

It turns out that it is *much* easier to multiply through this expression by c/c, where c is the speed of light. Then

$$k = \frac{2\pi}{hc}\sqrt{2mc^2(U_0 - E)}.$$

This method has the advantage of allowing us to express the mass of subatomic particles in terms of easier to use units. We've done this before, we'll do it again.

(a) For the proton $mc^2 = 938$ MeV, so

$$k = \frac{2\pi}{(1240\text{ MeV}\cdot\text{fm})}\sqrt{2(938\text{ MeV})(10\text{ MeV} - 3.0\text{ MeV})} = 0.581\text{ fm}^{-1}.$$

The transmission coefficient is then

$$T = 16\frac{(3.0\text{ MeV})}{(10\text{ MeV})}\left(1 - \frac{(3.0\text{ MeV})}{(10\text{ MeV})}\right)e^{-2(0.581\text{ fm}^{-1})(10\text{ fm})} = 3.0\times10^{-5}.$$

(b) For the deuteron $mc^2 = 2 \times 938$ MeV, so

$$k = \frac{2\pi}{(1240\text{ MeV}\cdot\text{fm})}\sqrt{2(2)(938\text{ MeV})(10\text{ MeV} - 3.0\text{ MeV})} = 0.821\text{ fm}^{-1}.$$

The transmission coefficient is then

$$T = 16\frac{(3.0\text{ MeV})}{(10\text{ MeV})}\left(1 - \frac{(3.0\text{ MeV})}{(10\text{ MeV})}\right)e^{-2(0.821\text{ fm}^{-1})(10\text{ fm})} = 2.5\times10^{-7}.$$

P46-1 We will interpret low energy to mean non-relativistic. Then

$$\lambda = \frac{h}{p} = \frac{h}{\sqrt{2m_\mathrm{n}K}}.$$

The diffraction pattern is then given by

$$d\sin\theta = m\lambda = mh/\sqrt{2m_\mathrm{n}K},$$

where m is diffraction order while m_n is the neutron mass. We want to investigate the spread by taking the derivative of θ with respect to K,

$$d\cos\theta\, d\theta = -\frac{mh}{2\sqrt{2m_n K^3}}\, dK.$$

Divide this by the original equation, and then

$$\frac{\cos\theta}{\sin\theta}\, d\theta = -\frac{dK}{2K}.$$

Rearrange, change the differential to a difference, and then

$$\Delta\theta = \tan\theta\,\frac{\Delta K}{2K}.$$

We dropped the negative sign out of laziness; but the angles are in radians, so we need to multiply by $180/\pi$ to convert to degrees.

P46-5 Maybe you can remember back to a high school analytic geometry class where your teacher derived the expression for the cosine or the sine of the sum of two angles. I expect that these memories may have been blocked out, like your memories of your first filling from the dentist.

Here is the derivation, but with only five lines, as opposed to the nasty diagram and fifty lines that you saw in high school. First, the rule for exponents

$$e^{i(a+b)} = e^{ia}\, e^{ib}.$$

Then apply Eq. 46-12, $e^{i\theta} = \cos\theta + i\sin\theta$,

$$\cos(a+b) + i\sin(a+b) = (\cos a + i\sin a)(\sin b + i\sin b).$$

Expand the right hand side, remembering that $i^2 = -1$,

$$\cos(a+b) + i\sin(a+b) = \cos a\cos b + i\cos a\sin b + i\sin a\cos b - \sin a\sin b.$$

Since the real part of the left hand side must equal the real part of the right and the imaginary part of the left hand side must equal the imaginary part of the right, we actually have *two* equations. They are

$$\cos(a+b) = \cos a\cos b - \sin a\sin b$$

and

$$\sin(a+b) = \cos a\sin b + \sin a\cos b.$$

Not bad, eh?

Chapter 47

Electrons in Potential Wells

E47-1 (a) The allowed energy levels for an infinitely deep well are given by Eq. 47-3,

$$E_n = n^2 \frac{h^2}{8mL^2},$$

so the ground state energy level will be given by

$$E_1 = \frac{h^2}{8mL^2} = \frac{(6.63 \times 10^{-34} \text{J} \cdot \text{s})^2}{8(9.11 \times 10^{-31} \text{kg})(1.4 \times 10^{-14} \text{m})^2} = 3.1 \times 10^{-10} \text{ J}.$$

The answer is correct, but the units make it almost useless. We can divide by the electron charge to express this in electron volts, and then $E = 1900$ MeV. Note that this is an extremely relativistic quantity, so the energy expression loses validity.

(b) We can repeat what we did above, or we can apply a "trick" that is often used in solving these problems. Multiplying the top and the bottom of the energy expression by c^2 we get

$$E_1 = \frac{(hc)^2}{8(mc^2)L^2}$$

This might not look simpler, but it is. Looking at the numerator,

$$hc = (4.14 \times 10^{-15} \text{ eV} \cdot \text{s})(3.00 \times 10^8 \text{ m/s}) = 1240 \text{ eV} \cdot \text{nm} = 1240 \text{ MeV} \cdot \text{fm}.$$

The first way, $hc = 1240$ eV·nm, is appropriate when discussing atomic energy levels, while the second, $hc = 1240$ MeV·fm, is appropriate for nuclear energy levels. Either number is fairly easy to remember, and mc^2 is a more "natural" way to deal with the mass of the particles.

Then

$$E_1 = \frac{(1240 \text{ MeV} \cdot \text{fm})^2}{8(940 \text{ MeV})(14 \text{ fm})^2} = 1.0 \text{ MeV}.$$

(c) Finding an neutron inside the nucleus seems reasonable; but finding the electron would not. The energy of such an electron is considerably larger than binding energies of the particles in the nucleus.

E47-7 (a) We will take advantage of the "trick" that was developed in part (b) of Exercise 47-1. Then

$$E_n = n^2 \frac{(hc)^2}{8mc^2\,L} = (15)^2 \frac{(1240 \text{ eV} \cdot \text{nm})^2}{8(0.511 \times 10^6 \text{ eV})(0.0985 \text{ nm})^2} = 8.72 \text{ keV}.$$

(b) The magnitude of the momentum is *exactly* known because $E = p^2/2m$. This momentum is given by

$$pc = \sqrt{2mc^2 E} = \sqrt{2(511 \text{ keV})(8.72 \text{ keV})} = 94.4 \text{ keV}.$$

Notice that once again we multiplied through by c in order to work with easier units.

What we don't know is in which direction the particle is moving. It is bouncing back and forth between the walls of the box, so the momentum could be directed toward the right or toward the left. The uncertainty in the momentum is then

$$\Delta p = p$$

which can be expressed in terms of the box size L by

$$\Delta p = p = \sqrt{2mE} = \sqrt{\frac{n^2 h^2}{4L^2}} = \frac{nh}{2L}.$$

(c) The uncertainty in the position is 98.5 pm; the electron could be *anywhere* inside the well.

E47-11 The ground state of hydrogen, as given by Eq. 47-21, is

$$E_1 = -\frac{me^4}{8\epsilon_0^2 h^2} = -\frac{(9.109 \times 10^{-31} \text{ kg})(1.602 \times 10^{-19} \text{ C})^4}{8(8.854 \times 10^{-12} \text{ F/m})^2(6.626 \times 10^{-34} \text{ J} \cdot \text{s})^2} = 2.179 \times 10^{-18} \text{ J}.$$

In terms of electron volts the ground state energy is

$$E_1 = -(2.179 \times 10^{-18} \text{ J})/(1.602 \times 10^{-19} \text{ C}) = -13.60 \text{ eV}.$$

E47-15 The binding energy is the energy required to remove the electron. If the energy of the electron is negative, then that negative energy is a measure of the energy required to set the electron free.

The first excited state is when $n = 2$ in Eq. 47-21. It is *not* necessary to re-evaluate the constants in this equation every time, instead, we start from

$$E_n = \frac{E_1}{n^2} \text{ where } E_1 = -13.60 \text{ eV}.$$

Then the first excited state has energy

$$E_2 = \frac{(-13.6 \text{ eV})}{(2)^2} = -3.4 \text{ eV}.$$

The binding energy is then 3.4 eV.

E47-17 (a) The energy of this photon is

$$E = \frac{hc}{\lambda} = \frac{(1240 \text{ eV} \cdot \text{nm})}{(1281.8 \text{ nm})} = 0.96739 \text{ eV}.$$

The final state of the hydrogen must have an energy of no more than -0.96739, so the largest possible n of the final state is

$$n < \sqrt{13.60 \text{ eV}/0.96739 \text{ eV}} = 3.75,$$

so the final n is 1, 2, or 3. The initial state is only slightly higher than the final state. The jump from $n = 2$ to $n = 1$ is *too* large (see Exercise 15), any other initial state would have a larger energy difference, so $n = 1$ is *not* the final state.

So what level might be above $n = 2$? We'll try

$$n = \sqrt{13.6 \text{ eV}/(3.4 \text{ eV} - 0.97 \text{ eV})} = 2.36,$$

which is *so* far from being an integer that we don't need to look farther. The $n = 3$ state has energy 13.6 eV/9 = 1.51 eV. Then the initial state could be

$$n = \sqrt{13.6 \text{ eV}/(1.51 \text{ eV} - 0.97 \text{ eV})} = 5.01,$$

which is close enough to 5 that we can assume the transition was $n = 5$ to $n = 3$.

(b) This belongs to the Paschen series.

E47-21 In order to have an inelastic collision with the 6.0 eV neutron there must exist a transition with an energy difference of less than 6.0 eV. For a hydrogen atom in the ground state $E_1 = -13.6$ eV the nearest state is

$$E_2 = (-13.6 \text{ eV})/(2)^2 = -3.4 \text{ eV}.$$

Since the difference is 10.2 eV, it will *not* be possible for the 6.0 eV neutron to have an inelastic collision with a ground state hydrogen atom.

Interestingly enough, even two collisions with two 6.0 eV neutrons will still be elastic, so long as the time between the collisions is greater than some minimum time. What would be the minimum time? It would be found by considering the energy uncertainty relationship. We would need the atom to "absorb" an amount of energy from the first neutron on the order of 6 eV and hold onto it until the second neutron strikes it. This can only happen if the energy of the atom is uncertain, but this level of uncertainty would last no more than about

$$\Delta t \approx \frac{h}{2\pi \Delta E} = \frac{(4 \times 10^{-15} \text{ eV} \cdot \text{s})}{2\pi(6 \text{ eV})} = 1 \times 10^{-16} \text{s}.$$

That's a short time. A pair of collisions within this time period is unlikely.

E47-25 The first Lyman line is the $n = 1$ to $n = 2$ transition. The second Lyman line is the $n = 1$ to $n = 3$ transition. The first Balmer line is the $n = 2$ to $n = 3$ transition. Since the photon frequency is proportional to the photon energy ($E = hf$) and the photon energy is the energy difference between the two levels, we have

$$f_{n \to m} = \frac{E_m - E_n}{h}$$

where the E_n is the hydrogen atom energy level. Then

$$
\begin{aligned}
f_{1 \to 3} &= \frac{E_3 - E_1}{h}, \\
&= \frac{E_3 - E_2 + E_2 - E_1}{h} = \frac{E_3 - E_2}{h} + \frac{E_2 - E_1}{h}, \\
&= f_{2 \to 3} + f_{1 \to 2}.
\end{aligned}
$$

E47-27 (a) The energy levels in the He$^+$ spectrum are given by

$$E_n = -Z^2(13.6 \text{ eV})/n^2,$$

where $Z = 2$, as is discussed in Sample Problem 47-6. The photon wavelengths for the $n = 4$ series are then

$$\lambda = \frac{hc}{E_n - E_4} = \frac{hc/E_4}{1 - E_n/E_4},$$

which can also be written as

$$
\begin{aligned}
\lambda &= \frac{16hc/(54.4 \text{ eV})}{1 - 16/n^2}, \\
&= \frac{16hcn^2/(54.4 \text{ eV})}{n^2 - 16}, \\
&= \frac{Cn^2}{n^2 - 16},
\end{aligned}
$$

361

where $C = hc/(3.4 \text{ eV}) = 365$ nm.

(b) The wavelength of the first line is the transition from $n = 5$,

$$\lambda = \frac{(365 \text{ nm})(5)^2}{(5)^2 - (4)^2} = 1014 \text{ nm}.$$

The series limit is the transition from $n = \infty$, so

$$\lambda = 365 \text{ nm}.$$

(c) The series starts in the infrared (1014 nm), and ends in the ultraviolet (365 nm). So it must also include some visible lines.

E47-33 (a) We'll use Eqs. 47-25 and 47-26. At $r = 0$

$$\psi^2(0) = \frac{1}{\pi a_0^3} e^{-2(0)/a_0} = \frac{1}{\pi a_0^3} = 2150 \text{ nm}^{-3},$$

while

$$P(0) = 4\pi(0)^2 \psi^2(0) = 0.$$

(b) At $r = a_0$ we have

$$\psi^2(a_0) = frac1{\pi a_0^3} e^{-2(a_0)/a_0} = \frac{e^{-2}}{\pi a_0^3} = 291 \text{ nm}^{-3},$$

and

$$P(a_0) = 4\pi(a_0)^2 \psi^2(a_0) = 10.2 \text{ nm}^{-1}.$$

E47-37 If $l = 3$ then m_l can be 0, ±1, ±2, or ±3.

(a) From Eq. 47-30, $L_z = m_l h/2\pi$.. So L_z can equal 0, $\pm h/2\pi$, $\pm h/\pi$, or $\pm 3h/2\pi$.

(b) From Eq. 47-31, $\theta = \arccos(m_l/\sqrt{l(l+1)})$, so θ can equal 90°, 73.2°, 54.7°, or 30.0°.

(b) The magnitude of $\vec{L}$ is given by Eq. 47-28,

$$L = \sqrt{l(l+1)} \frac{h}{2\pi} = \sqrt{3} h/\pi.$$

E47-45 If $m_l = 4$ then $l \geq 4$. But $n \geq l + 1$, so $n > 4$. We only know that $m_s = \pm 1/2$.

P47-1 We can simplify the energy expression as

$$E = E_0 \left(n_x^2 + n_y^2 + n_z^2\right) \text{ where } E_0 = \frac{h^2}{8mL^2}.$$

To find the lowest energy levels we need to focus on the values of n_x, n_y, and n_z.

It doesn't take much imagination to realize that the set $(1, 1, 1)$ will result in the smallest value for $n_x^2 + n_y^2 + n_z^2$. The next choice is to set one of the values equal to 2, and try the set $(2, 1, 1)$.

Then it starts to get harder, as the next lowest might be either $(2, 2, 1)$ or $(3, 1, 1)$. The only way to find out is to try. I'll tabulate the results for you:

n_x	n_y	n_z	$n_x^2 + n_y^2 + n_z^2$	Mult.	n_x	n_y	n_z	$n_x^2 + n_y^2 + n_z^2$	Mult.
1	1	1	3	1	3	2	1	14	6
2	1	1	6	3	3	2	2	17	3
2	2	1	9	3	4	1	1	18	3
3	1	1	11	3	3	3	1	19	3
2	2	2	12	1	4	2	1	21	6

Multiplicity refers to how many ways you can assign the quantum numbers. The problem didn't ask, so I suppose I shouldn't tell, but I did.

We are now in a position to state the five lowest energy levels. The fundamental quantity is

$$E_0 = \frac{(hc)^2}{8mc^2L^2} = \frac{(1240 \text{ eV} \cdot \text{nm})^2}{8(0.511 \times 10^6 \text{ eV})(250 \text{ nm})^2} = 6.02 \times 10^{-6} \text{ eV}.$$

The five lowest levels are found by multiplying this fundamental quantity by the numbers in the table above.

P47-5 We will want an expression for

$$\frac{d^2}{dx^2} \psi_0.$$

Doing the math one derivative at a time,

$$
\begin{aligned}
\frac{d^2}{dx^2} \psi_0 &= \frac{d}{dx}\left(\frac{d}{dx}\psi_0\right), \\[4pt]
&= \frac{d}{dx}\left(A_0(-2\pi m\omega x/h)e^{-\pi m\omega x^2/h}\right), \\[4pt]
&= A_0(-2\pi m\omega x/h)^2 e^{-\pi m\omega x^2/h} + A_0(-2\pi m\omega/h)e^{-\pi m\omega x^2/h}, \\[4pt]
&= \left((2\pi m\omega x/h)^2 - (2\pi m\omega/h)\right)A_0 e^{-\pi m\omega x^2/h}, \\[4pt]
&= \left((2\pi m\omega x/h)^2 - (2\pi m\omega/h)\right)\psi_0.
\end{aligned}
$$

In the last line we factored out ψ_0. This will make our lives easier later on.

Now we want to go to Schrödinger's equation, and make some substitutions.

$$-\frac{h^2}{8\pi^2 m}\frac{d^2}{dx^2}\psi_0 + U\psi_0 = E\psi_0,$$

$$-\frac{h^2}{8\pi^2 m}\left((2\pi m\omega x/h)^2 - (2\pi m\omega/h)\right)\psi_0 + U\psi_0 = E\psi_0,$$

$$-\frac{h^2}{8\pi^2 m}\left((2\pi m\omega x/h)^2 - (2\pi m\omega/h)\right) + U = E,$$

where in the last line we divided through by ψ_0. Now for some algebra,

$$U = E + \frac{h^2}{8\pi^2 m}\left((2\pi m\omega x/h)^2 - (2\pi m\omega/h)\right),$$

$$= E + \frac{m\omega^2 x^2}{2} - \frac{h\omega}{4\pi}.$$

But we are given that $E = h\omega/4\pi$, so this simplifies to

$$U = \frac{m\omega^2 x^2}{2}$$

which looks like a harmonic oscillator type potential.

P47-9 This problem isn't really that much of a problem. Start with the magnitude of a vector in terms of the components,

$$L_x^2 + L_y^2 + L_z^2 = L^2,$$

and then rearrange,

$$L_x^2 + L_y^2 = L^2 - L_z^2.$$

According to Eq. 47-28 $L^2 = l(l+1)h^2/4\pi^2$, while according to Eq. 47-30 $L_z = m_l h/2\pi$. Substitute that into the equation, and

$$L_x^2 + L_y^2 = l(l+1)h^2/4\pi^2 - m_l^2 h^2/4\pi^2 = \left(l(l+1) - m_l^2\right)\frac{h^2}{4\pi^2}.$$

Take the square root of both sides of this expression, and we are done.

What does the "note" mean? These two components, L_x and L_y, are not separately quantized, unlike L^2 and L_z. There does seem to be some favoritism here. Why is L_z special, and not L_x or L_y? L_z was a convenient choice, but we could have instead quantized either L_x or L_y. The point is that we can't quantize all three components of $\vec{L}$, we can only quantize one component at a time.

The maximum value for m_l is l, while the minimum value is 0. Consequently,

$$\sqrt{L_x^2 + L_y^2} = \sqrt{l(l+1) - m_l^2}\,h/2\pi \le \sqrt{l(l+1)}\,h/2\pi,$$

and

$$\sqrt{L_x^2 + L_y^2} = \sqrt{l(l+1) - m_l^2}\,h/2\pi \ge \sqrt{l}\,h/2\pi.$$

364

P47-13 We want to evaluate the difference between the values of P at $x = 2$ and $x = 2$. Then

$$
\begin{aligned}
P(2) - P(1) &= \left(1 - e^{-4}(1 + 2(2) + 2(2)^2)\right) - \left(1 - e^{-2}(1 + 2(1) + 2(1)^2)\right), \\
&= 5e^{-2} - 13e^{-4} = 0.439.
\end{aligned}
$$

Chapter 48

Atomic Structure

E48-1 The highest energy x-ray photon will have an energy equal to the bombarding electrons, as is shown in Eq. 48-1,

$$\lambda_{\min} = \frac{hc}{eV}$$

Insert the appropriate values into the above expression,

$$\lambda_{\min} = \frac{(4.14 \times 10^{-15} \text{ eV} \cdot \text{s})(3.00 \times 10^8 \text{ m/s})}{eV} = \frac{1240 \times 10^{-9} \text{ eV} \cdot \text{m}}{eV}.$$

We *can* cancel the "e" in the numerator with the "*e*" in the denominator, because the "e" in electron volt refers to the charge of one electron, which is what the "*e*" in the denominator represents. We *cannot* cancel the "*V*" in the numerator with the "*V*" in the denominator, because the numerator is the unit "Volt" while the denominator is the potential difference through which the electron is accelerated.

   The expression is then

$$\lambda_{\min} = \frac{1240 \times 10^{-9} \text{ V} \cdot \text{m}}{V} = \frac{1240 \text{ kV} \cdot \text{pm}}{V}.$$

So long as we are certain that the "*V*" will be measured in units of kilovolts, we can write this as

$$\lambda_{\min} = 1240 \text{ pm}/V.$$

E48-5 (a) Changing the accelerating potential of the x-ray tube will decrease $\lambda_{\min}$. The new value will be (using the results of Exercise 48-1)

$$\lambda_{\min} = 1240 \text{ pm}/(50.0) = 24.8 \text{ pm}.$$

   (b) λ_{K_β} doesn't change. It is a property of the atom, not a property of the accelerating potential of the x-ray tube. The only way in which the accelerating potential might make a difference is if $\lambda_{K_\beta} < \lambda_{\min}$ for which case there would not be a λ_{K_β} line.

(c) λ_{K_α} doesn't change. See part (b).

E48-9 The 50.0 keV electron makes a collision and loses half of its energy to a photon, then the photon has an energy of 25.0 keV. The electron is now a 25.0 keV electron, and on the next collision again loses loses half of its energy to a photon, then this photon has an energy of 12.5 keV. On the third collision the electron loses the remaining energy, so this photon has an energy of 12.5 keV. The wavelengths of these photons will be given by

$$\lambda = \frac{(1240 \text{ keV} \cdot \text{pm})}{E},$$

which is a variation of Exercise 45-1.

E48-15 (a) The ground state question is fairly easy. The $n = 1$ shell is completely occupied by the first two electrons. So the third electron will be in the $n = 2$ state. The lowest energy angular momentum state in any shell is the s sub-shell, corresponding to $l = 0$. There is only one choice for m_l in this case: $m_l = 0$. There is no way at this level of coverage to distinguish between the energy of either the spin up or spin down configuration, so we'll arbitrarily pick spin up.

(b) Determining the configuration for the first excited state will require some thought. We could assume that one of the K shell electrons ($n = 1$) is promoted to the L shell ($n = 2$). Or we could assume that the L shell electron is promoted to the M shell. Or we could assume that the L shell electron remains in the L shell, but that the angular momentum value is changed to $l = 1$. The question that we would need to answer is which of these possibilities has the lowest energy.

The answer is the last choice: increasing the l value results in a small increase in the energy of multi-electron atoms.

E48-17 We will assume that the ordering of the energy of the shells and sub-shells is the same. That ordering is

$$1s < 2s < 2p < 3s < 3p < 4s < 3d < 4p < 5s < 4d < 5p$$
$$< 6s < 4f < 5d < 6p < 7s < 5f < 6d < 7p < 8s.$$

If there is no spin the s sub-shell would hold 1 electron, the p sub-shell would hold 3, the d sub-shell 5, and the f sub-shell 7. inert gases occur when a p sub-shell has filled, so the first three inert gases would be element 1 (Hydrogen), element $1 + 1 + 3 = 5$ (Boron), and element $1 + 1 + 3 + 1 + 3 = 9$ (Fluorine).

Is there a pattern? Yes. The new inert gases have *half* of the atomic number of the original inert gases. The factor of one-half comes about because there are no longer two spin states for each set of n, l, m_l quantum numbers.

We can save time and simply divide the atomic numbers of the remaining inert gases in half: element 18 (Argon), element 27 (Cobalt), element 43 (Technetium), element 59 (Praseodymium). Would there be any more? Yes.

The pattern is

electron configuration	atomic number,
1	1
$1 + 4$	5
$1 + 4 + 4$	9
$1 + 4 + 9 + 4$	18
$1 + 4 + 9 + 9 + 4$	27
$1 + 4 + 9 + 16 + 9 + 4$	43
$1 + 4 + 9 + 16 + 16 + 9 + 4$	59
$1 + 4 + 9 + 16 + 25 + 16 + 9 + 4$	84
$1 + 4 + 9 + 16 + 25 + 25 + 16 + 9 + 4$	109

I suppose that's enough.

E48-21 (a) The Bohr orbits are circular orbits of radius $r_n = a_0 n^2$ (Eq. 47-20). The electron is orbiting where the force is

$$F_n = \frac{e^2}{4\pi\epsilon_0 r_n^2},$$

and this force is equal to the centripetal force, so

$$\frac{mv^2}{r_n} = \frac{e^2}{4\pi\epsilon_0 r_n^2}.$$

where v is the velocity of the electron. Rearranging,

$$v = \sqrt{\frac{e^2}{4\pi\epsilon_0 m r_n}}.$$

The time it takes for the electron to make one orbit can be used to calculate the current,

$$i = \frac{q}{t} = \frac{e}{2\pi r_n/v} = \frac{e}{2\pi r_n}\sqrt{\frac{e^2}{4\pi\epsilon_0 m r_n}}.$$

The magnetic moment of a current loop is the current times the area of the loop, so

$$\mu = iA = \frac{e}{2\pi r_n}\sqrt{\frac{e^2}{4\pi\epsilon_0 m r_n}}\pi r_n^2,$$

which can be simplified to

$$\mu = \frac{e}{2}\sqrt{\frac{e^2}{4\pi\epsilon_0 m r_n}}r_n.$$

But $r_n = a_0 n^2$, so

$$\mu = n\frac{e}{2}\sqrt{\frac{a_0 e^2}{4\pi\epsilon_0 m}}.$$

This might not look right, but $a_0 = \epsilon_0 h^2/\pi m e^2$, so the expression can simplify to

$$\mu = n\frac{e}{2}\sqrt{\frac{h^2}{4\pi^2 m^2}} = n\left(\frac{eh}{4\pi m}\right) = n\mu_B.$$

(b) In reality the magnetic moments depend on the angular momentum quantum number, not the principle quantum number. Although the Bohr theory correctly predicts the magnitudes, it does not correctly predict when these values would occur.

E48-25 The energy change can be derived from Eq. 48-13; we multiply by a factor of 2 because the spin is completely flipped. Then

$$\Delta E = 2\mu_z B_z = 2(9.27\times 10^{-24}\,\text{J/T})(0.190\,\text{T}) = 3.52\times 10^{-24}\text{J}.$$

The corresponding wavelength is

$$\lambda = \frac{hc}{E} = \frac{(6.63\times 10^{-34}\text{J}\cdot\text{s})(3.00\times 10^8\text{m/s})}{(3.52\times 10^{-24}\text{J})} = 5.65\times 10^{-2}\text{m}.$$

This is somewhere near the microwave range.

E48-29 We need to find out how many 10 MHz wide signals can fit between the two wavelengths. The lower frequency is

$$f_1 = \frac{c}{\lambda_1} = \frac{(3.00\times 10^8\,\text{m/s})}{700\times 10^{-9}\,\text{m})} = 4.29\times 10^{14}\,\text{Hz}.$$

The higher frequency is

$$f_1 = \frac{c}{\lambda_1} = \frac{(3.00\times 10^8\,\text{m/s})}{400\times 10^{-9}\,\text{m})} = 7.50\times 10^{14}\,\text{Hz}.$$

The number of signals that can be sent in this range is

$$\frac{f_2 - f_1}{(10\,\text{MHz})} = \frac{(7.50\times 10^{14}\,\text{Hz}) - (4.29\times 10^{14}\,\text{Hz})}{(10\times 10^6\,\text{Hz})} = 3.21\times 10^7.$$

That's quite a number of television channels.

E48-33 (a) At thermal equilibrium the population ratio is given by

$$\frac{N_2}{N_1} = \frac{e^{-E_2/kT}}{e^{-E_1/kT}} = e^{-\Delta E/kT}.$$

But ΔE can be written in terms of the transition photon wavelength, so this expression becomes

$$N_2 = N_1 e^{-hc/\lambda kT}.$$

Putting in the numbers,

$$N_2 = (4.0 \times 10^{20}) e^{-(1240 \text{ eV·nm})/(582 \text{ nm})(8.62 \times 10^{-5} \text{ eV/K})(300 \text{ K}))} = 6.62 \times 10^{-16}.$$

That's effectively *none*.

(b) If the population of the upper state were 7.0×10^{20}, then in a single laser pulse

$$E = N\frac{hc}{\lambda} = (7.0 \times 10^{20})\frac{(6.63 \times 10^{-34} \text{ J·s})(3.00 \times 10^8 \text{ m/s})}{(582 \times 10^{-9} \text{ m})} = 240 \text{ J}.$$

P48-1 Let λ_1 be the wavelength of the first photon. Then $\lambda_2 = \lambda_1 + 130$ pm. The total energy transfered to the two photons is then

$$E_1 + E_2 = \frac{hc}{\lambda_1} + \frac{hc}{\lambda_2} = 20.0 \text{ keV}.$$

We can solve this for λ_1,

$$\frac{20.0 \text{ keV}}{hc} = \frac{1}{\lambda_1} + \frac{1}{\lambda_1 + 130 \text{ pm}},$$

$$= \frac{2\lambda_1 + 130 \text{ pm}}{\lambda_1(\lambda_1 + 130 \text{ pm})},$$

which can also be written as

$$\lambda_1(\lambda_1 + 130 \text{ pm}) = (62 \text{ pm})(2\lambda_1 + 130 \text{ pm}),$$
$$\lambda_1^2 + (6 \text{ pm})\lambda_1 - (8060 \text{ pm}^2) = 0.$$

This equation has solutions

$$\lambda_1 = 86.8 \text{ pm and } -92.8 \text{ pm}.$$

Only the positive answer has physical meaning. The energy of this first photon is then

$$E_1 = \frac{(1240 \text{ keV·pm})}{(86.8 \text{ pm})} = 14.3 \text{ keV}.$$

(a) After this first photon is emitted the electron still has a kinetic energy of

$$20.0 \text{ keV} - 14.3 \text{ keV} = 5.7 \text{ keV}.$$

(b) We found the energy and wavelength of the first photon above. The energy of the second photon *must* be 5.7 keV, with wavelength

$$\lambda_2 = (86.8 \text{ pm}) + 130 \text{ pm} = 217 \text{ pm}.$$

P48-7 We assume in this crude model that one electron moves in a circular orbit attracted to the helium nucleus but repelled from the other electron. Look back to Sample Problem 47-6; we need to use some of the results from that Sample Problem to solve this problem.

The factor of e^2 in Eq. 47-20 (the expression for the Bohr radius) and the factor of $(e^2)^2$ in Eq. 47-21 (the expression for the Bohr energy levels) was from the Coulomb force between the single electron and the single proton in the nucleus. This force is

$$F = \frac{e^2}{4\pi\epsilon_0 r^2}.$$

In our approximation the force of attraction between the one electron and the helium nucleus is

$$F_1 = \frac{2e^2}{4\pi\epsilon_0 r^2}.$$

The factor of two is because there are two protons in the helium nucleus.

There is also a repulsive force between the one electron and the other electron,

$$F_2 = \frac{e^2}{4\pi\epsilon_0 (2r)^2},$$

where the factor of $2r$ is because the two electrons are on opposite side of the nucleus.

The net force on the first electron in our approximation is then

$$F_1 - F_2 = \frac{2e^2}{4\pi\epsilon_0 r^2} - \frac{e^2}{4\pi\epsilon_0 (2r)^2},$$

which can be rearranged to yield

$$F_{\text{net}} = \frac{e^2}{4\pi\epsilon_0 r^2}\left(2 - \frac{1}{4}\right) = \frac{e^2}{4\pi\epsilon_0 r^2}\left(\frac{7}{4}\right).$$

It is apparent (to me, at least) that we need to substitute $7e^2/4$ for every occurrence of e^2.

(a)   The ground state radius of the helium atom will then be given by Eq. 47-20 with the appropriate substitution,

$$r = \frac{\epsilon_0 h^2}{\pi m (7e^2/4)} = \frac{4}{7} a_0.$$

(b)   The energy of *one* electron in this ground state is given by Eq. 47-21 with the substitution of $7e^2/4$ for every occurrence of e^2, then

$$E = -\frac{m(7e^2/4)^2}{8\epsilon_0^4 h^2} = -\frac{49}{16} \frac{me^4}{8\epsilon_0^4 h^2}.$$

We already evaluated all of the constants to be 13.6 eV.

One last thing. There are *two* electrons, so we need to double the above expression. The ground state energy of a helium atom in this approximation is

$$E_0 = -2\frac{49}{16}(13.6 \text{ eV}) = -83.3 eV.$$

(c)   Removing one electron will allow the remaining electron to move closer to the nucleus. The energy of the remaining electron is given by the Bohr theory for He^+, and is

$$E_{He^+} = (4)(-13.60 \text{ eV}) = 54.4 \text{ eV},$$

so the ionization energy is 83.3 eV - 54.4 eV = 28.9 eV. This compares well with the accepted value.

Chapter 49

Electrical Conduction in Solids

E49-1 (a) Equation 49-2 is

$$n(E) = \frac{8\sqrt{2}\pi m^{3/2}}{h^3}E^{1/2} = \frac{8\sqrt{2}\pi (mc^2)^{3/2}}{(hc)^3}E^{1/2}.$$

We can evaluate this by substituting in all known quantities,

$$n(E) = \frac{8\sqrt{2}\pi (0.511 \times 10^6 \text{ eV})^{3/2}}{(1240 \times 10^{-9} \text{ eV} \cdot \text{m})^3}E^{1/2} = (6.81 \times 10^{27} \text{ m}^{-3} \cdot \text{eV}^{-3/2})E^{1/2}.$$

Once again, we simplified the expression by writing hc wherever we could, and then using $hc = 1240 \times 10^{-9}$ eV $\cdot$ m.

(b) Then, if $E = 5.00$ eV,

$$n(E) = (6.81 \times 10^{27} \text{ m}^{-3} \cdot \text{eV}^{-3/2})(5.00 \text{ eV})^{1/2} = 1.52 \times 10^{28} \text{ m}^{-3} \cdot \text{eV}^{-1}.$$

E49-5 (a) The approximate volume of a single sodium atom is

$$V_1 = \frac{(0.023 \text{ kg/mol})}{(6.02 \times 10^{23} \text{part/mol})(971 \text{ kg/m}^3)} = 3.93 \times 10^{-29} \text{m}^3.$$

The volume of the sodium ion sphere is

$$V_2 = \frac{4\pi}{3}(98 \times 10^{-12} \text{ m})^3 = 3.94 \times 10^{-30} \text{ m}^3.$$

The fractional volume available for conduction electrons is

$$\frac{V_1 - V_2}{V_1} = \frac{(3.93 \times 10^{-29}\text{m}^3) - (3.94 \times 10^{-30} \text{ m}^3)}{(3.93 \times 10^{-29}\text{m}^3)} = 90\%.$$

(b) The approximate volume of a single copper atom is

$$V_1 = \frac{(0.0635 \, \text{kg/mol})}{(6.02 \times 10^{23} \text{part/mol})(8960 \, \text{kg/m}^3)} = 1.18 \times 10^{-29} \text{m}^3.$$

The volume of the copper ion sphere is

$$V_2 = \frac{4\pi}{3} (96 \times 10^{-12} \, \text{m})^3 = 3.71 \times 10^{-30} \, \text{m}^3.$$

The fractional volume available for conduction electrons is

$$\frac{V_1 - V_2}{V_1} = \frac{(1.18 \times 10^{-29} \text{m}^3) - (3.71 \times 10^{-30} \, \text{m}^3)}{(1.18 \times 10^{-29} \text{m}^3)} = 69\%.$$

(c) Sodium, since more of the volume is available for the conduction electron.

E49-9 The Fermi energy is given by Eq. 49-5,

$$E_{\text{F}} = \frac{h^2}{8m} \left(\frac{3n}{\pi} \right)^{2/3},$$

where n is the density of conduction electrons. For gold we have

$$n = \frac{(19.3 \, \text{g/cm}^3)(6.02 \times 10^{23} \text{part/mol})}{(197 \, \text{g/mol})} = 5.90 \times 10^{22} \text{ elect./cm}^3 = 59 \text{ elect./nm}^3$$

The Fermi energy is then

$$E_{\text{F}} = \frac{(1240 \text{ eV} \cdot \text{nm})^2}{8(0.511 \times 10^6 \text{ eV})} \left(\frac{3(59 \text{ electrons/nm}^3)}{\pi} \right)^{2/3} = 5.53 \text{ eV}.$$

Once again, we did that $h^2/m = (hc)^2/(mc^2)$ trick.

E49-13 Equation 49-5 is

$$E_{\text{F}} = \frac{h^2}{8m} \left(\frac{3n}{\pi} \right)^{2/3},$$

and if we collect the constants,

$$E_{\text{F}} = \frac{h^2}{8m} \left(\frac{3}{\pi} \right)^{2/3} n^{3/2} = An^{3/2},$$

where, if we multiply the top and bottom by c^2

$$A = \frac{(hc)^2}{8mc^2} \left(\frac{3}{\pi} \right)^{2/3} = \frac{(1240 \times 10^{-9} \text{ eV} \cdot \text{m})^2}{8(0.511 \times 10^6 \text{ eV})} \left(\frac{3}{\pi} \right)^{2/3} = 3.65 \times 10^{-19} \text{ m}^2 \cdot \text{eV}.$$

E49-17   The steps to solve this exercise are equivalent to the steps for Exercise 49-9, except now the iron atoms each contribute 26 electrons and we have to find the density.

First, the density is

$$\rho = \frac{m}{4\pi r^3/3} = \frac{(1.99 \times 10^{30} \text{kg})}{4\pi (6.37 \times 10^6 \text{m})^3/3} = 1.84 \times 10^9 \text{kg/m}^3$$

Then

$$n = \frac{(26)(1.84 \times 10^6 \text{ g/cm}^3)(6.02 \times 10^{23} \text{part/mol})}{(56 \text{ g/mol})} = 5.1 \times 10^{29} \text{ elect./cm}^3,$$

$$= 5.1 \times 10^8 \text{ elect./nm}^3$$

The Fermi energy is then

$$E_\text{F} = \frac{(1240 \text{ eV} \cdot \text{nm})^2}{8(0.511 \times 10^6 \text{ eV})} \left(\frac{3(5.1 \times 10^8 \text{ elect./nm}^3)}{\pi} \right)^{2/3} = 230 \text{ keV}.$$

Once again, we did that $h^2/m = (hc)^2/(mc^2)$ trick.

E49-21   Using the results of Exercise 19,

$$T = \frac{2fE_\text{F}}{3k} = \frac{2(0.0130)(4.71 \text{ eV})}{3(8.62 \times 10^{-5} \text{eV} \cdot \text{K})} = 474 \text{ K}.$$

E49-25 (a)   Refer to Sample Problem 49-5 where we learn that the mean free path λ can be written in terms of Fermi speed v_F and mean time between collisions τ as

$$\lambda = v_\text{F} \tau.$$

The Fermi speed is

$$v_\text{F} = c\sqrt{2E_F/mc^2} = c\sqrt{2(5.51 \text{ eV})/(5.11 \times 10^5 \text{ eV})} = 4.64 \times 10^{-3} c.$$

The time between collisions is

$$\tau = \frac{m}{ne^2\rho} = \frac{(9.11 \times 10^{-31} \text{kg})}{(5.86 \times 10^{28} \text{m}^{-3})(1.60 \times 10^{-19} \text{C})^2(1.62 \times 10^{-8} \Omega \cdot \text{m})} = 3.74 \times 10^{-14} \text{s}.$$

We found n by looking up the answers from Exercise 49-23 in the back of the book. The mean free path is then

$$\lambda = (4.64 \times 10^{-3})(3.00 \times 10^8 \text{m/s})(3.74 \times 10^{-14} \text{s}) = 52 \text{ nm}.$$

(b) The spacing between the ion cores is approximated by the cube root of volume per atom. This atomic volume for silver is

$$V = \frac{(108 \text{ g/mol})}{(6.02 \times 10^{23} \text{part/mol})(10.5 \text{ g/cm}^3)} = 1.71 \times 10^{-23} \text{cm}^3.$$

The distance between the ions is then

$$l = \sqrt[3]{V} = 0.257 \text{ nm}.$$

The ratio is

$$\lambda/l = 190.$$

E49-29 The number of silicon atoms per unit volume is

$$n = \frac{(6.02 \times 10^{23} \text{part/mol})(2.33 \text{ g/cm}^3)}{(28.1 \text{ g/mol})} = 4.99 \times 10^{22} \text{ part./cm}^3.$$

If one out of $1.0eex7$ are replaced then there will be an additional charge carrier density of

$$4.99 \times 10^{22} \text{ part./cm}^3/1.0 \times 10^7 = 4.99 \times 10^{15} \text{ part./cm}^3 = 4.99 \times 10^{21} \text{m}^{-3}.$$

(b) The ratio is

$$(4.99 \times 10^{21} \text{m}^{-3})/(2 \times 1.5 \times 10^{16} \text{m}^{-3}) = 1.7 \times 10^5.$$

The extra factor of two is because *all* of the charge carriers in silicon (holes and electrons) are charge carriers.

E49-33 The first one is an insulator because the lower band is filled and band gap is so large; there is no impurity.

The second one is an extrinsic n-type semiconductor: it is a semiconductor because the lower band is filled and the band gap is small; it is extrinsic because there is an impurity; since the impurity level is close to the top of the band gap the impurity is a donor.

The third sample is an intrinsic semiconductor: it is a semiconductor because the lower band is filled and the band gap is small.

The fourth sample is a conductor; although the band gap is large, the lower band is *not* completely filled.

The fifth sample is a conductor: the Fermi level is above the bottom of the upper band.

The sixth one is an extrinsic p-type semiconductor: it is a semiconductor because the lower band is filled and the band gap is small; it is extrinsic because there is an impurity; since the impurity level is close to the bottom of the band gap the impurity is an acceptor.

E49-37 (a) Apply that ever so useful formula

$$\lambda = \frac{hc}{E} = \frac{(1240 \text{ eV} \cdot \text{nm})}{(5.5 \text{ eV})} = 225 \text{ nm}.$$

Why is this a *maximum?* Because longer wavelengths would have *lower* energy, and so not enough to cause an electron to jump across the band gap.

(b) Ultraviolet.

P49-1 We can calculate the electron density from Eq. 49-5,

$$
\begin{aligned}
n &= \frac{\pi}{3} \left(\frac{8mc^2 E_F}{(hc)^2} \right)^{3/2}, \\
&= \frac{\pi}{3} \left(\frac{8(0.511 \times 10^6 \text{ eV})(11.66 \text{ eV})}{(1240 \text{ eV} \cdot \text{nm})^2} \right)^{3/2}, \\
&= 181 \text{ electrons/nm}^3.
\end{aligned}
$$

From this we calculate the number of electrons per particle,

$$\frac{(181 \text{ electrons/nm}^3)(27.0 \text{ g/mol})}{(2.70 \text{ g/cm}^3)(6.02 \times 10^{23} \text{ particles/mol})} = 3.01,$$

which we can reasonably approximate as 3.

P49-5 (a) We want to use Eq. 49-6; although we don't know the Fermi energy, we do know the differences between the energies in question. In the un-doped silicon $E - E_F = 0.55$ eV for the bottom of the conduction band. The quantity

$$kT = (8.62 \times 10^{-5} \text{ eV/K})(290 \text{ K}) = 0.025 \text{ eV},$$

which is a good number to remember— at room temperature kT is 1/40 of an electron-volt.

Then

$$p = \frac{1}{e^{(0.55 \text{ eV})/(0.025 \text{ eV})} + 1} = 2.8 \times 10^{-10}.$$

In the doped silicon $E - E_F = 0.084$ eV for the bottom of the conduction band. Then

$$p = \frac{1}{e^{(0.084 \text{ eV})/(0.025 \text{ eV})} + 1} = 3.4 \times 10^{-2}.$$

(b) For the donor state $E - E_F = -0.066$ eV, so

$$p = \frac{1}{e^{(-0.066 \text{ eV})/(0.025 \text{ eV})} + 1} = 0.93.$$

Chapter 50

Nuclear Physics

E50-1 We want to follow the example set in Sample Problem 50-1. The distance of closest approach is given by

$$d = \frac{qQ}{4\pi\epsilon_0 K_\alpha},$$

$$= \frac{(2)(29)(1.60\times 10^{-19}\text{C})^2}{4\pi(8.85\times 10^{-12}\text{C}^2/\text{Nm}^2)(5.30\text{MeV})(1.60 \times 10^{-13}\text{J/MeV})},$$

$$= 1.57\times 10^{-14}\,\text{m}.$$

That's pretty close.

E50-5 We can make an estimate of the mass number A from Eq. 50-1,

$$R = R_0 A^{1/3},$$

where $R_0 = 1.2$ fm. If the measurements indicate a radius of 3.6 fm we would have

$$A = (R/R_0)^3 = ((3.6 \text{ fm})/(1.2 \text{ fm}))^3 = 27.$$

Aluminum, perhaps? We would need to make some measurements of the nuclear charge as well to know which element this is. Although the method works, it is much easier (and considerably more accurate) to find the mass number through other means, such as allowing the ions to travel through a uniform magnetic field.

E50-9 We'll follow the approach given in Sample Problem 50-3. The binding energy is the amount of energy required to disassemble the nucleus. ^{62}Ni is composed of 28 protons (because it is nickel) and $62 - 28 = 34$ neutrons. The combined mass of the free particles is

$$M = Zm_\text{p} + Nm_\text{n} = (28)(1.007825 \text{ u}) + (34)(1.008665 \text{ u}) = 62.513710 \text{ u}.$$

Note that we used the mass of hydrogen where we wrote m_p, because each proton will be accompanied by one electron, and although the binding energy of the electron is negligible, the mass energy isn't.

The binding energy is the difference

$$E_\text{B} = (62.513710 \text{ u} - 61.928349 \text{ u})(931.5 \text{ MeV/u}) = 545.3 \text{ MeV},$$

and the binding energy per nucleon is then

$$(545.3 \text{ MeV})/(62) = 8.795 \text{ MeV}.$$

E50-13 We've actually done this problem a fair number of times already! Look back to Ex. 46-27 for one such example. The neutron confined in a nucleus of radius R will have a position uncertainty on the order of $\Delta x \approx R$. The momentum uncertainty will then be no less than

$$\Delta p \geq \frac{h}{2\pi \Delta x} \approx \frac{h}{2\pi R}.$$

Assuming that $p \approx \Delta p$, we have

$$p \geq \frac{h}{2\pi R},$$

and then the neutron will have a (minimum) kinetic energy of

$$E \approx \frac{p^2}{2m} \approx \frac{h^2}{8\pi^2 m R^2}.$$

But $R = R_0 A^{1/3}$, and want to apply the trick of getting the mass in terms of c^2, so

$$E \approx \frac{(hc)^2}{8\pi^2 mc^2 R_0^2 A^{2/3}}.$$

For an atom with $A = 100$ we get

$$E \approx \frac{(1240 \text{ MeV} \cdot \text{fm})^2}{8\pi^2 (940 \text{ MeV})(1.2 \text{ fm})^2 (100)^{2/3}} = 0.668 \text{ MeV}.$$

This is about a factor of 5 or 10 less than the binding energy per nucleon.

E50-19 We will do this one the easy way because we can. This method won't work except when there is an integer number of half-lives. The activity of the sample will fall to one-half of the initial decay rate after one half-life; it will fall to one-half of one-half (one-fourth) after two half-lives. So two half-lives have elapsed, for a total of $(2)(140 \text{ d}) = 280 \text{ d}$.

E50-23 (a)   The decay constant for ^{67}Ga can be derived from Eq. 50-8,

$$\lambda = \frac{\ln 2}{t_{1/2}} = \frac{\ln 2}{(2.817 \times 10^5 \text{ s})} = 2.461 \times 10^{-6} \text{s}^{-1}.$$

The activity is given by $R = \lambda N$, so we want to know how many atoms are present. That can be found from

$$3.42 \text{ g} \left(\frac{1 \text{ u}}{1.6605 \times 10^{-24} \text{ g}} \right) \left(\frac{1 \text{ atom}}{66.93 \text{ u}} \right) = 3.077 \times 10^{22} \text{ atoms}.$$

So the activity is

$$R = (2.461 \times 10^{-6}/\text{s}^{-1})(3.077 \times 10^{22} \text{ atoms}) = 7.572 \times 10^{16} \text{ decays/s}.$$

(b)   After 1.728×10^5 s the activity would have decreased to

$$R = R_0 e^{-\lambda t} = (7.572 \times 10^{16} \text{ decays/s}) e^{-(2.461 \times 10^{-6}/\text{S}^{-1})(1.728 \times 10^5 \text{ S})} = 4.949 \times 10^{16} \text{ decays/s}.$$

E50-27 (a)   There is no easy way, we must apply Eq. 50-7,

$$R = R_0 e^{-\lambda t}.$$

We first need to know the decay constant from Eq. 50-8,

$$\lambda = \frac{\ln 2}{t_{1/2}} = \frac{\ln 2}{(1.234 \times 10^6 \text{ s})} = 5.618 \times 10^{-7} \text{s}^{-1}.$$

And the the time is found from

$$
\begin{aligned}
t &= -\frac{1}{\lambda} \ln \frac{R}{R_0}, \\
&= -\frac{1}{(5.618 \times 10^{-7} \text{s}^{-1})} \ln \frac{(170 \text{ counts/s})}{(3050 \text{ counts/s})}, \\
&= 5.139 \times 10^6 \text{ s} \approx 59.5 \text{ days}.
\end{aligned}
$$

Note that counts/s is *not* the same as decays/s. Not all decay events will be picked up by a detector and recorded as a count; we are assuming that whatever scaling factor which connects the initial count rate to the initial decay rate is valid at later times as well. Such an assumption is a reasonable assumption.

(b) The purpose of such an experiment would be to measure the amount of phosphorus that is taken up in a leaf. But the activity of the tracer decays with time, and so without a correction factor we would record the wrong amount of phosphorus in the leaf. That correction factor is R_0/R; we need to multiply the measured counts by this factor to correct for the decay.

In this case

$$\frac{R}{R_0} = e^{\lambda t} = e^{(5.618 \times 10^{-7} \text{S}^{-1})(3.007 \times 10^5 \text{ S})} = 1.184.$$

There is one other problem. The decay rate is *not* the same as the count rate. Some sort of correction will occur because not all decays are counted by a detector!

E50-33 The Q values are

$$
\begin{aligned}
Q_3 &= (235.043923 - 232.038050 - 3.016029)(931.5 \text{ MeV}) = -9.46 \text{ MeV}, \\
Q_4 &= (235.043923 - 231.036297 - 4.002603)(931.5 \text{ MeV}) = 4.68 \text{ MeV}, \\
Q_5 &= (235.043923 - 230.033127 - 5.012228)(931.5 \text{ MeV}) = -1.33 \text{ MeV}.
\end{aligned}
$$

Only reactions with positive Q values are energetically possible.

E50-37 (a) The kinetic energy of this electron is significant compared to the rest mass energy, so we *must* use relativity to find the momentum. The total energy of the electron is $E = K + mc^2$, the momentum will be given by

$$
\begin{aligned}
pc &= \sqrt{E^2 - m^2 c^4} = \sqrt{K^2 + 2Kmc^2}, \\
&= \sqrt{(1.00 \text{ MeV})^2 + 2(1.00 \text{ MeV})(0.511 \text{ MeV})} = 1.42 \text{ MeV}.
\end{aligned}
$$

The de Broglie wavelength is then

$$\lambda = \frac{hc}{pc} = \frac{(1240 \text{ MeV} \cdot \text{fm})}{(1.42 \text{ MeV})} = 873 \text{ fm}.$$

(b) The radius of the emitting nucleus is

$$R = R_0 A^{1/3} = (1.2 \text{ fm})(150)^{1/3} = 6.4 \text{ fm}.$$

(c) The longest wavelength standing wave on a string fixed at each end is twice the length of the string. Although the rules for standing waves in a box are slightly more complicated, it is a fair assumption that the electron could not exist as a standing a wave in the nucleus.

(d) See part (c).

E50-41 The radiation absorbed dose (rad) is related to the roentgen equivalent man (rem) by the quality factor, so for the chest x-ray

$$\frac{(25 \text{ mrem})}{(0.85)} = 29 \text{ mrad}.$$

This is well beneath the annual exposure average.

Each rad corresponds to the delivery of 10^{-5} J/g, so the energy absorbed by the patient is

$$(0.029)(10^{-5} \text{ J/g})\left(\frac{1}{2}\right)(88 \text{ kg}) = 1.28 \times 10^{-2} \text{ J}.$$

Don't forget to make sure that the mass units are properly converted!

E50-45 This one is a bit of a "trick" question, with a considerable amount of unnecessary, but still very distracting, information. The hospital uses a 6000 Ci source, and that is all the information we need to find the number of disintegrations per second:

$$(6000 \text{ Ci})(3.7 \times 10^{10} \text{ decays/s} \cdot \text{Ci}) = 2.22 \times 10^{14} \text{ decays/s}.$$

We are told the half life, but to find the number of radioactive nuclei present we want to know the decay constant. Then

$$\lambda = \frac{\ln 2}{t_{1/2}} = \frac{\ln 2}{(1.66 \times 10^8 \text{ s})} = 4.17 \times 10^{-9} \text{ s}^{-1}.$$

The number of ^{60}Co nuclei is then

$$N = \frac{R}{\lambda} = \frac{(2.22 \times 10^{14} \text{ decays/s})}{(4.17 \times 10^{-9} \text{ s}^{-1})} = 5.32 \times 10^{22}.$$

This might look like a large number, but it is only

$$(59.93 \text{ g/mol})\frac{(5.32 \times 10^{22})}{(6.02 \times 10^{23} \text{ mol}^{-1})} = 5.3 \text{ g}.$$

E50-49 We want to make a few assumptions before we start on this problem. Uranium doesn't decay directly into lead, so we need to consider the half-lives of the intermediate nuclei. Checking through the decay chain we find that two intermediate nuclei have half-lives on the order of 10^4 and 10^5 years, but this is so much smaller than 10^9 that we can reasonably assume that only the half-life of ^{238}U needs to be considered here.

To further simplify our work we will assume that the mass of ^{238}U is 238 g/mol, and the mass of ^{206}Pb is 206 g/mol. Now we can apply Eq. 50-18 to find the age of the rock,

$$
\begin{aligned}
t &= \frac{t_{1/2}}{\ln 2} \ln\left(1 + \frac{N_F}{N_I}\right), \\
&= \frac{(4.47 \times 10^9 \mathrm{y})}{\ln 2} \ln\left(1 + \frac{(2.00 \times 10^{-3}\mathrm{g})/(206 \mathrm{\ g/mol})}{(4.20 \times 10^{-3}\mathrm{g})/(238 \mathrm{\ g/mol})}\right), \\
&= 2.83 \times 10^9 \mathrm{y}.
\end{aligned}
$$

E50-55 We will write these reactions in the same way as Eq. 50-20 represents the reaction of Eq. 50-19. It is helpful to work backwards before proceeding by asking the following question: what nuclei will we have if we subtract one of the allowed projectiles?

The goal is ^{60}Co, which has 27 protons and $60 - 27 = 33$ neutrons.

1. Removing a proton will leave 26 protons and 33 neutrons, which is ^{59}Fe; but that nuclide is unstable.

2. Removing a neutron will leave 27 protons and 32 neutrons, which is ^{59}Co; and that nuclide is stable.

3. Removing a deuteron will leave 26 protons and 32 neutrons, which is ^{58}Fe; and that nuclide is stable.

It looks as if only ^{59}Co(n)^{60}Co and ^{58}Fe(d)^{60}Co are possible. If, however, we allow for the possibility of other daughter particles we should also consider some of the following reactions.

1. Swapping a neutron for a proton: ^{60}Ni(n,p)^{60}Co.

2. Using a neutron to knock out a deuteron: ^{61}Ni(n,d)^{60}Co.

3. Using a neutron to knock out an alpha particle: ^{63}Cu(n,α)^{60}Co.

4. Using a deuteron to knock out an alpha particle: ^{62}Ni(d,α)^{60}Co.

E50-61 (a) The binding energy of this neutron can be found by considering the Q value of the reaction ^{90}Zr(n)^{91}Zr which is

$$(89.904704 + 1.008665 - 90.905645)(931.5 \mathrm{\ MeV}) = 7.19 \mathrm{\ MeV}.$$

(b) The binding energy of this neutron can be found by considering the Q value of the reaction ^{89}Zr(n)^{90}Zr which is

$$(88.908889 + 1.008665 - 89.904704)(931.5 \mathrm{\ MeV}) = 12.0 \mathrm{\ MeV}.$$

This neutron is bound more tightly that the one in part (a).

(c) The binding energy per nucleon is found by dividing the binding energy by the number of nucleons:

$$\frac{(40 \times 1.007825 + 51 \times 1.008665 - 90.905645)(931.5 \text{ MeV})}{91} = 8.69 \text{ MeV}.$$

The neutron in the outside shell of ^{91}Zr is less tightly bound than the average nucleon in ^{91}Zr.

P50-1 Before doing anything we need to know whether or not the motion is relativistic. The rest mass energy of an α particle is

$$mc^2 = (4.00)(931.5 \text{ MeV}) = 3.73 \text{ GeV},$$

and since this is much greater than the kinetic energy we can assume the motion is non-relativistic, and we can apply non-relativistic momentum and energy conservation principles. The initial velocity of the α particle is then

$$v = \sqrt{2K/m} = c\sqrt{2K/mc^2} = c\sqrt{2(5.00 \text{ MeV})/(3.73 \text{ GeV})} = 5.18 \times 10^{-2}c.$$

For an elastic collision where the second particle is at originally at rest we have the final velocity of the first particle as

$$v_{1,\text{f}} = v_{1,\text{i}}\frac{m_2 - m_1}{m_2 + m_1} = (5.18 \times 10^{-2}c)\frac{(4.00\text{u}) - (197\text{u})}{(4.00\text{u}) + (197\text{u})} = -4.97 \times 10^{-2}c,$$

while the final velocity of the second particle is

$$v_{2,\text{f}} = v_{1,\text{i}}\frac{2m_1}{m_2 + m_1} = (5.18 \times 10^{-2}c)\frac{2(4.00\text{u})}{(4.00\text{u}) + (197\text{u})} = 2.06 \times 10^{-3}c.$$

(a) The kinetic energy of the recoiling nucleus is

$$K = \frac{1}{2}mv^2 = \frac{1}{2}m(2.06 \times 10^{-3}c)^2 = (2.12 \times 10^{-6})mc^2$$
$$= (2.12 \times 10^{-6})(197)(931.5 \text{ MeV}) = 0.389 \text{ MeV}.$$

(b) Energy conservation is the fastest way to answer this question, since it is an elastic collision. Then

$$(5.00 \text{ MeV}) - (0.389 \text{ MeV}) = 4.61 \text{ MeV}.$$

P50-5   The decay rate is given by $R = \lambda N$, where N is the number of radioactive nuclei present. If R exceeds P then nuclei will decay faster than they are produced; but this will cause N to decrease, which means R will decrease until it is equal to P. If R is less than P then nuclei will be produced faster than they are decaying; but this will cause N to increase, which means R will increase until it is equal to P. In either case equilibrium occurs when $R = P$, and it is a stable equilibrium because it is approached no matter which side is larger. Then

$$P = R = \lambda N$$

at equilibrium, so $N = P/\lambda$.

P50-9   As was done in Problem 1 we need to address the relativity issue. And as was found in Problem 1 this is also a non-relativistic problem. Assuming the ^{238}U nucleus is originally at rest the total initial momentum is zero, which means the magnitudes of the final momenta of the α particle and the ^{234}Th nucleus are equal.

The α particle has a final velocity of

$$v = \sqrt{2K/m} = c\sqrt{2K/mc^2} = c\sqrt{2(4.196 \text{ MeV})/(4.0026 \times 931.5 \text{ MeV})} = 4.744{\times}10^{-2}c.$$

Since the magnitudes of the final momenta are the same, the ^{234}Th nucleus has a final velocity of

$$(4.744 {\times} 10^{-2}c)\left(\frac{(4.0026 \text{ u})}{(234.04 \text{ u})}\right) = 8.113 \times 10^{-4}c.$$

The kinetic energy of the ^{234}Th nucleus is

$$K = \frac{1}{2}mv^2 = \frac{1}{2}m(8.113{\times}10^{-4}c)^2 = (3.291 \times 10^{-7})mc^2$$
$$= (3.291 \times 10^{-7})(234.04)(931.5 \text{ MeV}) = 71.75 \text{ keV}.$$

The Q value for the reaction is then

$$(4.196 \text{ MeV}) + (71.75 \text{ keV}) = 4.268 \text{ MeV},$$

which agrees well with the Sample Problem.

P50-13   The decay constant for ^{90}Sr is

$$\lambda = \frac{\ln 2}{t_{1/2}} = \frac{\ln 2}{(9.15 \times 10^8 \text{ s})} = 7.58 \times 10^{-10} \text{s}^{-1}.$$

The number of nuclei present in 400 g of ^{90}Sr is

$$N = (400 \text{ g})\frac{(6.02 \times 10^{23}/\text{mol})}{(89.9 \text{ g/mol})} = 2.68 \times 10^{24},$$

so the overall activity of the 400 g of ^{90}Sr is

$$R = \lambda N = (7.58 \times 10^{-10} \text{s}^{-1})(2.68 \times 10^{24})/(3.7 \times 10^{10}/\text{Ci} \cdot \text{s}) = 5.49 \times 10^4 \text{ Ci}.$$

This is spread out over a 2000 km^2 area, so the "activity surface density" (I just made that term up) is

$$\frac{(5.49 \times 10^4 \text{ Ci})}{(20006 \text{ m}^2)} = 2.74 \times 10^{-5} \text{ Ci/m}^2.$$

If the allowable limit is 0.002 mCi, then the area of land that would contain this activity is

$$\frac{(0.002 \times 10^{-3} \text{ Ci})}{(2.74 \times 10^{-5} \text{ Ci/m}^2)} = 7.30 \times 10^{-2} \text{m}^2.$$

This is not very much land, and since the half life of ^{90}Sr is 29 years one should expect that this problem will be around for longer than a generation. But is it really a problem?

My (quick) research on the web indicates that one cow needs about one of acre of land for sustained grazing. Assuming that the cow eats the entire acre (which grows back, but is then clean) and *all* of the ^{90}Sr which lands on the soil is consumed by the cow, then in the course of the year the cow will absorb enough ^{90}Sr to have an activity of

$$(4 \times 10^3 \text{m}^2)(2.74 \times 10^{-5} \text{ Ci/m}^2) = 0.11 \text{ Ci}.$$

Much of the ^{90}Sr would go into the bones of the cow, but we will assume that it all passes into the milk. During the course of the year the cow will produce some 5,000 liters of milk (again, found from the web, so don't cite me) so the activity per liter is about 0.02 mCi.

According to this, a small glass of milk from such a contaminated cow would result in an amount of ^{90}Sr which would easily exceed the allowed bone burden for one person.

Chapter 51

Energy from the Nucleus

E51-3 (a) There are

$$\frac{(1.00\,\text{kg})(6.02\times10^{23}\text{mol}^{-1})}{(235\text{g/mol})} = 2.56\times10^{24}$$

atoms in 1.00 kg of ^{235}U.

(b) If each atom releases 200 MeV, then

$$(200\times10^6\text{ eV})(1.6\times10^{-19}\text{J/ eV})(2.56\times10^{24}) = 81.9\times10^{13}\text{ J}$$

of energy could be released from 1.00 kg of ^{235}U.

(c) This amount of energy would keep a 100-W lamp lit for

$$t = \frac{(81.9\times10^{13}\text{ J})}{(100\text{ W})} = 8.19\times10^{11}\text{s} \approx 26,000\text{ y!}$$

E51-7 When the ^{233}U nucleus absorbs a neutron we are given a total of 92 protons and 142 neutrons. Gallium has 31 protons and around 39 neutrons; chromium has 24 protons and around 28 neutrons. There are then 37 protons and around 75 neutrons left over. This would be rubidium, but the number of neutrons is *very* wrong. Although the elemental identification is correct, because we must conserve proton number, the isotopes are *wrong* in our above choices for neutron numbers.

E51-13 The energy released is

$$(235.043923 - 140.920044 - 91.919726 - 2\times1.008665)(931.5\text{ MeV}) = 174\text{ MeV}.$$

E51-17 Since ^{239}Pu is one nucleon heavier than ^{238}U only one neutron capture is required. The atomic number of Pu is two *more* than U, so two beta decays will be required. The reaction series is then

$$^{238}\text{U} + \text{n} \rightarrow \ ^{239}\text{U},$$
$$^{239}\text{U} \rightarrow \ ^{239}\text{Np} + \beta^- + \bar{\nu},$$
$$^{239}\text{Np} \rightarrow \ ^{239}\text{Pu} + \beta^- + \bar{\nu}.$$

E51-21 Let the energy released by one fission be E_1. If the average time to the next fission event is t_{gen}, then the "average" power output from the one fission is $P_1 = E_1/t_{\text{gen}}$. If every fission event results in the release of k neutrons, each of which cause a later fission event, then after every time period t_{gen} the number of fission events, and hence the average power output from *all* of the fission events, will increase by a factor of k.

For long enough times we can write

$$P(t) = P_0 k^{t/t_{\text{gen}}}.$$

E51-25 The time constant for this decay is

$$\lambda = \frac{\ln 2}{(2.77 \times 10^9 \, \text{s})} = 2.50 \times 10^{-10} \text{s}^{-1}.$$

The number of nuclei present in 1.00 kg is

$$N = \frac{(1.00 \, \text{kg})(6.02 \times 10^{23} \, \text{mol}^{-1})}{(238 \, \text{g/mol})} = 2.53 \times 10^{24}.$$

The decay rate is then

$$R = \lambda N = (2.50 \times 10^{-10} \text{s}^{-1})(2.53 \times 10^{24}) = 6.33 \times 10^{14} \text{s}^{-1}.$$

The power generated is the decay rate times the energy released per decay,

$$P = (6.33 \times 10^{14} \text{s}^{-1})(5.59 \times 10^6 \, \text{eV})(1.6 \times 10^{-19} \, \text{J/eV}) = 566 \, \text{W}.$$

E51-29 If ^{238}U absorbs a neutron it becomes ^{239}U, which will decay by beta decay to first ^{239}Np and then ^{239}Pu; we looked at this in Exercise 51-17. This can decay by alpha emission according to

$$^{239}\text{Pu} \rightarrow ^{235}\text{U} + \alpha.$$

E51-35 The energy released is

$$(3 \times 4.002603 - 12.0000000)(931.5 \, \text{MeV}) = 7.27 \, \text{MeV}.$$

E51-39 The rate of consumption is 6.2×10^{11} kg/s, the core has 1/8 the mass but only 35% is hydrogen, so the time remaining is

$$t = (0.35)(1/8)(2.0 \times 10^{30} \text{kg})/(6.2 \times 10^{11} \text{kg/s}) = 1.4 \times 10^{17} \text{s},$$

or about 4.5×10^9 years.

E51-45 Assume momentum conservation, then

$$p_\alpha = p_\text{n} \text{ or } v_\text{n}/v_\alpha = m_\alpha/m_\text{n}.$$

The ratio of the kinetic energies is then

$$\frac{K_\text{n}}{K_\alpha} = \frac{m_\text{n} v_\text{n}^2}{m_\alpha v_\alpha^2} = \frac{m_\alpha}{m_\text{n}} \approx 4.$$

Then $K_\text{n} = 4Q/5 = 14.07$ MeV while $K_\alpha = Q/5 = 3.52$ MeV.

P51-1 (a) Equation 50-1 is
$$R = R_0 A^{1/3},$$

where $R_0 = 1.2$ fm. The distance between the two nuclei will be the sum of the radii, or
$$R_0 \left((140)^{1/3} + (94)^{1/3} \right).$$

The potential energy will be

$$
\begin{aligned}
U &= \frac{1}{4\pi\epsilon_0} \frac{q_1 q_2}{r}, \\
&= \frac{e^2}{4\pi\epsilon_0 R_0} \frac{(54)(38)}{((140)^{1/3} + (94)^{1/3})}, \\
&= \frac{(1.60 \times 10^{-19} \text{C})^2}{4\pi(8.85 \times 10^{-12} \text{C}^2/\text{Nm}^2)(1.2 \text{ fm})} 211, \\
&= 253 \text{ MeV}.
\end{aligned}
$$

(b) The energy will eventually appear as thermal energy.

P51-7 (a) Demonstrating the consistency of this expression is considerably easier than deriving it from first principles. From Problem 50-4 we have that a uniform sphere of charge Q and radius R has potential energy

$$U = \frac{3Q^2}{20\pi\epsilon_0 R}.$$

This expression was derived from the fundamental expression

$$dU = \frac{1}{4\pi\epsilon_0}\frac{q\,dq}{r}.$$

For gravity the fundamental expression is

$$dU = \frac{Gm\,dm}{r},$$

so we replace $1/4\pi\epsilon_0$ with G and Q with M. But like charges repel while all masses attract, so we pick up a negative sign.

(b) The initial energy would be zero if $R = \infty$, so the energy released is

$$U = \frac{3GM^2}{5R} = \frac{3(6.7\times10^{-11}\text{N}\cdot\text{m}^2/\text{kg}^2)(2.0\times10^{30}\text{kg})^2}{5(7.0\times10^8\text{m})} = 2.3\times10^{41}\text{J}.$$

At the current rate (see Sample Problem 51-6), the sun would be

$$t = \frac{(2.3\times10^{41}\text{J})}{(3.9\times10^{26}\text{W})} = 5.9\times10^{14}\text{ s},$$

or 187 million years old.

Chapter 52

Particle Physics and Cosmology

E52-1 (a) The gravitational force is given by Gm^2/r^2, while the electrostatic force is given by $q^2/4\pi\epsilon_0 r^2$. The ratio is

$$\frac{4\pi\epsilon_0 Gm^2}{q^2} = \frac{4\pi(8.85\times10^{-12}\mathrm{C}^2/\mathrm{Nm}^2)(6.67\times10^{-11}\mathrm{Nm}^2/\mathrm{kg}^2)(9.11\times10^{-31}\mathrm{kg})^2}{(1.60\times10^{-19}\mathrm{C})^2},$$

$$= 2.4\times10^{-43}.$$

Gravitational effects would be swamped by electrostatic effects at *any* separation.

(b) The ratio is

$$\frac{4\pi\epsilon_0 Gm^2}{q^2} = \frac{4\pi(8.85\times10^{-12}\mathrm{C}^2/\mathrm{Nm}^2)(6.67\times10^{-11}\mathrm{Nm}^2/\mathrm{kg}^2)(1.67\times10^{-27}\mathrm{kg})^2}{(1.60\times10^{-19}\mathrm{C})^2},$$

$$= 8.1\times10^{-37}.$$

E52-5 The energy of one of the pions will be

$$E = \sqrt{(pc)^2 + (mc^2)^2} = \sqrt{(358.3 \text{ MeV})^2 + (140 \text{ MeV})^2} = 385 \text{ MeV}.$$

There are two of these pions, so the rest mass energy of the ρ_0 is 770 MeV.

E52-9 (a) Baryon number is conserved by having two "p" on one side and a "p" and a Δ^0 on the other. Charge will only be conserved if the particle x is positive. Strangeness will only be conserved if x is strange. Since it can't be a baryon it must be a meson. Then x is K^+.

(b) Baryon number on the left is 0, so x must be an anti-baryon. Charge on the left is zero, so x must be neutral because "n" is neutral. Strangeness is everywhere zero, so the particle must be $\bar{n}$.

(c) There is one baryon on the left and one on the right, so x has baryon number 0. The charge on the left adds to zero, so x is neutral. The strangeness of x must also be 0, so it must be a π^0.

E52-17 A strangeness of $+1$ corresponds to the existence of an $\bar{s}$ anti-quark, which has a charge of $+1/3$. The only quarks that can combine with this anti-quark to form a meson will have charges of $-1/3$ or $+2/3$. It is only possible to have a net charge of 0 or $+1$. The reverse is true for strangeness -1.

E52-25 The minimum energy required to produce the pairs is through the collision of two 140 MeV photons. This corresponds to a temperature of

$$T = (140 \text{ MeV})/(8.62 \times 10^{-5} \text{eV/K}) = 1.62 \times 10^{12} \text{K}.$$

This temperature existed at a time

$$t = \frac{(1.5 \times 10^{10} \text{s}^{1/2} \text{K})^2}{(1.62 \times 10^{12} \text{K})^2} = 86 \, \mu\text{s}.$$

P52-1 The total energy of the pion is $135 + 80 = 215$ MeV. The gamma factor of relativity is

$$\gamma = E/mc^2 = (215 \text{ MeV})/(135 \text{ MeV}) = 1.59,$$

so the velocity parameter is

$$\beta = \sqrt{1 - 1/\gamma^2} = 0.777.$$

The lifetime of the pion as measured in the laboratory is

$$t = (8.4 \times 10^{-17} \text{ s})(1.59) = 1.34 \times 10^{-16} \text{s},$$

so the distance traveled is

$$d = vt = (0.777)(3.00 \times 10^8 \text{m/s})(1.34 \times 10^{-16} \text{s}) = 31 \text{ nm}.$$

P52-7 (a) The force on a particle in a spherical distribution of matter depends only on the matter contained in a sphere of radius *smaller* than the distance to the center of the spherical distribution. And then we can treat all that relevant matter as being concentrated at the center. If M is the total mass, then

$$M' = M\frac{r^3}{R^3},$$

is the fraction of matter contained in the sphere of radius $r < R$. The force on a star of mass m a distance r from the center is

$$F = GmM'/r^2 = GmMr/R^3.$$

This force is the source of the centripetal force, so the velocity is

$$v = \sqrt{ar} = \sqrt{Fr/m} = r\sqrt{GM/R^3}.$$

The time required to make a revolution is then

$$T = \frac{2\pi r}{v} = 2\pi\sqrt{R^3/GM}.$$

Note that this means that the system rotates as if it were a solid body!

We would expect a different result if the the mass distribution were a disk instead of a sphere.

(b)   If, instead, all of the mass were concentrated at the center, then the centripetal force would be

$$F = GmM/r^2,$$

so

$$v = \sqrt{ar} = \sqrt{Fr/m} = \sqrt{GM/r},$$

and the period would be

$$T = \frac{2\pi r}{v} = 2\pi\sqrt{r^3/GM}.$$